Osiris, Isis & Planet X

Chasing The Centuries

By

Rob Solàrion

Bloomington, IN

Milton Keynes, UK

AuthorHouse™
1663 Liberty Drive, Suite 200
Bloomington, IN 47403
www.authorhouse.com
Phone: 1-800-839-8640

AuthorHouse™ UK Ltd.
500 Avebury Boulevard
Central Milton Keynes, MK9 2BE
www.authorhouse.co.uk
Phone: 08001974150

First published by AuthorHouse 4/18/2006

ISBN: 1-4259-2621-5 (sc)
ISBN: 1-4259-2622-3 (dj)

Library of Congress Control Number: 2006902854

Printed in the United States of America
Bloomington, Indiana

This book is printed on acid-free paper.

Cover Design By Rob Solàrion

Rob Solàrion
UFO Museum Library
Roswell, New Mexico, USA
13 October 2005

See Chapter 16.

Dedicated To
The Memories Of
Joey, Mimi, Minh &
Robert Dwayne

"The reptiles verily descend. The Earth is resplendent as a well-watered garden. At that time Enki and Eridu had not appeared. Daylight did not shine. Moonlight had not emerged." Fragment, Oldest Known Sumerian Tablet

"I shall establish a savage. Man will be his name. Verily, savage-man I shall create. He will be charged with the service of the gods, that they might be at ease." Sumerian Creation Epic

"Human beings appear to be a slave race, languishing on an isolated planet in a small galaxy. As such, the human race was once a source of labor for an extraterrestrial civilization and still remains a possession today. To keep control over its possession and to maintain Earth as something of a prison, that other civilization has bred never-ending conflict between human beings, has promoted human spiritual decay, and has erected on Earth conditions of unremitting physical hardship. This situation has existed for thousands of years, and it continues today." William Bramley, *The Gods Of Eden*

Table Of Contents

Part I

Osiris, Isis & Planet X

Chapter 1

Osiris & Isis
Galactic Origin Of Planet X Nibiru

*

Whence did Planet X Nibiru originate? That is the question to be answered in this book.

After the discovery of the Planets Uranus and Neptune, astronomers noticed gravitational anomalies in their orbits. Let me quote from the 2006 *World Almanac*, page 341:

“About a century ago, a hypothetical planet was believed to lie beyond Neptune and Uranus because neither planet followed paths predicted by astronomers when all known gravitational influences were considered. In little more than a guess, a mass equal to Earth was assigned to the mysterious body, and mathematical searches were begun. Amid some controversy about the validity of the predictive process, Pluto was discovered nearly where it had been predicted to lie, by Claude Tombaugh at the Lowell Observatory in Flagstaff, Arizona, in 1930. ... It is now clear that Pluto could not have influenced Neptune and Uranus to go astray. Besides being the smallest planet, Pluto is actually smaller than at least 7 of the solar system’s moons, including Earth’s moon. Although there may be additional solar system planets to be discovered and confirmed, astronomers no longer believe there are unexplained perturbations in the orbits of Uranus or Neptune that might be caused by another planet.”

Working today from the assumption that Planet X Nibiru's existence is an astronomical probability, even though it has not been officially "discovered" yet, we now know that the planet exerting those gravitational influences upon Uranus and Neptune was in fact Planet X, our Solar System's tenth planet, *The Twelfth Planet* of Zecharia Sitchin, who referenced the Sumerian system which counted the Sun and Earth's Moon amongst the "planets". Was Planet X a part of this Solar System from the very beginning? That is debatable, and some modern "Nibiruologists" (if I may coin a new word!) like Andy Lloyd might answer in the affirmative, based upon early stellar formation patterns. (See Appendix A.) On the other hand, in his acclaimed series *The Earth Chronicles*, Zecharia Sitchin has stated categorically that Planet X Nibiru was "captured" by this Solar System at some point in the distant past, perhaps even a million or more years ago. But Sitchin does not attempt to determine whence this "Planet of the Gods" originated.

Because of Planet X's elongated, comet-like, oval orbit around the Sun, practically perpendicular to the ecliptical orbits of the other nine planets and their moons, I, like Sitchin, believe that Planet X Nibiru was not an original member of this Solar System but was captured by our Sun's gravity at a much later time. And when, exactly, could that have been?

Sirius A, commonly known simply as Sirius, is the brightest star in our nighttime sky and one of the closest stars at about 8.3 light-years. In ancient Egyptian legends, this star is also called "Sothis" and is considered to be an equivalent of the Goddess Isis. The Sirius System is a double-star system, perhaps even a triple-star system, with the main components being Sirius A and a smaller star known as Sirius B, which is invisible to the naked eye. Sirius B was not observed until the 20th Century and then only with powerful telescopes, because the brightness of Sirius A made observation of Sirius B impossible with the less powerful telescopes of previous centuries.

Sirius B is a "white-dwarf" or "neutron-dwarf" (the term that I shall use). It is at the end of its period of stellar evolution and life. Previously, or so our astronomers instruct us, such "neutron-dwarfs" were "red-giant" stars that imploded and collapsed in on themselves as denser dwarf stars. Thus, prior to its metamorphosis into a neutron-dwarf, Sirius B was a red-giant. We are told that these processes of stellar evolution take millions of years. But paradoxically, right here on Earth, we still possess legends from antiquity that "Sirius" used to be "redder than Mars"; and some of these legends, such as those recorded by Roman historian Seneca, are fewer than 2,000 years old.

Yet we Cro-Magnon Sapien humans go back no more than 100,000 years at the very maximum, probably no farther back than only about 40,000 years, even if that far; this is by no means certain in our science. If stellar evolution takes millions of years, then how could we Cro-Magnons, *before we even came into existence*, retain a legend of such a shift from red-giant to neutron-dwarf for any star, Sirius B or otherwise? We humans barely retain any history or legends of any kind earlier than about 7,000 years ago, and that is probably stretching the timeframe a bit. Did ancient astronomer-priests actually witness the implosion of a nearby red-giant and leave behind records still in existence today?

As will be shown in subsequent chapters, the primitive Dogon tribe of Mali in Africa have "myths" that "gods" arrived here on Earth from "Digitaria" which can be identified with Sirius B. The Sumerians recorded that Earth was settled by a group of amphibian or "saurian-like" beings from Sirius, under the leadership of a Sirian man named "Oannes", who first came ashore in what is now modern-day Kuwait. Zecharia Sitchin has written that we Cro-Magnons are a crossbreed between the "foreign" Nibiruans and the "local" Neanderthals. R.A. Boulay has taken this idea further and provided compelling evidence that these "aliens" were "reptilian" or "saurian" in nature, although Sitchin thinks that they were more "human-like". All anthropologists will tell you that Cro-Magnon Sapiens is a "new" species that did not evolve directly from Neanderthal. The precise origin of Cro-Magnon is still a mystery to modern scientists, unless one accepts the explanations put forth by Sitchin and Boulay.

In times long passed, when humans would have gazed towards the Sirius System, they would have seen two bright stars: a red-giant star and a smaller blue-white star. The Ancient Egyptians named these two stars, respectively, Osiris and Isis, who had a child named Horus. The following is a synopsis of what we know from Ancient Egypt.

*

New Larousse Encyclopedia Of Mythology
(Hamlyn Publishing, London & New York, 1978 Edition)

PART 2, Egyptian Mythology (By J. Viaud)

Pages 16-21 (Excerpted, Not Necessarily In Order)

Inserted **[COMMENTs]** By Rob Solàrion (and throughout the book)

Osiris, which is the Greek rendering of the Egyptian Ousir, was identified by the Greeks with several of their own gods, but principally with Dionysus and Hades. At first Osiris was a nature god and embodied the spirit of vegetation which dies with the harvest to be reborn when the grain sprouts. Afterwards he was worshipped throughout Egypt as a god of the dead, and in this capacity reached first rank in the Egyptian pantheon.

[COMMENT: Zecharia Sitchin has identified Osiris with the Greek Zeus, and I agree. R.A. Boulay has identified Osiris also with the Hebrew Yahweh, and I again agree. Both have identified Osiris with Nibiruan Crown-Prince Enlil. The Greek Dionysus is the equivalent of Nibiruan Baron Nabu, the Sumerian Nabo, the Egyptian Bakha and the Roman Bacchus. The Greek Hades is the equivalent of Nibiruan Duke Nergal, the Sumerian Irrigal or Gaga, the Egyptian Anubis and the Roman Pluto. See Chapter 12 ("Names Of The Gods"). This legend that Osiris was reborn with the harvest resembles that of the rebirth of the so-called "Phoenix Bird" and probably alludes to the periodic return of Planet X Nibiru, the "rebirth" of Planet X. RS]

Hieroglyphic texts contain numerous allusions to the life and death of Osiris during his sojourn on earth; but it is above all thanks to Plutarch that we know his legend so well. The first son of Geb and Nut, he was born in Thebes in Upper Egypt.

[COMMENT: The Egyptian Geb (Seb) and Nut (Neith) are none other than Nibiruan Emperor Anu and Empress Antu. They were known to the Greeks as Kronos and Rhea, to the Romans as Saturn and Ops, and to the Levantines as Leviathan and Apas. RS]

Isis (a Greek rendering of Aset, Eset) was identified by the Greeks with Demeter, Hera, Selene and even — because of a late confusion between Isis and Hathor — with Aphrodite. In later days the popularity of Isis became such that she finally absorbed the qualities of all the other goddesses; but originally she seems to have been a modest divinity of the Delta, the protective deity of Perehbet, north of Busiris, where she always retained a renowned temple. Very soon she was given as wife to Osiris, the god of the neighbouring town. She bore him a son, Horus, who formed the third member of the trinity. Her popularity grew rapidly with that of her husband and son. This is her legend as Plutarch tells it to us:

The first daughter of Geb and Nut was born in the swamps of the Delta on the fourth intercalary day. Osiris, her eldest brother, chose her as his consort and she mounted to the throne with him. She helped him in his great work of civilising Egypt by teaching women to grind corn, spin flax and weave cloth. She also taught men the art of curing disease and, by instituting marriage, accustomed them to domestic life. When her husband departed on his pacific conquest of the world, she remained in Egypt as regent. She governed wisely while awaiting his return.

[**COMMENT:** The Egyptian Isis is the equivalent of Nibiruan Queen Ninkhursag, the Sumerian Ninmah, the Greek Hera and the Roman Juno. She was not technically "married" to Osiris, since the "official wife" of Enlil / Osiris / Zeus was Nibiruan Crown-Princess Ninlil, the Egyptian Ma'at, the Greek Maia and the Roman Majesta. Enlil / Osiris / Zeus had children by both Ninlil / Ma'at / Maia and Ninkhursag / Isis / Hera. According to Sitchin, Queen Ninkhursag / Isis / Hera was/is the "Chief Medical Officer" and "Geneticist" of Planet X Nibiru. She created the original "Adamu" from a combination of Nibiruan and Neanderthal genetics. Thus, it is not surprising to read that Isis taught men the art of curing disease. Whether Isis was also identified with the Greek Demeter or Selene is unknown (probably not), but she was certainly not the same "goddess" as Aphrodite, who is the equivalent of Nibiruan Princess-Royal Inanna, the Sumerian Ishtar, the Levantine Astarte, the Egyptian Hathor and the Roman Venus. RS]

Set (Seth, Sutekh), whom the Greeks called Typhon, was the name of Osiris' evil brother who finally became the incarnation of the spirit of evil, in eternal opposition to the spirit of good. The son of Geb and Nut, he was, Plutarch tells us, prematurely born on the third intercalary day. He tore himself violently from his mother's womb. He was rough and wild, his skin was white and his hair was red — an abomination to the Egyptians, who compared it to the pelt of an ass.

[**COMMENT:** The Egyptian Set, or Seth, is the equivalent of Nibiruan Prince Enki. He is also the equivalent of the Sumerian Ea or Urki, the Greek Poseidon and the Roman Neptune. He was not the Greco-Roman Typhon, who is the Nibiruan Baron Ninurta, the Sumerian Ningirsu, the Egyptian Ptah, the Greek Hephaestus or the Roman Vulcan. Enki/Set is the Biblical "fallen angel" known as Satan, enemy of Yahweh. The idea that he is "evil" comes from the fact that there is an ongoing battle for "Planet Tiamat" (Earth) between half-brothers Enlil and Enki, and it is in the "propaganda" of Enlil that we find this idea of an "evil" Enki/Set. RS]

Set was jealous of Osiris, his elder brother, and secretly aspired to the throne. In order to seize it he availed himself of the great festivals which were celebrated at Memphis on the occasion of Osiris' victorious return to his kingdom. Having first assured himself of the presence of seventy-two accomplices, he invited his brother to a banquet during the course of which he gave orders that a marvellously fashioned coffer should be brought in. This chest, he explained jokingly, would belong to whomsoever fitted it exactly. Osiris, falling in with the pleasantry, lay down in the coffer without suspicion. The conspirators at once rushed forward, closed the lid and nailed it solidly down. They threw it into the Nile, whence it was carried to the sea and borne across the waves to the Phoenician coast where it came to rest at the base of a tamarisk tree. The tree grew with such astonishing rapidity that the chest was entirely enclosed within its trunk.

[COMMENT: Note the number 72 in the above paragraph. This implies that the 72 Hermetic Archons, 72 Greek Titans or 72 Hebrew Lemegeton Demon Spirits — to wit, the ruling élite nobility of Planet X Nibiru — were in on this plot, much as the Roman Senate conspired to assassinate Julius Caesar. This idea will be further discussed later in this book. RS]

Now Malcandre, the king of Byblos, gave orders that the tamarisk should be cut down in order to serve as a prop for the roof of his palace. When this was done, the marvellous tree gave off so exquisite a scent that its reputation reached the ears of Isis, who immediately understood its significance. Without delay she went to Phoenicia. There the queen, Astarte, confided to her the care of her newly born son. Isis adopted the baby and would have conferred immortality upon it had its mother not broken the charm by her cries of terror upon seeing the goddess bathe the baby in purificatory flames. In order to reassure her, Isis revealed her true name and the reason for her presence. Then, having been presented with the trunk of the miraculous tree, she drew forth the coffer of her husband, bathed it in tears, and bore it back in haste to Egypt where, to deceive Set, she hid it in the swamps of Buto. Set, however, regained possession of his brother's body by chance and in order to annihilate it forever cut it into fourteen pieces which he scattered far and wide.

[COMMENT: This idea of a gigantic, marvellous tree obviously refers to Hyperborea, The Cosmic Tree. RS]

Isis, undiscouraged, searched for the precious fragments and found them all except the phallus which had been greedily devoured by a Nile crab, the Oxyrhynchid, forever accursed for this crime. The goddess reconstituted the body of Osiris, cunningly joining the fragments together. She then performed, for the first time in history, the rites of embalmment which restored the murdered god to eternal life. In this she was assisted by her sister Nephthys, her nephew Anubis, Osiris' grand vizier Thoth and by Horus, the posthumous son whom she had conceived by union with her husband's corpse, miraculously re-animated by her charms.

Afterwards she retired to the swamps of Buto to escape the wrath of Set and to bring up her son Horus until the day when he should be of an age to avenge his father. Thanks to her magic powers, Horus was able to overcome every danger which threatened him.

Resurrected and from thenceforward secure from the threat of death, Osiris could have regained his throne and continued to reign over the living. But he preferred to depart from this earth and retire to the 'Elysian Fields' where he warmly welcomed the souls of the just and reigned over the dead. Such is the legend of Osiris.

[COMMENT: The "Elysian Fields" (Champs Elysées) were the "partially lighted" or "twilight" zones between the boundary where one could still view "The Cosmic Tree" in the Northern Hemisphere and the non-tropical regions of the Southern Hemisphere, where the "Night Sun" was not visible in the sky. For additional details see Chapter 7 ("Heaven & Hell") of my book *Planet X Nibiru: Slow-Motion Doomsday.* RS]

Here we can only indicate briefly the many cosmic interpretations which the myth of Osiris has been given. As a vegetation spirit that dies and is ceaselessly reborn, Osiris represents the corn, the vine and trees. He is also the Nile which rises and falls each year; the light of the sun which vanishes in the shadows every evening to reappear more brilliantly at dawn. The struggle between the two brothers is the war between the desert and the fertile earth, between the drying wind and vegetation, aridity and fecundity, darkness and light. It was as god of the dead that Osiris enjoyed his greatest popularity; for he gave his devotees the hope of an eternally happy life in another world ruled over by a just and good king.

[COMMENT: And perhaps it is from the legend of Osiris that we derived our Judeo-Christian idea of a "Heaven" or "Afterlife" ruled by a benevolent "Lord God On High", sitting on a throne of gold, condemning all the "sinners" to a "Hell" which was governed by his rival brother "Satan", who like the Poseidon of Greece is often depicted carrying a "trident" or "pitchfork". The "god" Shiva of India is also shown with a trident and is undoubtedly the Indian equivalent of Enki / Set / Poseidon / Satan. RS]

Osiris is represented sometimes standing, sometimes seated on his throne, as a man tightly swathed in white mummy wrappings. His greenish face is surmounted by the high white mitre flanked by two ostrich feathers which is called 'Atef', the crown of Upper Egypt. Around his neck he wears a kind of cravat. His two hands, freed from the winding sheet, are folded across his breast and hold the whip and the sceptre in the form of a crook, emblems of supreme power.

Isis, indeed, was a potent magician and even the gods were not immune from her sorcery. It was told how, when she was still only a simple woman in the service of Ra, she persuaded the great god to confide to her his secret name. She had taken advantage of the fact that the sun-god was now an old man with shaking head and dribbling mouth. With earth moistened with his divine spittle she fashioned a venomous snake which bit Ra cruelly. Ra was incapable of curing himself of a wound whose origin he did not understand, and had recourse to the spells of Isis. But Isis refused to conjure away the poison until Ra, overcome with pain and hiding himself from the other gods, consented to reveal his true name, which he caused to pass directly from his own bosom into that of Isis.

[COMMENT: These "secret names" probably have something to do with all the peculiar "72 Names Of God" or "72 Goetian Spirit Sigils". But this is totally speculative. However, it is certainly *not* "speculative" that all these groups of 72 entities are identical. RS]

Isis is normally represented as a woman who bears on her head a throne, the ideogram of her name. Occasionally, but later, her head-dress is a disk, set between cow's horns, occasionally flanked with two feathers. Finally we sometimes find her represented with a cow's head set on a human body. These horns and the cow's head merely prove that Isis was by then identified with Hathor; but Plutarch, though he says he does not believe it, gives us another explanation. Isis, he tells us, wished to intervene on behalf of Set who, though her husband's murderer, was also her own brother. She tried to cheat her son Horus of his just vengeance; but Horus turned in rage against his mother and cut off her head. Thoth then transformed it by enchantment and gave her the head of a cow.

[COMMENT: The Egyptian Thoth is the equivalent of Nibiruan Prince Nannar, the Greek Hermes and the Roman Mercury, "god of magic" in whose worship or "Hermetic School" we later find Apollonius of Tyana, amongst others. RS]

As we have already said, Set, in Osirian myth, figures as the eternal adversary, a personification of the arid desert, of drought and darkness in opposition to the fertile earth, life-bringing water and light. All that is creation and blessing comes from Osiris; all that is destruction and perversity arises from Set. In primitive times, however, the evil character of Set was not so accentuated. The old pyramid texts make him not the brother of Osiris, but the brother of Horus the Elder, and speak of terrible struggles between them which were terminated by the judgment of the gods, who proclaimed Horus the victor and banished Set to the desert. It was only later, when the Osirian myth had grown and when the two Horuses had become confused, that Set was made the uncle of Horus and the eternal enemy of Osiris.

Set is represented as having the features of a fantastic beast with a thin, curved snout, straight, square-cut ears and a stiff forked tail. This creature cannot with certainly be identified and is commonly called the 'Typhonian animal'. Sometimes Set is depicted as a man with the head of this strange quadruped.

[COMMENT: And "Satan" is also often represented with a forked tail. Linguistically the two words "Set" and "Satan", as well as the word "Poseidon", are closely related. RS]

Horus is the Latin rendering of the Greek Horos and the Egyptian Hor. He was a solar god constantly identified with Apollo and represented by a falcon or a falconheaded god. Under the name Hor — which in Egyptian sounds like a word meaning 'sky' — the Egyptians referred to the falcon which they saw soaring high above their heads, and many thought of the sky as a divine falcon whose two eyes were the sun and the moon. The worshippers of this bird must have been numerous and powerful; for it was carried as a totem on prehistoric standards and from the earliest times was considered the pre-eminent divine being. The hieroglyph which represents the idea of 'god' was a falcon on its perch.

[**COMMENT:** Horus was not Apollo. Horus was Nibiruan Prince Ishkur, the Sumerian Adad, the Greek Ares and the Roman Mars. He was also the "god" known from the Levant to India as "Mithras", whose cult following found its way to Rome by way of the Roman occupying armies in Asia. Roman Emperor Theodosius, who in the 380s CE officially implemented the Catholicism proclaimed by the Council of Nicaea in 325 CE, decreed that the worship of Mithras would be punishable by death. The birthdate of Mithras, December 25, was usurped by the Catholic Church as the birth of "The Jesus Christ" in order to dilute its significance within Mithraism. The tactic seems to have worked. "Sapiens" in general doesn't remember a thing about Mithras. Next Christmas Day, try wishing someone you love a "Merry Mithrasmas". They'll look at you as if you have lost your mind. And for the record, the Greek Apollo was the equivalent of Nibiruan Prince Utu, the Sumerian Shamash and the Roman Helios. Utu/Apollo was the Commander of the Nibiruan Spaceport in the Sinai, described by Sitchin. See Chapters 11-12. And the "bird" or "winged disk" of Horus atop the "totem pole" undoubtedly refers to Planet X Nibiru atop The Cosmic Tree. See the Illustrations which accompany my previous book. RS]

*

All of these ancient stories have been embellished over the centuries, distorting their original meanings. But if Sirius A equals Isis, then what were Osiris and Horus? Let me propose a revolutionary new idea which I shall attempt to support in the chapters that follow.

Originally, the Sirius System contained two large stars, Sirius A and Sirius B. Both the white and the red stars were visible from Earth as an adjacent pair of stars, which were named Osiris and Isis. If Sirius A equals Isis, then it would follow that Sirius B would equal Osiris, the red-giant, the larger of the two. Circling these stars was a large planet, the "offspring" of these stars: Horus, or "Sorghum-Female" (see Chapter 4). This planet was inhabited by "Saurians" who possessed an advanced technology, including space travel.

As Osiris approached the moment of implosion, these Saurians realized that their own planet might be destroyed or at least damaged by this momentous event. So they began searching the nearby region of the

Galaxy for a potential place of refuge and inevitably arrived to explore the Earth. Then finally Osiris imploded, sending shockwaves throughout the entire Sirius System. This event is reflected in Egyptian mythology as the dismemberment of Osiris and the scattering of his body-parts; and as Sirius B shrank in size to subsequently evolve into a neutron-dwarf, it faded from view by those on Earth, leading to the assumption that Osiris had hidden himself in an "underworld".

The Saurians had sought to protect themselves from this shockwave of radiation and magnetism by constructing a gigantic golden heat-shield to contain their planet's atmosphere. But their planet Horus, or Sorghum-Female, was ejected from that system by the sheer force of the implosion and was hurtled into space in the direction of the Earth. How much time passed, we cannot know; but Horus was subsequently captured by the gravity of our Sun and became a new member, a tenth planet, of this Solar System; and as such, it began to influence the gravitational movements of Uranus and Neptune.

During this process of implosion, ejection, transit and capture, their atmospheric heat-shield suffered damage. How much damage, we cannot estimate; but apparently it was not immediately life-threatening to the inhabitants. Once established in our Solar System, these Saurians began to mine gold for use in repairing their protective shield. And this necessity eventually culminated in the creation of Cro-Magnon Sapiens, that is, us modern humans, their primitive gold-mining savage-men. See Chapter 11.

Horus, or Sorghum-Female, became known as a "Planet of the Crossing" because it had "crossed over" from the Sirius System to our own Solar System. In the early Sumerian language, which was most likely derived from the Nibiruan language (much as Italian evolved from Latin), the word for "crossing" was "nibiru". According to the *Enuma Elish*, "Nibiru shall hold the crossings of heaven and earth. ... Let 'Crossing' be his name. ..."

Over time, Planet X Nibiru stabilized into a most peculiar orbit, which includes the periodic "tethering" by means of an electromagnetic beam from the South Pole of that planet to the North Pole of our own, once every Nibiruan orbital period of 3,600 Earth Years, or 1 Nibiruan *shar*. For one-fourth of this period, 900 years, they sit as a Winged Disk atop a Cosmic Tree — "a golden pole with a golden cage on top", as it was described in an old Siberian legend from the Irtysh River Valley.

Horus himself has also been identified with The Winged Disk, as we read in a short quote from this reliable website.

http://thunderbolts.info/tpod/2005/arch05/050519wingdisk.htm

"The Winged Disk", Contributed by Rens van der Sluijs
19 May 2005
Thunderbolts Of The Gods by David Talbott & Wallace Thornhill
(Mikamar Publishing, Portland, Oregon, 2004)

"Hardly any emblem from ancient Near-Eastern art and architecture is more widely known than the winged disc. The winged disc is essentially a circle to which two outstretched wings are attached. It is a religious symbol representing a deity the ancients regarded as the supreme god of the sky. The symbol is found in Egypt, Mesopotamia, Anatolia, and northern Syria and comes in a number of local variations; the Egyptians often attached two coiling *uraeus* — serpents — to its sides or placed two little horns on top of the disc, whereas the Assyrians typically portrayed the disc as a ring containing the upper part of the god in human form. ...

"Why exactly was this deity represented as a winged disc? The short answer is that this particular god was envisioned as a giant bird. Egyptologists have long acknowledged that the Egyptian form of the image could represent Horus the falcon or hawk god. Most specialists would also concur that the winged disc signified some astral god, whose presence in the heavens was symbolized by the flight of a bird. In Egypt, the winged disc was certainly regarded as a symbol of the sun in the first millennium BCE, but there are indications that the myth of this 'sun god' had not originally been based on the behavior of the sun. The archaic sun god of the earlier strata of myth appears to have been some other sun-like object observed in the sky that had afterwards vanished."

Indeed "some other sun-like object observed in the sky that had afterwards vanished"! For 900 years, Planet X Nibiru is our "Second Sun" or "Night Sun", only to "vanish" as they depart for 2,700 years, returning to their orbital aphelion, the exact location of which we do not yet know. Either it could be at the farthest reaches of our Solar System, in what is called the Oort Cloud, or it could even be back at the Sirius System itself, where they may also remain stationary for another period of 900 years. In the latter case, then this "Planet of the Crossing" would "cross over" between our system and the Sirius System twice every *shar.*

In conclusion, let me state here that *Osiris, Isis & Planet X* is intended to be read as a companion volume, perhaps even as a prequel, to my earlier *Planet X Nibiru: Slow-Motion Doomsday* (and parts of it might also be considered as a sequel to my *Apollonius Of Tyana & The Shroud Of Turin*). Planet X Nibiru is a most complex subject, and it is quite impossible to touch upon every detail in a single volume. After all, Zecharia Sitchin alone has written ten books about it. A complete list of his and other books can be found at the end of Chapter 11, which concerns the more recent history of Planet X Nibiru.

Chapter 2

Cosmos

*

COSMOS: A Sketch Of A Physical Description Of The Universe
By Alexander Von Humboldt
Translated From The German By E.C. Otté
(London, 1851)

[COMMENT: In my library is a complete 5-volume first edition of Von Humboldt's pioneering research. I inherited this set of books from my Great-Uncle Whit, and they are still in amazingly good condition. *Cosmos* was written well before any of the subsequent books to be analyzed, and it is obvious that "controversy" has always surrounded Sirius. The following brief excerpt illustrates this fact quite well. RS]

Volume III, Pages 175-182

A difference of colour in the proper light of the fixed stars, as well as in the reflected light of the planets, was recognized at a very early period; but our knowledge of this remarkable phenomenon has been greatly extended by the aid of telescopic vision, more especially since attention has been so especially directed to the double stars. We do not here allude to the change of colour which, as already observed, accompanied scintillation even in the whitest stars, and still less to the transient and generally red colour exhibited by stellar light near the horizon (a phenomenon owing to the character of the atmospheric medium through which we see it), but to the white or coloured stellar light radiated from each cosmical body, in consequence of its peculiar luminous process, and the different constitution of its surface. The Greek astronomers were acquainted with red stars only, while modern science has discovered, by the aid of the telescope, in the radiant fields of the starry heaven, as in the blossoms of the phanerogamia, and in the metallic oxides, almost all the gradations of the prismatic spectrum between the extremes of refrangibility of the red and violet ray.

Ptolemy [lived *circa* 110-180 CE, RS] enumerates in his catalogue of the fixed stars six (*upokirroi*) *fiery red* stars, *viz.*: Arcturus, Aldebaran, Pollux, Antares, *alpha* Orionis (in the right shoulder), and Sirius. Cleomedes even compared Antares in Scorpio with the fiery red Mars, which is called both *purros* and *puroeides*.

[FOOTNOTE: The expression *upokirros*, which Ptolemy employs indiscriminately to designate the six stars named in his catalogue, implies a slightly marked transition from *fiery-yellow* to *fiery-red*; it therefore refers, strictly speaking, to a *fiery-reddish* colour. He seems to attach the general predicate *zanthos*, *fiery-yellow*, to all the other fixed stars. (*Almag.*, viii. 3 ed. Hamla, tom. ii. p.94.) *Kirros* is, according to Galen {lived *circa* 130-201 CE, RS} (*Meth. med.* 12), a pale fiery-red inclining to yellow. Gellius compares the word with *melinus*, which, according to Servius, has the same meaning as "gilvus" and "fulvus". As Sirius is said by Seneca {lived *circa* 5 BCE - 65 CE, RS} (*Nat. Quaest.*, i. 1) to be *redder than Mars*, and belongs to the stars called in the Almagest *upokirroi*, there can be no doubt that the word implies the predominance or, at all events, a certain proportion of red rays. The assertion that the affix *poikilos*, which Aratus {lived *circa* 300-240 BCE, RS}, V. 327, attaches to Sirius, has been *translated* by Cicero as "rutilus", is erroneous. Cicero {lived 106-43 BCE, RS} says, indeed, v. 348: "Namque pedes subter rutilo cum lumine claret, Fervidus ille Canis stellarum luce refulgens"; but "rutilo cum lunine" is not a *translation* of *poikilos*, but the mere addition of a free translation. (From letters addressed to me by Professor Franz.) "If," as Arago observes (*Annuaire*, 1842, p. 351), "the Roman orator, in using the term *rutilus*, purposely departs from the strict rendering of the Greek of Aratus, we must suppose that he recognized the reddish character of the light of Sirius."]

Of the six above named stars, five still retain a red or reddish light. Pollux is still indicated as a reddish, but Castor as a greenish star. Sirius therefore affords the only example of an historically proved change of colour, for it has at present a perfectly white light. A great physical revolution must therefore have occurred at the surface or in the photosphere of this fixed star (or *remote sun*, as Aristarchus of Samos called the fixed stars), before the process could have been disturbed by means of which the less refrangible red rays had obtained the preponderance, through the abstraction or absorption of other complementary rays, either in the photosphere of the star itself, or in the moving cosmical clouds by which it is surrounded.

[FOOTNOTE: Sir John Herschel, in the *Edinb. Review*, vol. 87, 1848, p. 189, and in Schum. *Astr. Nachr.*, 1839, no. 372: "It seems much more likely that in Sirius a red colour should be the effect of a medium interfered, than that in the short space of 2000 years so vast a body should have actually undergone such a material change in its physical constitution. It may be supposed owing to the existence of some sort of *cosmical cloudiness*, subject to internal movements, depending on causes of which we are ignorant." (Compare Arago in the *Annuaire pour 1842*, pp. 350-353.)]

[COMMENT: There would have been absolutely no reason for Ptolemy, Seneca and other respected ancient scholars deliberately to distort the information that Sirius was "redder than Mars". Thus, we are faced with two possibilities: (1) after the red-giant Osiris, or Sirius B, imploded into a neutron-dwarf, red light was still visible into the time of the First Century CE; or (2) these historians were simply reporting this red color as having existed at some time in the past. It is indeed regrettable that no "epoch of the disappearance of the red colour of Sirius" has been preserved, as Von Humboldt next laments. RS]

It is to be wished that the epoch of the disappearance of the red colour of Sirius had been recorded by a definite reference to the time, as this subject has excited a vivid interest in the minds of astronomers since the great advance made in modern optics. At the time of Tycho Brahe the light of Sirius was undoubtedly already white, for when the new star which appeared in Cassiopeia, in 1572, was observed in the month of March, 1573, to change from its previous dazzling white colour to a reddish hue, and again became white in January, 1574, the red appearance of the star was compared to the colour of Mars and Aldebaran, but not to that of Sirius. M. Sédillot, or other philologists conversant with Arabic and Persian astronomy, may perhaps some day succeed in discovering evidence of the earlier colour of Sirius, in the periods intervening from El-Batani (Albategnius) and El-Fergani (Alfraganus) to Abdur-rahman Sufi and Ebn-

Junis (that is, from 880 to 1007), and from Ebn-Junis to Nassir-Eddin and Ulugh-Beg (from 1007 to 1437).

[COMMENT: It is most intriguing that this star in the North Polar Constellation of Cassiopeia seemed to change from white to red and back to white within the space of only two years. Certainly Von Humboldt was not misrepresenting this information. Perhaps these changes can occur more quickly than we have previously considered. Maybe such stellar evolution doesn't take "millions" of years to occur, as our modern astronomers would have us to believe.

[On 31 December 1990 the following "Science Note" by Jennifer Nagorka appeared in *The Dallas Morning News*: "SEEING RED — A new reading of an ancient Chinese text confirms that the star Sirius, now a bright white object in the constellation Canis Major, has shifted colors. Scientists have been debating whether Sirius changed colors for several years, write French astronomers C. Gry and J.M. Bonnet-Bidaud in the journal *Nature*. Some Greek, Roman, Babylonian and medieval writings suggest that the star glowed reddish about 2,000 years ago. But researchers could not explain how the star could change color in a few centuries — a relatively short time on the galactic scale. A Chinese text, the Han dynasty history book, supports the idea that Sirius once blushed red, the French scientists contend. The text, written by astronomer Sima Quin, indicates that Chinese astronomers noted Sirius' reddening tint more than 2,000 years ago. The astronomers believe they can explain why Sirius shifted color — Sirius may have passed behind an interstellar cloud. Like a sheet of tinted plastic, the cloud would have altered the star's color, making it appear red from Earth, the scientists wrote." No "interstellar cloud" was involved in this color-changing process! RS]

El-Fergani (properly Mohammed Ebn-Kethir El-Fergani), who conducted astronomical observations in the middle of the tenth century at Rakka (Aracte) on the Euphrates, indicates as red stars (*stellae ruffie* of the old Latin translation of 1590) Aldebaran, and, singularly enough, Capella, which is now yellow and has scarcely a tinge of red, but he does not mention Sirius. If at this period Sirius had been no longer red, it would certainly be a striking fact that El-Fergani, who invariably follows Ptolemy, should not here indicate the change of colour in so celebrated a star. Negative proofs are however not often conclusive, and indeed El-Fergani makes no reference in the same passage to the colour of Betelguex (*alpha* Orionis), which is now red, as it was in the age of Ptolemy.

It has long been acknowledged that of all the brightest luminous fixed stars of heaven, Sirius takes the first and most important place, no less in a chronological point of view, than through its historical association with the earliest development of human civilization in the valley of the Nile. The era of Sothis — the heliacal rising of Sothis (Sirius) — on which Biot has written an admirable treatise, indicates, according to the most recent investigations of Lepsius, the complete arrangements of the Egyptian calendar into those ancient epochs, including nearly 3300 years before our era, "when not only the summer solstice, and consequently the beginning of the rise of the Nile, but also the heliacal rising of Sothis, fell on the day of the first water-month (or the first Pachon)." I will collect in a note the most recent, and hitherto unpublished, etymological researches on Sothis or Sirius from the Coptic, Zend, Sanscrit, and Greek, which may perhaps be acceptable to those who, from love for the history of astronomy, seek in languages and their affinities, monuments of the earlier conditions of knowledge.

[FOOTNOTE: I have extracted the following observations from letters addressed to me by Professor Lepsius (February, 1850). "The Egyptian name of Sirius is *Sothis*, designated as a female star; hence, *E Sothis* is identified in Greek with the goddess *Sote* (more frequently *Sit* in hieroglyphics), and in the temple of the great Ramses at Thebes with Isis-Sothis (Lepsius, *Chron. der Aegypter*, bd. i. s. 119, 136). The signification of the root is found in Coptic, and is allied with a numerous family of words, the members of which, although they apparently differ very widely from each other, admit of being arranged somewhat in the following order. By the threefold transference of the verbal signification, we obtain from the original meaning, to throw out — *projicere (sagittum, telum)* — first, *seminare, to sow*; next, *extendere, to extend or spread* (as spun threads); and lastly, what is here most important, *to radiate light* and *to shine* (as stars and fire). From this series of ideas we may deduce the names of the divinities, *Satis* (the female archer); *Sothis*, the radiating, and *Seth*, the fiery. We may also hieroglyphically explain *sit* or *seti*, the arrows as well as the ray; *seta*, to spin; *setu*, scattered seeds. *Sothis* is especially the *brightly radiating*, the star regulating the seasons of the year and periods of time. The small triangle, always represented yellow, which is a symbolical sign for Sothis, is used to designate the radiating sun when arranged in numerous triple rows issuing in a downward direction from the sun's disk. *Seth* is the fiery scorching god, in contradistinction to the warming, fructifying water of the Nile, the goddess *Satis* who inundates the soil. She is also the goddess of the cataracts, because the overflowing of the Nile began with the appearance of Sothis in the heavens at the summer solstice. In Vettius Valens the star itself is called *Seth* instead of Sothis; but neither the name nor the subject admits of our identifying *Thoth* with Seth or Sothis, as Ideler has done. (*Handbuch der Chronologie*, bd. i. s. 126)." (Lepsius, bd. i. s. 136.)

[COMMENT: Note the idea that Sothis was connected with the "throwing out" or "sowing" of something. This could likely refer to the fact that Planet X Nibiru was "thrown out" of the Sirius System, to be later captured by our Sun. And the fact that there is a "triangle" of objects could imply Osiris (Sirius B), Isis (Sirius A) and Horus (Planet X Nibiru). RS]

[Footnote Continued: I will close these observations taken from the early Egyptian periods with some Hellenic, Zend, and Sanscrit etymologies: "*Seir*, the sun," says Professor Franz, "is an old root, differing only in pronunciation from *ther, theros, heat, summer,* in which we meet with the same change in the vowel sound as in *teiros* and *teros* or *teras*. The correctness of these assigned relations of the radicals *seir* and *ther, theros*, is proved not only by the employment of *thereitatos* in Aratus, v. 149 (Ideler, *Sternnamen*, s. 241), but also by the later use of the forms *seiros, seirios,* and *seirinos, hot, burning*, derived from *seir.* It is worthy of notice that *seira* or *theirina imatia* is used the same as *therina imatia, light summer clothing.* The form *seirios* seems, however, to have had a wider application; for it constitutes the ordinary term appended to all stars influencing the summer heat: hence, according to the version of the poet, Archilochus, the sun was *seirios aster*, while Ibycus calls the stars generally *seirios katauanei uzus ellampon.*

["According to Hesychius and Suidas, *Seirios* does indeed signify both the sun and the Dog-star; but I fully coincide with M. Martin, the new editor of Theon of Smyrna, in believing that the passage of Hesiod (*Opera et Dies*, v. 417), refers to the sun, as maintained by Tzetzes and Proclus, and not to the Dog-star. From the adjective *seirios*, which has established itself as the '*epitheton perpetuum*' of the Dog-star, we derive the verb *seirion*, which may be translated 'to sparkle'. Aratus, v. 331, says of Sirius, *ozea seiriaei*, 'it sparkles strongly'. When standing alone, the word *Seiren*, the Siren, has a totally different etymology; and your conjecture, that it has merely an accidental similarity of sound with the brightly shining star Sirius, is perfectly well-founded. The opinion of those who, according to Theon Smyrnaeus (*Liber de Astronomia*, 1850, p. 202), derive *Seiren* from *seiriazein* (a moreover unaccredited form of *seirian*) is likewise entirely erroneous. While the motion of heat and light is implied by the expression *seirios*, the radical of the word *Seiren* represents the flowing tones of this phenomenon of nature. It appears to me probable, that *Seiren* is connected with *eirein* (Plato, *Cratyl.* 398 D, *po gar eirein legein esti*), in which the original sharp aspiration passed into a hissing sound." (From letters of Prof. Franz to me, January 1850.)

[The Greek *Seir*, the sun, easily admits, according to Bopp, "of being associated with the Sanscrit word *svar*, which does not indeed signify the sun itself, but the heavens (as something shining). The ordinary Sanscrit denomination for the sun is *surya*, a contraction of *svarya*, which is not used. The root *svar* signifies in general *to shine.* The Zend designation for the sun is *hvare*, with the *h* instead of the *s*. The Greek *ther, theros* and *thermos* comes from the Sanscrit word *gharma* (Nom. *gharmas*), warmth heat."

[The acute editor of the Rigveda, Max Müller, observes, that "the special Indian astronomical name of the Dog-star, *Lubdhaka*, which signifies *a hunter*, when considered in reference to the neighbouring constellation Orion, seems to indicate an ancient Arian community of ideas regarding these groups of stars." He is moreover principally inclined "to derive *Seirios* from the root *sri*, to go, to wander; so that the sun and the brightest of the stars, Sirius, were originally called wandering stars." (Compare also Pott, *Etymologische Forschungen*, 1833, s. 130.)]

[COMMENT: This reference to a "wandering star" connected with Sirius/Sothis might imply the "wandering" nature of its castaway Planet X Nibiru. And as Von Humboldt indicates, some of these words denote another sort of "Sun" than our regular daytime Sun. As the Finnish creation epic *Kalevala* relates, "When my Sun and Moon are absent, in the air no joy remaineth." RS]

Chapter 3

The Dawn Of Astronomy

*

The Dawn Of Astronomy
A Study Of The Temple Worship &
Mythology Of The Ancient Egyptians
By J. Norman Lockyer (1894; MIT, Cambridge, 1973)

Preface to 1973 Edition (Excerpt)
By Giorgio de Santillana (*Hamlet's Mill*)

[**COMMENT:** For additional information regarding the milestone book *Hamlet's Mill, An Essay On Myth & The Frame Of Time* by Giorgio de Santillana and Hertha von Dechend (Boston, 1969), the reader is referred to my earlier book *Planet X Nibiru: Slow-Motion Doomsday.* RS]

As this important but almost forgotten book is brought back into print after seventy years, it would be fair first to remind the reader of the author's credentials. Sir Norman Lockyer (1836-1920) was one of the major English astronomers of his time. Born in Rugby, he completed his education on the Continent of Europe, and came to astronomy by way of private study and a clerkship in the War Office. In 1870 he was appointed secretary to the Duke of Devonshire's Royal Commission on science; a few years later, on the foundation of the Royal College of Science in London, he became director of the solar physics observatory and professor of astronomical physics. From 1866, he had been a pioneer in sun and star spectroscopy. He inaugurated *Nature* in 1869 [presumably the scientific journal still published today, RS] and edited it until his death. His interests went far afield, as we shall see. Always something of a maverick, he suddenly abandoned the Royal College and removed the instruments to his estate at Sidmouth, where he founded his own observatory. It is now the "Lockyer's Observatory" and contains a mass of his papers still unpublished.

Lockyer's fame is solidly based on his study of the sun. In 1868 he described the flares and prominences of the sun as located in a layer he called the chromosphere, and applied the Doppler principle to its movements. In 1868 Lockyer and Janssen, working independently, discovered a spectroscopic method whereby the solar prominences could be studied in daylight, whereas previously they were observable only during a total eclipse. To commemorate this discovery, a medal bearing the names of both astronomers was struck by the French government in 1872. Lockyer received the Rumford medal in 1874, and was vice-president of the Royal Society in 1892-1893. Among his most important discoveries is that of a new element in the solar atmosphere that he called "helium" and that was found later among the rare gases on earth.

Pages 246-249

But now comes in a most interesting and important point. If observations of the sun at solstice or equinox had been alone made use of, the true length of the year would have been determined in a few years. But the next scene in Egyptian history shows us that the true length of the year was not determined, but only an approximation to it.

How was this? The astronomical answer is very simple.

I have already referred to the common practice of all ancient peoples that we know of to make sacrifices at dawn, and have shown how, in order to do this, they took their time from a star rising before the sun.

[COMMENT: Apollonius of Tyana, for one, said prayers and made sacrifices at dawn. RS]

An observation of this so-called "heliacal rising" of a star [in particular, Sirius in early July, RS] — if the star were properly chosen — would give them the interval necessary for their preparations before the sun itself appeared; and, as the highest festival of all was that of New Year's Day, it was especially important that the work should be well done then.

Now, if the stars had no precessional movement, the sun and stars, after each interval of a true year, would be in exactly the same position; but in consequence of the stars having the precessional movement to which I have before referred, the star so observed and the sun will *not* be in exactly the same position after the interval of a true year. On this account, then, the difference of time between the heliacal risings will not represent the length of a true year. But, further, the heliacal rising of a star will not take place on the same day for the whole of Egypt, the difference between Thebes and Memphis, depending upon their latitudes, amounting to about four days; and, further still, the almost constant mists in the mornings in the Nile valley prevent accurate observations of the moment of rising.

Still, as a matter of fact, the Egyptians defined their new year by the rising of a star, and the length of it by the interval separating two heliacal risings. Such a year could not be accurate; and again, as a matter of fact, their correction was not accurate, for the year was defined now as consisting of 365 days. It seems clear from this that the correction was made before the solar temples were in use.

In any case the year of 360 days had naturally to give way, and it ultimately did so, in favour of one of 365. The precise date of the change is, as we have seen, not known. The five days were added as epacts or epagomena; the original months were not altered, but a "little month" of five days was interpolated at the end of the year between Mesori of one year and Thoth of the next, as already stated.

[**COMMENT:** See Chapter 17 ("Introduction To Galactic Mathematics"). RS]

When the year of 365 days was established, it was evidently imagined that finality had been reached; and, mindful of the confusion which, as we have shown, must have resulted from the attempt to keep up a year of 360 days by intercalations, each Egyptian king, on his accession to the throne, bound himself by oath before the priest of Isis, in the temple of Ptah at Memphis, not to intercalate either days or months, but to retain the year of 365 days as established by the Antiqui. The text of the Latin translation preserved by Nigidius Figulus cannot be accurately restored; only thus much can be seen with certainty.

To retain this year of 365 days, then, became the first law for the king, and, indeed, the Pharaohs thenceforth throughout the whole course of Egyptian history adhered to it, in spite of their being subsequently convinced, as we shall see, of its inadequacy. It was a Macedonian king [Alexander The Great, RS] who later made an attempt to replace it by a better one.

We may reckon upon the conservatism of the priests of the temples retaining the tradition of the old rejected year in every case. Thus even at Philae in late times, in the temple of Osiris, there wee 360 bowls for sacrifice, which were filled daily with milk by a specified rotation of priests. At Acanthus there was a perforated cask into which one of the 360 priests poured water from the Nile daily. Indeed, these temple ceremonials are an evidence of their antiquity, and the further we put back the change from the 360 to 365 days, the greater the antiquity we must assign to them, and therefore to the temples themselves.

[COMMENT: In *Worlds In Collision* Dr. Immanuel Velikovsky collected copious evidence that prior to the "cataclysm" of the 8th century BCE the year consisted of exactly 360 days. Thus, we might conclude that when Planet X Nibiru is tethered "beyond the north" as our Cosmic Tree, its presence results in an Earth Year of 360 days; and therein lies this ancient confusion as to the precise length of the year. RS]

During three thousand years of Egyptian history the beginning of the year was marked by the rising of Sirius, which rising took place nearly coincidently with the rise of the Nile and the Summer Solstice. I have insisted upon the regularity of the rise of the Nile affording the ancient Egyptians, so soon as this regularity had been established, a moderately good way of determining the length of the year, but we have seen they did not so employ it.

It is also clear that so soon as the greatest northing and southing of the sun rising or setting at the solstices had been recognised, and the intervals between them in days had been counted, a still more accurate way would be open to them. The solstice *must* have occurred with greater regularity than the rise of the river, so that as accuracy of definition became more necessary the solstice would be preferred. The solstice was common to all Egypt; the commencement of the inundation was later as the place of observation was nearer the mouth of the river. This means they also did not employ [it], at all events in the first instance. Of the three coincident, or nearly coincident, phenomena, the rise of the Nile, the Summer Solstice, and the rising of Sirius, they at first chose the last.

According to Biot the heliacal rising of Sirius *at the solstice* took place on July 20 (Julian) in the year 3285 B.C.; and according to Oppolzer it took place on July 18 (Julian), in the year 3000 B.C.

[**COMMENT:** Then there follows a lengthy description and discussion of the "Egyptian Sothis Period" of 1,461 years. In Chapter 17, I have analyzed the Sothis Period and will not go into that here. As Lockyer has stated, the flooding of the Nile would only have served to *approximate* the New Year's Day. This flooding is entirely dependent upon the commencement of the rainy season in Ethiopia, source of the Blue Nile River; and this rainy season does not start on exactly the same day of each year, nor is the daily rainfall consistent from year to year. RS]

Pages 287-299

We gather, then, that the wonderful old-world myth of Isis and Osiris is astronomical from beginning to end, although Osiris in this case is not the sun, but the moon. But I have not yet finished with the mummy form; the waning moon is also Osiris. It is supposed to be dying from the time of full moon to new moon. The Egyptians in their mythology were nothing if not consistent; the moon was called Osiris from the moment it began to wane, as the sun was Osiris so soon as it began to set. A constellation paling at sunrise was also Osiris.

I have previously noted the symbolism of Sirius-Hathor as a cow in a boat associated with the constellation of Orion. There is a point connected with this which I did not then refer to, but which is of extreme importance for a complete discussion of the question now occupying us. We get associated with the cow in the boat, Orion (Sah) as Horus, but in other inscriptions we get Orion as a mummy — that is to say, in the course of Egyptian history the same constellation is symbolised as a rising sun at one time and a setting sun at another. Now, that must have been so if the Egyptian mythology were consistent and rested on an astronomical basis, because Sah rose in the dawn in the one case and faded at dawn in the other. From the table giving a generalised statement with regard to Osiris, similar to that we have already considered for Isis, it looks as if the mythology connected with Osiris is simply the mythology connected with any celestial body becoming invisible. We have the sun setting, the moon waning, a planet setting, stars setting, constellations fading at dawn. We see, therefore, that the Egyptian mythology was absolutely and completely consistent with the astronomical conditions by which they were surrounded; that, although it is wonderfully poetical, in no case is the poetry allowed to interfere with the strictest and most accurate reference to the astronomical phenomena which had to be dealt with.

The argument, then, for the use of Isis as a generic name is greatly strengthened by the similar way in which the term Osiris, which is acknowledged to be a generic name, is employed.

[**COMMENT:** If Osiris is "simply the mythology connected with any celestial body becoming invisible", then this would also encompass the sudden invisibility of Osiris, as Sirius B, following the implosion of the previous red-giant into its presently invisible, to the naked eye, neutron-dwarf. RS]

Chapter 4

The Sirius Mystery

*

The Sirius Mystery By Robert Kyle Grenville Temple
(St. Martin's Press, New York, 1976)

[COMMENT: *The Sirius Mystery* is considered to be *the* definitive book on the connection of the Dogon people with Sirius B, or *Digitaria*. I have included a number of excerpts, and I refer the reader to the whole book for additional information. RS]

Page 3

The most secret traditions of the Dogon all concern the star which the Dogon call after the tiniest seed known to them, the botanical name for which is *Digitaria*, and which is thus used in the [aforementioned] article [by Marcel Griaule and Germaine Dieterlen, RS] as the name of the star instead of the actual Dogon name, *po*. However, even in this [anthropological] article [published in the book *African Worlds*, edited by Daryll Forde, Oxford, 1954, RS] which deals exclusively with this subject, Griaule and Dieterlen only mention the actual existence of a star which really exists and does what the Dogon say Digitaria does, in a passing footnote and in this brief remark: 'The question has not been solved, nor even asked, of how men with no instruments at their disposal could know the movements and certain characteristics of stars which are scarcely visible.' But even in saying this, the anthropologists were indicating their own lack of astronomical expertise, for the star, Sirius B which revolves around Sirius [A], is by no means 'scarcely visible'. It is *totally invisible* and was only discovered in the last century with the use of the telescope. As Arthur [C.] Clarke put it to me in a letter of 17 July 1968, after he had suggested he would check the facts: 'By the way, Sirius B is about magnitude 8 — quite invisible even if Sirius A didn't completely obliterate it.' Only in 1970 was a photograph of Sirius B successfully taken by Irving Lindenblad of the U.S. Naval Observatory; this photograph is reproduced in Plate 1.

In the article which I had obtained from the Royal Anthropological Institute, Griaule and Dieterlen recorded that the Dogon said the star Digitaria revolved around Sirius [A] every fifty years. It didn't take me long to research Sirius B and discover that its orbital period around Sirius [A] was indeed fifty years. I now knew that I was really on to something. And from that moment I have been immersed in trying to get to the bottom of the mystery.

Pages 15-16

There is no need for me to continue marshalling quotations from distinguished scientists and astronomers in support of the possibility of intelligent life in space, as the situation is by now obvious. The odds against intelligent life occurring fairly frequently within our galaxy are impossible ones. Since this is established, we are faced with yet another factor: in our own history, technological development has been rapid within a short space of time. When civilizations all over the universe reach 'take-off points', they have a technological explosion. It is familiar to older members of our species today that when they were young there were no airplanes, automobiles, rockets, satellites, electricity, radio, or atom bombs. People were dying of diseases which today we do not take seriously, no one with a toothache could obtain modern dental treatment, the concept of elementary hygiene was a novelty. I am not reciting all these wonders merely as a ritual incantation to our new god of progress. The point to be grasped is the sudden combustible nature of progress of this kind. In the lifetime of a single person all this can come about.

'Take-off point' is probably a universal phenomenon. Intelligent societies all over the universe will probably have experienced it, or are due to. Now the lifetime of a single person is of no consequence on the great universal time scale for the development of civilizations, not to mention the formation of planets. Therefore any society in advance of our own is certain to be very much in advance of ours. Once intelligent societies reach take-off point, they rush so quickly upward in technological competence that a comparison between them and non-technological societies is almost absurd. It would be foolish for us to suppose that any society more advanced than ours would be just a few years ahead of us. It would more likely be just a few tens of thousands of years ahead of us. And the technology and nature of such a society are beyond our abilities to imagine. The intelligent societies existing in the universe, then, are going to be of two kinds: less advanced than ourselves, 'primitive'; and fantastically more advanced than ourselves, 'magical'. To be at the point where we are now, at the watershed between 'primitive' and 'magical', is such a rare event in the universal history, that we may be the only intelligent society in the entire galaxy which is at this moment experiencing such a stage in our evolution. We therefore should feel privileged to be witnesses of it. Of course, the nature of time comes in again with the impossibility of talking sensibly about simultaneity in the galaxy at all. But that is another subject, and one which we may ignore here.

A further thought follows upon the above observations. Granted that there are two forms of society in the universe aside from our own bizarre transition stage, the 'primitive' societies are obviously only of interest to those more advanced than themselves, for they are incapable of communicating with anybody else. They are like we were as little as a hundred years ago: provincial, quiet, probably quite murderous, and smug, with the occasional visionary who is burned at the stake or crucified causing a moral ripple. But they cannot send or receive messages between the stars. In our transition stage, aptly enough, we can receive such messages with existing equipment, but could not send any unless we constructed expensive and special means to do so. Now that means that the only societies carrying on an interstellar dialogue of any kind are the 'magical' societies. These societies will be so advanced that they probably have emerging primitives like ourselves 'taped'. They certainly have standard sets of procedures for dealing with the likes of us, and may already have commenced their operations with the long-range intent of bringing us into their club. But just as no London gentlemen's club wishes to have a savage in a g-string waving his spear and poisoned arrows about in the members' lounge, so the interstellar club is unlikely to plug us straight into the circuits as a fully-fledged member.

But what I am getting at is not merely to impress upon the reader that a pecking order is likely to exist in the interstellar club of any galaxy, at least to the extent of having restrictions on novices, but to make the point which emerges from this. And the point is, that such highly advanced societies have possibly developed to such a pitch of technological expertise that interstellar travel has become possible for them, whereby they can physically transport themselves over at least modest interstellar distances of a few light-years to their near neighbours. And if that is the case, then our own planet, which any half-witted extraterrestrial astronomer in the neighbourhood could assume as a likely place for life to exist, has almost certainly been physically visited by extraterrestrials in their travels. This could have happened at any time in our lengthy history as a planet. No doubt, at the very least, our distant ancestors the cave-men would have been observed by extraterrestrial probes, who would have made a note that something was happening on this planet — slowly happening, but nevertheless actually happening. And as [Carl] Sagan and [I.S.] Shklovskii said in the quotation from their book [*Intelligent Life In The Universe*, New York, 1966, RS]: 'It is reasonable that ... the rate of sampling of our planet should have increased, perhaps to once every ten thousand years. ... But if the interval between sampling is only several thousand years, there is then a possibility that contact with an extra-terrestrial civilization has occurred within historical times.'

If this were so, it would certainly have left some impact upon man and been incorporated somehow into his traditions. But if several thousand years had elapsed between that time and the present, the traces of the impact on man's culture would have been mostly dissipated and, it would seem, nearly impossible to elucidate. Unless some specific and unmistakable survival were found to exist, in circumstances which would probably be unusual, it seems that the hope of reconstructing scattered clues and fragments of the original tradition would be futile.

[COMMENT: Well stated, indeed! RS]

Pages 31-32

The Dogon know perfectly well that it is the turning of the Earth on its axis which makes the sky seem to turn round. They speak of ' ... the apparent movement of the stars from east to west, as men see them'. The Dogon are thus free from the illusions of our European ancestors, who thought the sky and stars wheeled round the Earth (though there was an exception to such primitive notions in Europe which no historian of science has ever reported, at least as far as I have been able to discover after a great deal of searching. I have summarized this 'secret' tradition in Appendix I, and pointed out its connection with the Sirius mystery).

[**COMMENT:** Appendix I begins with these words: "**The Moons of the Planets, the Planets around Stars, and Revolutions and Rotations of Bodies in Space — Described by the Neoplatonic Philosopher Proclus.** ' ... In each of the planetary spheres there are invisible stars which revolve together with their spheres ... ' So said Proclus the Platonic successor in A.D.438." This brief appendix discusses the life and work of Greek philosopher Proclus, who lived 410-485 CE. Temple's attempt to connect the writings of Proclus to Sirius B are tenuous at best, in my opinion. Temple quotes Professor A.C. Lloyd of the University of Liverpool, who was given the task of discussing Proclus as part of his contribution to the *Cambridge History of Later Greek and Early Mediaeval Philosophy*, a compendium which did not exist before 1967 and which was reprinted with corrections in 1970. Professor Lloyd's description of Proclus is reminiscent of the life of Apollonius of Tyana: "Proclus moved in important political circles, but like other leading Platonists he was a champion of pagan worship against imperial policy and found himself more than once in trouble. There is no doubt of his personal faith in religious practices. A vegetarian diet, prayers to the sun, the rites of a Chaldaean initiate, even the observance of Egyptian holy days were scrupulously practised. He is said to have got his practical knowledge of theurgy from a daughter of Plutarch {the Platonist, not the author of the *Lives*, RS}, and according to his own claim he could conjure up luminous phantoms of Hecate. Nor is there any doubt that he put theurgy, as liberation of the soul, above philosophy. But while his philosophy is full of abstract processions and reversions, philosophy was nothing for him if not itself a reversion, a return to the One, though achieving only an incomplete union. Its place can be seen in an almost fantastically elaborated metaphysical system: but although this system would not have been created had there not been a religion to justify, its validity does not depend and was not thought by Proclus to depend on the religion." In Volume III of his *Dictionary of Greek and Roman Biography and Mythology*, Professor William Smith devotes several pages to a discussion of Proclus. RS]

The placenta is used by the Dogon as a symbol of a 'system' of a group of stars of planets. Our own solar system seems to be referred to as 'Ogo's placenta', whereas the system of the star Sirius and its companion star and satellites, *etc.*, is referred to as 'Nommo's placenta'. Nommo is the collective name for the great culture-hero and founder of civilization who came from the Sirius system to set up society on the Earth. Nommo — or, to be more precise, the Nommos — were amphibious creatures, and are to be seen in the two tribal drawings in Figure 32 and Figure 34 in this book.

Ancient Dogon Figurine

Yannis (see Chapter 15) had a book in Greek about the Dogons and the Nommos. This is a scan from a page of that book, depicting the Nommo Insignia on a UFO over Greece and as a rock sculpture on a Greek mountainside. See also the graphic preceding Parts I and II of this book, as well as on my Postscript page at the end.

This is another scan from the same Greek book, the title of which I do not recall. This bas-relief depicts two "giant" gods, with two smaller humans in front of them and a symbolic serpent behind them. A race of reptilian "saurians" would undoubtedly prefer snakes as pets, just as we mammalian humans prefer mammals like dogs and cats. These humans are only as tall as the knees of their "gods".

Insignia of Nibiruan Crown-Prince Enlil
Note the similarity in design to a "millwheel".

These Nommos are more or less equivalent with the Sumerian and Babylonian tradition of Oannes. All of this subject is discussed in Chapter Eight [of Temple's book, RS], where it is necessary to consider details of what kind of creatures may live on a planet in the Sirius system. For the moment we are really more concerned with the Dogon astronomical and other scientific knowledge. Their descriptions of 'spacemen' and landings of 'spaceships' — or at least what seem to be such — are left to Chapter Eight.

Here is the way in which Griaule and Dieterlen record the Dogon beliefs about the two cosmic placentas I have just mentioned: 'Two systems, that are sometimes linked together, intervene, and are at the origin of various calendars, giving a rhythm to the life and activities of man. ... One of them, nearest to the Earth, will have the sun as an axis, the sun is the testament to the rest of Ogo's placenta, and another, further away, Sirius, testament to the placenta of the Nommo, monitor of the Universe.'

The movements of the bodies within these 'placentas' are likened to the circulation of blood in the actual placenta, and the bodies in space are likened to coagulations of blood into lumps. This principle is also applied to larger systems: 'In the formation of the stars, we recall that the "path of the blood" is represented by the Milky Way ... ', ' ... the planets and satellites (and companions) are associated to the circulating blood and to the "seeds" ... that flow with the blood.' The system of Sirius, which is known as 'land of the fish', and is the placenta of Nommo, is specifically called the 'double placenta in the sky', referring to the fact that it is a binary star system. The 'earth' which is in the Sirius system is 'pure earth', whereas the 'earth' which is in our solar system is 'impure earth'.

[COMMENT: And, of course, this "placenta" could actually refer to the electromagnetic tethering beam which periodically connects the negative South Magnetic Pole of Planet X Nibiru to the positive North Magnetic Pole of the Earth. If Planet X originated within the Sirius System, then it would not be illogical for this "placenta" also to be related to Sirius. RS]

The landing of Nommo on our Earth is called 'the day of the fish', and the planet he came from in the Sirius system is known as the '(pure) earth of the day of the fish ... not (our) impure earth ...' In our own solar system all the planets emerged from the placenta of our sun. This is said of the planet Jupiter, which 'emerged from the blood which fell on the placenta'. The planet Venus was also formed from blood which fell on the placenta. (Venus 'was blood red when she was created, her colour fading progressively'.) Mars, too, was created from a coagulation of 'blood'. Our solar system is, as we have noted, called the placenta of Ogo, the Fox, who is impure. Our own planet Earth is, significantly, 'the place where Ogo's umbilical cord was attached to his placenta ... and recalls his first descent'. In other words, the Earth is where Ogo 'plugged in', as it were, to this system of planets. What Ogo the Fox seems to represent is man himself, an imperfect intelligent species who 'descended' or originated on this planet, which is the planet in our solar system to which the great umbilical cord is attached.

[**COMMENT:** Temple is mistaken about the meaning of Ogo, which undoubtedly refers to Planet X from which the "umbilical cord" descends to Earth. And this idea of a "fox" is also reminiscent of the Scandinavian legend of a "Fenris Wolf" doing battle with a "Midgard Serpent", as well as those "myths" of a tomcat, squirrel or monkey climbing up and down the "totem pole" or "golden pole with a golden cage on top" — "Nail of the North" — discussed at length in my earlier *Planet X Nibiru: Slow-Motion Doomsday.* RS]

Ogo is ourselves, in all our cosmic impurity. It comes as a shock to realize that we are Ogo, the imperfect, the meddler, the outcast. Ogo rebelled at his creation and remained unfinished. He is the equivalent of Lucifer in our own tradition in the Christian West. And in order to atone for our impurity it is said over and over by the Dogon that the Nommo dies and is resurrected, acting as a a sacrifice for us, to purify and cleanse the Earth. The parallels with Christ are extraordinary, even extending to Nommo being resurrected. But these religious elements are not the subject with which I propose to deal. Let each reader pursue them as he sees fit, on his own initiative. I only raise the subject that, as Ogo, we may be cosmic pariahs, because I only hope that we must not always remain so. The Dogon seem to hold out hope of 'redemption' just as Jesus Christ did in his great message to the world. Redemption can mean what you want it to mean. But perhaps it would be more sensible to view 'sin' less as a sort of infraction of social rules and more as a form of impurity such as Ogo represents. The perversions of Christianity have always seemed to me to incorporate a perversion of the notion of 'sin' and the means by which 'sin' can be exploited as a means of temporal blackmail over other human beings. To rid ourselves of some impurity may be closer to what is needed, and those writers who have speculated that we suffer from a genetic fault may even be correct. If so, are we actually in cosmic quarantine at this moment?

We are told that the Nommo will come again. A certain 'star' in the sky will appear once more and will be the 'testament to the Nommo's resurrection'. When the Nommo originally landed on Earth, he 'crushed the Fox, thus marking his future domination over the Earth which the Fox had made'. So perhaps man's brutish nature has already been sufficiently subdued in our distant past. Perhaps it was those visitors whom the Dogon call the Nommos who really did 'crush the Fox' in us, who all but destroyed Ogo, and have given us all the best elements of civilization which we possess. We remain as a curious mixture of the brute and the civilized, struggling against the Ogo within us.

[COMMENT: If we are "struggling against the Ogo within us", then we may be struggling against our own "Saurian" genetics, inherited from Prince Enki of Planet X Nibiru. The idea of death and resurrection connected with the Nommo undoubtedly refers to the periodic Crossover and Cosmic Tree ("Sampo" in the *Kalevala*) of Planet X Nibiru. RS]

Pages 42-45

THE ORIGINS AND FEATURES OF DIGITARIA

The eighth [Dogon ruling] Hogon instructed his people in the features of the star, and, more generally, of the Sirius system.

Sirius appears red to the eye, Digitaria white. The latter lies at the origin of things. 'God created *Digitaria* before any other star.' It is the 'egg of the world', *aduna tal*, the infinitely tiny and, as it developed, it gave birth to everything that exists, visible or invisible. It is made up of three of the four basic elements: air, fire and water. The element earth is replaced by metal. To start with, it was just a seed of *Digitaria exilis, põ*, called euphemistically *kize uzi*, 'the little thing', consisting of a central nucleus which ejected ever larger seeds or shoots in a conical spiral motion. The first seven seeds or shoots are represented graphically by seven lines, increasing in length, within the sac formed in turn by an oval symbolizing the egg of the world.

[COMMENT: That "Sirius appears red to the eye" is a curious statement by Temple, and he offers no explanation. The idea that the element earth was replaced by the element metal could strongly suggest an artificial world or gigantic "spaceship" of sorts. It could even be said that our own International Space Station is a "world" constructed of air, fire, water and metal! RS]

The entire work of Digitaria is summarized in a drawing whose various parts are carried out in the following order: a vertical line issues from the oval — the first shoot to emerge from the sac; another segment, the second shoot, takes up a crosswise position, and thus supplies the four cardinal points: the stage of the world. The straightness of these two segments symbolizes the continuity of things, their perseverance in one state. Last, a third shoot, taking the place of the first, gives it the form of an oval which is open in its lower section, and surrounds the base of the vertical segment. The curved form, as opposed to the straight, suggests the transformation and progress of things. The personage thus obtained, called the 'life of the world', is the created being, the agent, the microcosm summarizing the universe.

In its capacity as the heavy embryo of a world issued each year, Digitaria is represented in Wazouba either by a dot or by a sac enveloping a concentric circle of ten dots (the eight ancestral Nommos and the initial couple of Nommo). Its continual movement produces beings whose souls emerge at intervals from the dots and are guided towards the star Sorghum which sends them on to Nommo. This movement is copied by the rhombus which disperses the creation of the Yourougou in space. Six figures are arranged around the circle, as if ejected from it:

a two-pronged fork: trees;
a stem with four diagonal lines: small millet;
four dots arranged as a trapezium:
    cow with its head marked by a short line;
four diverging lines starting from the base of a bent stem:
    domestic animals;
four dots and a line: wild animals;
an axis flanked by four dots: plants and their foliage.

The original work is likewise symbolized by a filter-basket made of straw called *nun goro*, 'bean cap'. This utensil consists of a sheath in the form of a continuous helical spiral, the centre of which starts at the bottom. The spiral supports a network of double radii. The spiral and the helix are the initial vortical motion of the world; the radii represent the inner vibration of things.

[COMMENT: Here one can fashion a peculiar correlation of ideas. The European "fairy tale" of Jack and the Beanstalk obviously has implications relevant to the tethered Cosmic Tree, a vine that stretches into the heavens to a land of giants. Thus, rhetorically, why else would the Dogon also have such a "bean cap" analogy as this, amongst their own local "fairy tales"? RS]

Originally, then, Digitaria is a materialized, productive motion. Its first product was an extremely heavy substance which was deposited outside the cage of movement represented by the filter-basket. The mass thus formed brought to mind a mortar twice as big as the ordinary utensil used by women. According to the version told to the men, this mortar has three compartments: the first contains the aquatic beings, the second, terrestrial beings, and the third, the creatures of the air. In reality the star is conceived of as a thick oval forming a backcloth from which issues a spiral with three whorls (the three compartments).

[COMMENT: At my Internet forum we have had long discussions about the exact "species" of these beings from Planet X Nibiru. Some of us insist that they are "saurian" or "reptilian", whilst others tow the Sitchin line that they are merely gigantic "humanoids" of some sort. Temple asserts that Oannes and his entourage seemed "amphibian". If they are reptilian, then they could not survive at sub-freezing temperatures; or at best they would immediately go into hibernation — hence their need for a planetary "heat-shield" to keep them warm? Most dinosaurs were reptilian, and our scientists tell us that there is a definite link between the disappearance of the dinosaurs and the evolution of birds, which are considered to be "descendants" of the dinosaurs. The offspring of both hatch out from eggs laid by the females, and many birds can easily withstand extremely low temperatures. We have "myths" of a Thunderbird, Owlman, Mothman and Dracula, none of which contains "reptilian" details. Moreover, in our distant legends, "angels" are often depicted as having wings. So perhaps these Nibiruans are more "avian" in nature than "reptilian". Or they could be an otherworldly high-tech crossbreed of all three species! RS]

According to the version instructed to the women, the compartments are four in number and contain grain, metal, vegetables and water. Each compartment is in turn made up of twenty compartments; the whole contains the eight fundamental elements.

The star is the reservoir and the source of everything: 'It is the granary for everything in the world.' The contents of the star-receptacle are ejected by centrifugal force, in the form of infinitesimals comparable to the seeds of *Digitaria exilis* which undergo rapid development: 'The thing which goes (which) emerges outside (the star) becomes as large as it every day.' In other words, what issues from the star increases each day by a volume equal to itself.

Because of this role, the star which is considered to be the smallest thing in the sky is also the heaviest: 'Digitaria is the smallest thing there is. It is the heaviest star.' It consists of a metal called *sagala*, which is a little brighter than iron and so heavy 'that all earthly beings combined cannot lift it'. In effect the star weighs the equivalent of 480 donkey loads (about 38,000 kg. — 85,000 lb.), the equivalent of all seeds, or of all the iron on earth, although, in theory, it is the size of a stretched ox-skin or a mortar.

[**COMMENT:** Reading the foregoing passage again for inclusion in this book, I couldn't help but think that this "creation process" could just as easily be a primitive description of the formation of the electromagnetic tethering beam that Planet X puts out towards our North Pole, establishing another Cosmic Tree at the commencement of each new *shar* of 3,600 Earth Years. And this idea of "480 donkey loads" goes back to something which I have mentioned before, *i.e.*, that a Middle-Eastern reference to "40 somethings" such as "it rained for 40 days and 40 nights" or "Ali Baba and the 40 Thieves" simply means that "it rained for a long time" or "Ali Baba and his Big Gang of Thieves"; and since 480 is 12 X 40, we could further infer that "480 somethings" indicate a very, very long time (such as "480 years" between the Exodus and Solomon's Temple, when in fact it was 592 years) or a very, very heavy object (such as "480 donkey loads" for the weight of Digitaria). RS]

THE POSITION OF DIGITARIA

The orbit of Digitaria is situated at the centre of the world, 'Digitaria is the axis of the whole world', and without its movement no other star could hold its course. This means that it is the master of ceremonies of the celestial positions; in particular it governs the position of Sirius, the most unruly star; it separates it from the other stars by encompassing it with its trajectory.

OTHER STARS IN THE SIRIUS SYSTEM

But Digitaria is not Sirius's only companion: the star *emme ya*, Sorghum-Female, is larger than it, four times as light (in weight), and travels along a greater trajectory in the same direction and in the same time as it (fifty years). Their respective positions are such that the angle of the radii is at right angles. The positions of this star determine various rites at Yougo Dogorou. Sorghum-Female is the seat of the female souls of all living or future beings. It is euphemism that describes them as being in the waters of family pools; the star throws out two pairs of radii (beams) (a female figure) which, on reaching the surface of the waters, catch the souls.

It is the only star which emits these beams which have the quality of solar rays because it is the 'sun of women', *nyãn nay*, 'a little sun', *nay dagi*. In fact it is accompanied by a satellite which is called the 'star of Women', *nyãn tolo*, or Goatherd, *enegirin* (literally: goat-guide), a term which is a pun on *emme girin* (literally: sorghum-guide). Nominally then it would be more important as the guide of Sorghum-Female. Furthermore, there is some confusion with the major star, the Goatherd, which is familiar to everyone.

The star of women is represented by a cross, a dynamic sign which calls to mind the movement of the whole Sirius system. Sorghum-Female is outlined by three points, a male symbol of authority, surrounded by seven dots, or four (female) plus three (male) which are the female soul and the male soul.

Taken as a whole, the Sorghum-Female system is represented by a circle containing a cross (the four cardinal directions), whose centre consists of a round spot (the star itself) and whose arms serve as a receptacle for the male and female souls of all beings. This figure, called the 'Sorghum-Female pattern', *emme ya tõnu*, occupies one of the centres of an ellipse called 'the pattern of men', *anam tõnu*, consisting of a full line called the "goatherd's course', *enegirin ozu*, flanked by two dotted lines, the outside of which is the path of the male souls, and the inside the path of the female souls.

The Sirius-Digitaria-Sorghum system is represented by a 'pattern of the Sigui', *sigi tõnu*, consisting of an oval (the world) in which one of the centres is Sirius. The two alternate positions of Digitaria at the time of the Sigui are marked and the positions at the same moment of Sorghum-Female are marked on two concentric circles encompassing Sirius.

[COMMENT: This Sorghum-Female pattern of the cross within the circle is identical to other representations of the disk atop The Cosmic Tree. The cross would represent four "spokes" within the cosmic "millwheel" which is also portrayed with more than four "spokes". Please refer back to the Insignia of Enlil included earlier in this chapter. Again, to quote the Siberian Irtysh River Valley legend, "There is a mill which grinds by itself, swings of itself, and scatters the dust a hundred versts away. And there is a golden pole with a golden cage on top which is also the Nail of the North." The Dogon obviously knew that Planet X Nibiru originated in the Sirius System, but then their legends of its origin became confused with their legends of The Cosmic Tree. RS]

Pages 58-59

The heliacal rising of Sirius [in July] was also important to other ancient peoples [besides the Egyptians, RS]. Here is a dramatic description by the ancient Greek poet Aratus of Soli of the rising of Sirius (often known as the Dog Star as it is in the constellation Canis, or 'Dog'): 'The tip of his [the Dog's] terrible jaw is marked by a star that keenest of all blazes with a searing flame and him men call Sirius. When he rises with the Sun [his heliacal rising], no longer do the trees deceive him by the feeble freshness of their leaves. For easily with his keen glance he pierces their ranks, and to some he gives strength but of others he blights the bark utterly.'

We see that this dramatic description of the rising of the star indicates an event which was certainly noticed by ancient peoples. Throughout Latin literature there are many references to 'the Dog Days' which followed the heliacal rising of Sirius in the summer. These hot, parched days were thought by that time to derive some of their ferocity and dryness from the 'searing' of Sirius. Traditions arose of Sirius being 'red' because it was in fact red at its heliacal rising, just as any other body at the horizon is red. When making rhetorical allusion to the Dog Days, the Latins would often speak of Sirius being red at that time, which it was.

[COMMENT: As noted earlier, perhaps Sirius B still emitted a visible reddish glow when these ancient descriptions were written, before it totally disappeared from view as a neutron-dwarf about 1,500 years ago, at the very latest. RS]

We tend to be unaware that stars rise and set at all. This is not entirely due to our living in cities ablaze with electric lights which reflect back at us from our fumes, smoke, and artificial haze. When I discussed the stars with a well-known naturalist, I was surprised to learn that even a man such as he, who has spent his entire lifetime observing wildlife and nature, was totally unaware of the movements of the stars. And he is no prisoner of smog-bound cities. He had no inkling, for instance, that the Little Bear could serve as a reliable night clock as it revolves in tight circles around the Pole Star (and acts as a celestial hour-hand at half speed — that is, it takes 24 hours rather than 12 for a single revolution).

I wondered what could be wrong. Our modern civilization does not ignore the stars only because most of us can no longer see them. There are definitely deeper reasons. For even if we leave the sulphurous vapours of our Gomorrahs to venture into a natural landscape, the stars do not enter into any of our back-to-nature schemes. They simply have no place in our outlook anymore. We look at them, our heads flung back in awe, and wonder that they can exist in such profusion. But that is as far as it goes, except for the poets. This is simply a 'gee whiz' reaction. The rise in interest in astrology today does not result in much actual star-gazing. And as for the space programme's impact on our view of the sky, many people will attentively follow the motions of a visible satellite against a backdrop of stars whose positions are absolutely meaningless to them. The ancient mythological figures sketched in the sky were taught us as children to be quaint 'shepherds' fantasies' unworthy of the attention of adult minds. We are interested in the satellite because we made it, but the stars are alien and untouched by human hands — therefore vapid. To such a level has our technological mania, like a bacterial solution in which we have been stewed from birth, reduced us.

It is only the integral part of the landscape which can relate to the stars. Man has ceased to be that. He inhabits a world which is more and more his own fantasy. Farmers relate to the skies, as well as sailors, camel caravans, and aerial navigators. For theirs are all integral functions involving the fundamental principle — now all but forgotten — of orientation. But in an almost totally secular and artificial world, orientation is thought to be unnecessary. And the numbers of people in insane asylums or living at home doped on tranquilizers testifies to our aimless, drifting metaphysic. And to our having forgotten orientation either to seasons (except to turn on the air-conditioning if we sweat or the heating system if we shiver) or to direction (our one token acceptance of cosmic direction being the wearing of sun-glasses because the sun is 'over there').

We have debased what was once the integral nature of life channelled by cosmic orientations — a wholeness — to the enervated tepidity of skin sensations and retinal discomfort. Our interior body clocks, known as circadian rhythms, continue to operate inside us, but find no contact with the outside world. They therefore become ingrown and frustrated cycles which never interlock with our environment. We are causing ourselves to become meaningless body machines programmed to what looks, in its isolation, to be an arbitrary set of cycles. But by tearing ourselves from our context, like the still-beating heart ripped out of the body of an Aztec victim, we inevitably do violence to our psyches. I would call the new disease, with its side effect of 'alienation of the young', *dementia temporalis.*

When I tried to remedy my own total ignorance of this subject originally, I found it an extremely difficult process. I discovered that I was reading coherent explanatory matter which I 'understood' but did not comprehend. For comprehension consists of understanding from the inside as well as understanding from the outside. Things that do not really matter to us, or into which we do not imaginatively project our own consciousness, remain strange to us; we understand them outside (like a man feeling the skin of an orange) but we have no inherent relation with the thing, and hence are ultimately divorced from its reality. This increasing isolation and alienation, a cultural blight of which there is almost universal complaint in the 'civilized' world, is yet another consequence of *dementia temporalis.* For how can you get inside anything in the end if you have ceased to be inside your own local universe with its cycles and natural events? To be outside nature is to be an outsider in all things.

[**COMMENT:** And to think that these ideas were written thirty years ago! This disconnection of humanity from the natural Cosmos is even worse today, in light of our 24-hour preoccupation with television, movies, sports and the Internet. As I was preparing this book, I mentioned its theme to an email friend, who replied, "I don't really know much about the stars and astronomy. I can find the Big Dipper and the Little Dipper. But I do well to anchor myself here on Earth!" *Dementia temporalis* indeed! RS]

Pages 82-83

We must return to the treatise 'The Virgin of the World'. This treatise is quite explicit in saying that Isis and Osiris were sent to help the Earth by giving primitive mankind the arts of civilization. ... And in the treatise Isis claims that the 'Black Rite' honours her and 'gives perfection'. It is also concerned with the mysterious thing called 'Night' — 'who weaves her web with rapid light though it be less than Sun's'. It is made plain that 'Night' is not the night sky because it moves in the Heaven along with 'the other mysteries in turn that move in Heaven, with ordered motions and with periods of times, with certain hidden influences bestowing order on the things below and co-increasing them'.

We must scrutinize the description of what is labelled 'Night' in this treatise. This description makes it perfectly clear that 'Night' is not 'night', but a code word. For it is said to have 'light though it be less than Sun's'. The dark companion of Sirius [A] is a star and has light, though less than the sun. Also 'Night' is said 'to weave her web with rapid light' which specifically describes the object as being in motion. Since Sirius B orbits Sirius A in fifty years, it moves more rapidly even than three of our sun's planets in our own solar system — Pluto, Neptune, and Uranus. Of these three, Uranus is the most rapid, and its orbit about the sun takes eighty-four years. So here is a star orbiting more rapidly than a planet! That may indeed be said to constitute 'weaving a web with rapid light'!

[COMMENT: This matter of "Night" is discussed at some length in *Planet X Nibiru: Slow-Motion Doomsday* in the chapters titled "The Night Sun" and "Heaven & Hell". Suffice it to say here that during periods of The Cosmic Tree, "Night" is experienced only in those lower extremities of the Southern Hemisphere, the "underworld", where the northern "Night Sun" is not visible. Here again Temple seems to be misinterpreting the Dogon legends. RS]

Page 85

The Egyptian star clocks date from at least the reigns of Seti I (1303-1290 B.C.) and Ramses IV (1158-1152 B.C.) of the XIXth and XXth Dynasties respectively, on the walls of whose tombs they are found. Therefore these star clocks are at least as old as 1300 B.C. and seem to go back to the very origins of Egyptian culture.

[**COMMENT:** These are traditional, "old-style" dates, which have been "reconstructed" by Dr. Immanuel Velikovsky in his *Ages In Chaos* series. See Chapter 18 for full details. RS]

By the first millennium B.C. they had been changed and a fifteen-day week substituted for the ten-day week. Other innovations took place as well at later dates, and the system fell into a considerable decay and became, it seems, a relic. I should imagine that a rise in the popularity of the sun god Ra made stars and especially Sirius seem less important. In any case, the innate integrity of the Sirius system in Egypt began to rot away and be ignored by the first millennium B.C., as it was superseded by ideas more obvious and less esoteric to impatient priests. Perhaps when this began to happen some purists may have gone off to other places where they hoped to retain the traditions without interference from decadent Pharaohs. We shall return much later to this idea, with some surprising information.

But let us return to Sumer and continue in hot pursuit. In Tablet VI of the *Enuma elish* we find an interesting passage. In it are mentioned the Anunnaki, who were the sons of An (An means 'heaven'), also known as Anu the great god. These Anunnaki were fifty in number and were called 'the fifty great gods'. Nearly always these Anunnaki were anonymous, the emphasis being on their number and their greatness and their control over fate. No certain identification of any important Sumerian god with any one of the Anunnaki exists except peripherally (as I shall describe later). In fact, all Sumerologists have been puzzled by the Anunnaki. They have not been 'identified' and no one knows exactly what is meant by them. They recur often throughout the texts, which makes it all the more annoying that nowhere are they explicitly explained. But their apparent importance to the Sumerians cannot be questioned.

[**COMMENT:** Temple published *The Sirius Mystery* in 1976, the same year that Zecharia Sitchin first published *The Twelfth Planet*, to be followed in subsequent years by his other nine books. By now, of course, we are well aware that the Anunnaki are residents of Planet X Nibiru. RS]

Pages 95-96

Now let us try to think of what we know is connected with the celestial Anunnaki and Sirius which also fits into this idea of there being seven Anunnaki-gods in the underworld. Remember that in both Sumer and Egypt each god of significance in astronomical terms has his own ten-day period or 'week'. If we multiply seven (gods) times ten days we get seventy days. Is there any basis for this length of time being of significance for the underworld in either Sumer or Egypt? Yes! In Egypt the underworld is called the Duat (or Tuat) and the seventy-day period is very significant there and relates intimately to Sirius, as we have seen.

[COMMENT: The "Duat" or "Underworld", as I have noted earlier, is discussed at length in my *Planet X Nibiru: Slow-Motion Doomsday.* See also Chapter 9 for other ideas on the possible location for the "Duat" as a "Stairway to Heaven". RS]

Parker and Neugebauer say: 'It is here made clear that Sirius (Sothis) gives the pattern for all the other decanal stars.' Sirius was, astronomically, the foundation of the entire Egyptian religious system. Its celestial movements determined the Egyptian calendar, which is even known as the Sothic Calendar. Its heliacal rising marked the beginning of the Egyptian year and roughly coincided with the flooding of the Nile. (Plutarch says the Nile itself was sometimes called Sirius.) This heliacal rising was the occasion of an important feast. One can imagine a kind of New Year-*cum*-Easter. The heliacal rising was the occasion when Sirius again rose into visibility in the sky after a period of seventy days of being out of sight, during which time it was conceived of as being in the Duat, or underworld. A further connection with Anubis comes in here, as Anubis was conceived of as embalming Sothis for these seventy days in the Duat. But as we all know, an embalmed mummy is supposed to come alive again. And this is what happens to the mummy of Sothis. Sothis is reborn on the occasion of her heliacal rising. Parker and Neugebauer also say: 'During the entire time of its purification it (Sothis, the star) was considered dead and it was only with its rising again out of the Duat that it could once more be considered as living.'

[COMMENT: One could also interpret this as meaning that when Planet X Nibiru is not present as The Cosmic Tree, during its 2,700-year period of invisible transit to aphelion and back, it is also in the "Duat". RS]

The Egyptians stubbornly clung to the traditional seventy days as the prototype of an underworld experience, despite its inconvenience, and, as we have already seen, 'Sirius gives the pattern for all the other decanal stars'. In fact, it was the practice through all of Egyptian history for there to be a period of precisely seventy days for the embalming of a human mummy — in imitation of Sirius. Even during the late Ptolemaic period, the embalming process invariably lasted the precise period of seventy days.

Thus we find the explanation of the seven Anunnaki of the underworld! It is also interesting to note that in ancient Mexico the underworld was thought to have seven caves.

It is worth noting that in the story *Etana*, about the legendary King Etana not long after the Great Flood, who had to ascend to heaven in order to have something done about his inability to have children (and thereby managed to have a son and heir), mentions 'the divine Seven' and describes them as Igigi, emphasizing the apparent interchangeability of the terms Igigi and Anunnaki. Also 'the great Anunnaki' are described as 'They who created the regions, who set up the establishments'.

[COMMENT: The terms Igigi, Anunnaki and Nefilim all denote various classes of inhabitants of Planet X Nibiru. Humans on Earth are referred to in the Nibiruan language as "Lulu" (male) and "Lulawa" (female). RS]

In the 'Descent of Ishtar to the Nether World' the Anunnaki are described as being brought forth (they are referred to as if they were stuffed animals being brought out of a closet, dusted off, and displayed in a taxidermists' contest) and seated on thrones of gold. Once more the throne concept appears. It seems all the Anunnaki ever do is sit and be symbolic.

Good little Anunnaki, like poodles, sit and smile at Anu. They are never given personalities, poor fellows. I might mention that in this story the nether world is described as having seven gates leading to seven successive rooms (or caves). It is obvious that the period of seventy days during which Sirius was 'in the underworld' to the Egyptians led to a breaking down of the seventy days into ten-day weeks, each with a god, giving seven gods. But these seven gods of the underworld must not have personalities lest there be the distraction of personal qualities to detract from the purely numerical significance of the concept. And of course the seven rooms of the seven gods are successive, leading from 'week' to 'week' until Sirius again rises.

So we see yet another essential link between the early Sumerian concepts and the Egyptian concepts. When will Professor Neugebauer take notice of this and cease ruminating among the late Babylonians and Persians?

Pages 117-118

The reader is by now presumably immune to shock at the endless 'surprises' which arise in the course of this enquiry. Hence he will no doubt be prepared to learn that if we shorten the 'e' (from eta to epsilon) in Greek, we have the *heru*-derived word (which has dropped the aspirate, probably in connection with the shortening of the vowel) *erion*, which means — 'woollen fleece'!

[COMMENT: This idea of a "fleece" refers to the "golden fleece" of Jason and the Argonauts, who were fifty in number, and are discussed by Temple in connection with the significance of the number fifty as it relates to Digitaria. Even Sitchin mentions the number fifty in connection with the Anunnaki, when one of their important spacecrafts crashed into Saturn, killing all fifty on board.

[And there is another significance to this number fifty. In our historical records, we find references to 72 Archons of Destiny, 72 Lemegeton Demon Spirits of Solomon, 72 Goetian Spirits/Sigils, 72 Greek Titans, and 72 Names of God. The ruling élite of Planet X Nibiru is a "Council of Twelve" who have 12 paired "consorts". After former Emperor Alalu and Empress Lilitu were ousted in a *coup d'état* by Anu and Antu, this number shrank to 22. If one subtracts the new Council of Twelve, then one is left with 10 of the original consorts. In the Tarot card deck, there are 22 "Major Arcana". If these 22 are added to the 50 Anunnaki (their "Parliament" if you will), one gets a total of 72 high-ranking Nibiruans, who are clearly the same 72 as those mentioned in various other traditions, as has been noted. For more information, see Chapter 11 ("Brief History Of Planet X Nibiru") and Chapter 12 ("Names Of The Gods"). RS]

There is a possibility that Herakles ('the glory of Hera'), the original captain of the *Argo* according to [Robert] Graves, and his protectress the goddess Hera (wife to Zeus and Queen of the gods) are derived from *heru* and they are known to be related to the word Seirios, which gave us Sirius and the Sanskrit *svar*, *suryas*, *etc*. In Sanskrit *Sura* means 'hero', indicating that these words may relate also. Liddell and Scott believe this complex of words to be separate from the Helios-complex, but their opinion is only an opinion. *Surana* means 'fiery', just as Seirios can in the sense of 'scorching' (due to the supposed 'scorching' of the Dog Star, *etc.*).

Back to our fleece. We find that the Greek word for a woollen fleece is related to the Egyptian word for Horus, the Greek word for sun, *etc.*, *etc.* So much for the puzzling nature of that now moot question: Why a fleece? Back to sacred puns again, which besiege us endlessly.

Let us not forget the Sumerians. Let us look again at that list of the fifty names of Marduk. One of them is the name Nebiru. It is commonly taken to be the name of the planet Jupiter, but there is confusion there, and the word is discussed in *Hamlet's Mill* [by Giorgio de Santillana and Hertha von Dechend, RS] and many other places as one of the infuriating Sumerian words which we would like to understand. Where did it come from? What does it mean? Why is it one of the fifty names?

[COMMENT: These remarks seem rather quaint in retrospect, because we now know exactly what the word "Nebiru" or "Nibiru" refers to: Planet X. RS]

Immediately after this forty-ninth name, Marduk is called 'Lord of the Lands' (its Akkadian form, which has no significance for us, is Bel Matati; I do not know the Sumerian form, which might be of interest to us). Then, after this supposed fiftieth name comes another name, namely Ea (Enki). Then Marduk is described as being of fifty names. It seems not to make perfect sense, since he has just been given fifty-one names. One way in which to make it sensible is to treat 'Lord of the Lands' (which is given in English in Speiser and Heidel, unlike all the other names) as a synonym of Nebiru. If we do this, then Ea is the fiftieth name and everything is all right.

Now, let us look at the Egyptian language once again. We find that the word *Neb* is extremely common and is used in many combinations and means 'Lord'. Without further ado, let me make clear that I believe the Sumerian Nebiru to be derived from the Egyptian Neb-Heru. If we treat Heru in its older Egyptian sense as the sun, then the descriptions of Nebiru in the Babylonian *Enuma elish* could read as a perfect description of Neb-Heru — 'the Lord the sun': 'Nebiru shall hold the crossings of heaven and earth. ... He who the midst of the Sea restlessly crosses, Let "Crossing" be his name, who controls its midst,' *etc.*, though overlaid with this, as with the traditional Horus, is a strictly stellar element which is behind the more obvious solar element. However, I do not wish unduly to confuse the issue by peeling off too many layers at once. Suffice it to recall the previously mentioned associations of Horus with the Sirius system and note that there is a Heru-ami-Sept-t 'Horus of Sothis' and Heru-Sept 'Horus the Dog Star' and then to note, again in association with Nebiru which is supposed to have been Jupiter, that there is in Egyptian a Heru-sba-res 'Horus, star of the south, *i.e.* Jupiter', and Heru-up-Shet, 'the planet Jupiter'; also in the *Enuma elish* Nebiru is clearly described as 'a star'. Horus also exists as Heru-ami-u which is 'a hawk-headed crocodile with a tail terminating in a dog's head'. The dog is related to Sirius. Heru-ur-shefit is a jackal form of Horus; *heru* is also the name of a sceptre and of a jackal-*headed* standard in the other world. A form of Horus using the common word Neb is Heru-Neb-urr-t, meaning 'Horus as possessor of the supreme crown'. Another of several is Heru-Neb-pat, meaning 'Horus, lord of men'. Heru-Neb-taui is 'Horus, Lord ot the Two Lands'. Recall our synonym for Nebiru — 'Lord of the Lands'!

[COMMENT: Clearly the words "Neb-Heru" and "Nebiru" are linguistically related to "Nibiru". There is no denying that. And here Temple makes a comparison of Nibiru with the Egyptian "Horus", son of Osiris and Isis. Thus, just as Horus was a child of Osiris and Isis, so also was Planet X Nibiru an "offspring" of Sirius B and Sirius A! RS]

We are getting deeper and deeper into the legend of the golden fleece, of origins of Greek and Middle Eastern ideas in Egypt, along with key words and names, *etc.* All these centre round the curious Sirius complex. What more will we uncover? Perhaps we need a break from all these Egyptian words. There are many other aspects of our subject, and it leads us ever closer to the solution of our mystery — which is the origin of the subject.

Pages 133-134

In the Homeric *Hymn* quoted we find it specifically stated that Minoan Cretans (contemporaneous with ancient Egypt, of course, and who traded with the Egyptians) from Knossos took Apollo to Delphi, the site of an *omphalos.* And these Knossians are stated to respect oracles. And near Knossos is a site called Omphalos which is one degree of latitude south of the site of Kythera, which is one degree south of Delos, which is one degree south of Delphi.

Parke gives further information. He mentions the connections well known to have existed *between Delos and dodona* through what are known as 'the Hyperborean gifts', which were sent to Delos by way of Dodona from the mysterious northern Hyperboreans, whose land has never been located with any certainty at all, but which is thought by many to have been Britain. In Book II of Diodorus Siculus one finds a description of the Hyperboreans observing celestial objects through what sounds to me and some other scholars distinctly like a telescope. The description should be consulted by the interested reader.

[COMMENT: This passage is followed by a drawing depicting a battle between two serpents, one larger than the other. The caption reads: "Figure 18. Detail of mural from Pompeii reproduced by W.H. Roscher. The *omphalos* is identical to the one at Delos. Here the friendly omphalos-serpent is being harassed by a python." Furthermore, mysterious Hyperborea ("land beyond the mountains where the North Wind rises") is nothing more than Planet X Nibiru atop The Cosmic Tree, The World Tree, The Sacred Tree, Mount Olympus, Mount Zion, Mount Meru. It is certainly not Britain, as Temple and others have suggested. RS]

Pages 163-165

The father of Orthrus the Sirius-dog and his brother Cerberus the fifty-headed dog was the monster Typhon whom we mentioned a moment ago. And it is worthwhile for us to see what Liddell and Scott's *Greek Lexicon* have to say about the meaning of the name Typhon and also related forms of this word.

One meaning of *Typhon*, curiously enough, is 'a kind of comet' — in other words, a moving star! Another form is either Typhoeus or Typhos and specifically refers to the youngest son of Gaia, who was mother also of the three fifty-headed monsters and of Garamas. *Typhos* means 'smoke, vapour', and also 'conceit, vanity (because it clouds or darkens a man's intellect)'. *Typhos* means 'blind' and specifically 'in the sense of misty, darkened'. The verb *Typhloo* means 'to blind, make blind' or 'to blind, baffle'. It also means 'to wrap in smoke'.

Since Typhon is specifically said to be the father of Sirius (Orthrus) and one of its unexplained definitions is a description of a moving star, and its son has fifty heads, I take all the references to obscurity and invisibility to mean that Typhon represents Sirius B which is the dark companion of

Sirius [A] and is invisible to us. In other words, we are *typhlos* (blind) to Typhon because it seems as if it were obscured or *typhloo*'d by *typhos* (vapour, smoke), and we are baffled, blind (*typhlos*) in the sense of the subject being darkened (*typhloo*).

[COMMENT: In Chapter 12 I have equated the Mayan god of the "smoking mirror", Tezcatlipoca, with Nibiruan Baron Marduk. However, if there is also "smoke" connected with Typhon, whom I have equated with Baron Ninurta, then perhaps Tezcatlipoca should be equated with Ninurta, not Marduk; however, I do not plan to modify my charts in Chapter 12 to reflect this possibility. I leave such details to later research by specialists in these fields. And I might add that Dr. Velikovsky also tried to analyze the name "Typhon" in his *Worlds In Collision*. This is an *extremely* complex subject, in case anybody still needs to be reminded of that. RS]

A possible origin of the word Typhon may be the Egyptian word *tephit* or *teph-t*, both of which have the meaning of 'cave, cavern, hole in the ground'. This Egyptian word describes perfectly the chasm at Delphi in which Python was supposed to lie rotting, his corpse giving off the fumes out of the earth. And, as we have seen, Python was equated with Typhon in early times.

If we take the Egyptian word *tep* we discover that it means 'mouth' and in the form *tep ra* it means 'mouth of the god' literally, but in fact the real meaning of this is 'divine oracle'. *Tep* is an unaspirated *teph*. Hence the *tep* of Delphi has a *tephit*, or cavernous abyss beneath it. Later I shall consider the Egyptian word *tep* in its further ramifications. But for the moment it is sufficient to see that Typhon almost certainly originates from the Egyptian word describing a cavern or hole in the earth, as the Egyptians founded the *tep* or oracle at Delphi and naturally used their own word to describe the cavern. As Delphi passed into Greek culture and the Egyptians became forgotten in all but vague legends such as the famous visit of the Canopic Herakles to Delphi, *etc*., the original word to describe Delphi's cavern would have been retained through the natural conservative inclinations of religious organizations who retain antique words and language for notoriously long periods of time, forgetting their origins. Hence a Greek who had no knowledge of Egyptian culture or that it had ever penetrated to his homeland in earlier days would nevertheless call the cavern at Delphi which produced the sulphurous fumes the den of Typhon after its original Egyptian designation of *tephit*. It has been noted by people other than myself and with greater knowledge that the Sumerian word for cavern, *abzu*, survived in Greek as *abyssos*, leading to our English 'abyss'.

The fumes arising from the Delphic cavern obviously gave rise to the usage of forms of the word for 'obscuring with smoke, dark', *etc.* And the fact that the personified Typhon became closely associated with Sirius was obviously due to the fact that this word which had entered Greek usage and been extended to considerations of 'darkness, obscurity', was useful in the traditional Sirius lore as adopted in Greece. The other meanings for the word then developed from there, except for the obvious popular usages, such as applying the word to a description of 'vanity' because vanity clouds a man's intellect — a really superb extension of the meaning for use in poetic and common expression.

It is probably considerations such as the Typhonic in the sense of Sirius B's association with darkness and obscurity, and hence with cavernous blackness, that some of the Sirius-related divinities were reputed to live in the dark underworld in later times. The prototype of these is quite specifically Anubis, the embalmer of mummies. Anubis was not originally meant to be a death god *per se* and his association with mummies and the underworld has been previously explained. Egyptian mummies were, as I have said, embalmed over a period of seventy days, to correspond with the number of days each year when the star Sirius was 'in the Duat, or Underworld' and was not visible in the night sky. Hence the seventy-day 'death' of Sirius each year was the fundamental and earliest underworld aspect of the Sirius lore. Of course, Anubis, as the expression of the orbit of Sirius B, was invisible all the time, and not only for seventy days a year. Hence the permanent Typhonic darkness could be even further extended in later lore and a heightened sense of the importance of the underworld aspects could arise. This concept of invisibility and darkness must have become more and more important as time went on, and the grasp of the nature of the mysteries became weakened by successive generations of initiates who were further and further from the original sources of information, though the Dogon even down to our time have maintained the information in a remarkably pure state. So there developed the underworld nature of the fifty-headed Cerberus-Anubis in Greek times. With the earlier Egyptians, as always with them, the underworld concept had been on more than one level. To the public the underworld aspect seemed to be entirely explicable by the disappearance of Sirius for seventy days — a fact which anyone could notice — and its reappearance following that period at dawn on the occasion of the star's heliacal rising. But the priests knew that the dark companion of Sirius was never visible.

It would be worthwhile now to look a little more closely at the dog Orthrus, who was Sirius. Orthrus is the dog of the herdsman Eurytion. [Robert] Graves interestingly compares this Eurytion with the Sumerian Enkidu, the companion of Gilgamesh who was hairy and wild and came from the steppes and was imbued with incredible strength: 'Eurytion is the "interloper", a stock character ... The earliest mythical example of the interloper is the same Enkidu: he interrupted Gilgamesh's sacred marriage with the Goddess of Erech [Uruk], and challenged him to battle.' It is particularly interesting to find the Greek companion of Sirius compared by Graves to the Sumerian Enkidu, whom I also have identified with the companion of Sirius [A]. For 'companion of Sirius' is precisely what Eurytion is; if Orthrus is Sirius [A] and Eurytion the herdsman accompanies him, then Eurytion is the 'companion of Sirius'. And Enkidu is the strong hairy wild man who endured a trial of strength against Gilgamesh and became his companion after their wrestling match. Both Eurytion and Enkidu are hairy and rustic characters, and they seem to be related also to the god Pan, whose hairy and rustic nature classes him with them.

Pages 182-183

[Sir E.A.] Wallis Budge says another form of *tchens*, 'weight', is *tens*, which also means 'weight, heavy'. And the very next word in the giant dictionary is *teng* which means 'dwarf'! We thus see an apparent variation of the same word meaning 'heavy' and 'dwarf', and this word is specifically applied to the Sirius system.

But just in case there are any sceptics left (and there always are), a look at the Egyptian word *shenit* will be helpful. This word means 'the divine court of Osiris'. The same word *shenit* means 'circle, circuit', and *shent* means 'a circuiting, a going round, revolution'. *Shenu* means 'circuit, circle, periphery, circumference, orbit, revolution', ... which Wallis Budge gives, and which means 'the two circuits' — and twice fifty is a hundred, giving us the Great Year. *Shen ur* means 'the Great Circle' or 'the circuit of the Great Circle' or 'the islands of Shen-ur', which last is interesting in that it indicates that this place of the Great Circle is not only 'the divine court of Osiris', who is the husband of Sothis (Sirius), but is also a place with islands (stars or planets) where one can presumably live. It does seem that the Egyptians had quite as clear a conception of the Sirius system as the Dogon have.

The verb *shenu* means 'to go round, to encircle', but the verb *shen* means 'to hover over', and presumably the great orbit is above us in the sky, hovering over us in space.

[**COMMENT:** This "going round" or "circling" could also refer to the comet-like orbit of Planet X Nibiru, whilst in that period when it is invisible at its cosmic "Duat" during aphelion. As for the idea of "hovering", that is easily explainable by the fact that Planet X Nibiru, as Cosmic Tree or Winged Disk, "hovers over" our North Pole — "land beyond the mountains where the North Wind rises"! RS]

The Egyptian word *khemut* means 'hot parching winds, the khamasin, or khamsin, *i.e.* winds of the "fifty" hot days'. This is rather interesting. (Arabic *khamsin*, 'fifty' and Hebrew *khamshin*, 'fifty', are obviously derived from this Egyptian source.) In late times 'the dog days' about the time of the rising of Sirius and called 'dog days' from 'the Dog Star' were supposed to be hot and scorching. There are many references to this in writers like Pliny and Virgil. Here is an earlier tradition of hot days incorporating the Sirian number *fifty*. This same word *khemut* has familiar meanings in its related forms. *Khemiu-urfu* means 'the stars that rest not'. *Khemiu-hepu* means 'a class of stars'. *Khemiu-hemu* also means 'a class of stars'. In short, *khemiu* means 'stars'. So *khem* (though apparently not used on its own in surviving texts) really means 'star', as well as referring to fifty days. *Khem* also has the meanings 'shrine, holy of holies, sanctuary', and 'little, small', also 'he whose name is unknown, *i.e.* God', also 'god of procreation and generative power', also 'to be hot', and 'unknown'. All these meanings are relevant to the Sirius mysteries. The Sirius system was held to be the source of generative and procreative power as we have already seen; Sirius B was of course 'unknown', and was 'little, small', and was a star that rests not (that is, it is always orbiting, which is not at all usual for a star). And what is a star that rests not unless it be Sirius B? For only the planets, which were well known and differentiated by the ancient Egyptians, 'rested not' with the remarkable exception of Sirius B. Comets and meteors apart, and they too were well classed to themselves.

[**COMMENT:** When I served in the Peace Corps in Massaua, Eritrea (then Ethiopia), a Red Sea port city, we experienced these *khamsin* every summer. Vast windstorms over the Saudi Arabian desert would hurl sand into the air, sand which would then be blown westwardly over the ultra-humid Red Sea, picking up moisture and turning it into airborne "mud". This mud would then rain down inland onto Eritrea/Ethiopia and subsequently be blown northwards into Egypt. In a nutshell, the *khamsin* (so-called because there are about 50 of these storms each year during the summer months) is an extremely messy event; one's house is repeatedly covered by a film of mud, both inside and out. It is a "supreme nuisance" if nothing else, and that is an understatement. RS]

Pages 190-191

Turning to the Egyptian word *henti*, one finds that it is a name for Osiris and it also is 'a crocodile-headed god in the Tuat', which is the Egyptian underworld, and it also means simply 'crocodile gods'.

Hent is specifically 'the crocodile of Set'; *hen-t* is, interestingly, a specific locality of the underworld and means 'a district in the Tuat'. But, more widely, *hen-t* is 'a mythological locality' which is not necessarily in the underworld. It would seem that the fabulous Hen-t was a locality which had an underworld counterpart and obviously is somehow connected closely with both Osiris and crocodiles.

The name of this region, Hen-t, when taken as a common noun rather than as a name, means 'dual'. This is a strong clue as to the nature of the fabulous region. A region intimately connected with Osiris and whose name means 'dual' is reminiscent of Plutarch's description of that circle or ellipse with its dual aspect of separating the light from the darkness.

[COMMENT: This "twilight boundary" between the Northern Hemisphere with its "Night Sun" and the Southern Hemisphere with its true "Night" is discussed in greater detail in Chapter 7 ("Heaven & Hell") of my previous volume, as I have stated earlier. RS]

Lest the reader think this far-fetched I must hasten to add a further meaning on *hen-t* which is 'border, boundary', and another which is 'the two ends of heaven' — which all appear to refer to a circle and have the *hen-t* ('dual') nature of outside and inside and the two extremes connected by a diameter. *Hen-t* also means 'end, limit', and a *henti* is a specific period of time lasting for 120 years. Remember that the Sigui of the Dogon was every sixty-years and two Dogon Sigui make one Egyptian *henti*. In fact, a *hen-t* would be a Sigui or, perhaps, vice versa, depending upon one's grammatical reference.

(The use of the word 'dual' can be rather ambivalent and be construed as either halving or doubling by the context.) And this dual time period is also rather like the two fifty-month periods which make the hundred-month period of that sacred Great Year connected with Sirius, which has a dual aspect.

Henti also has the meaning endless — and the endless circling of Sirius B around Sirius A could be referred to here. Some such idea must be at work, otherwise how can the same word have the meaning of 'endless' and also of '120 years'? It must be a reference to an 'endless' cycling of perhaps the orbit of Sirius B or of the Sigui cycle's own basis. In any case, it signifies that the 120-year period was arrived at as an endlessly recurring cycle, and for that to have been the case, the 120-year period must have been quite important, which is exactly what one would anticipate. ...

Considering that *henti* means all this and also means 'crocodile gods', *etc.*, it is surprising to see that *henn* means 'to plough' and a *hennti* is 'a ploughman'. One immediately thinks of Jason *ploughing* the field for the dragon's (crocodile's?) teeth. It may well be that the 'serpent's teeth' motif which was a pun for 'the goddess Sirius' was extended in another layer of pun to 'dragon's teeth' as a reference to crocodiles.

[**COMMENT:** Needless to say, all of these references to "serpents" and "dragons" symbolize the "saurian" nature of the inhabitants of Planet X Nibiru. For additional information, the reader can go to my website and read *Flying Serpents & Dragons* by R.A. Boulay, in either English or French, with my comments interspersed throughout. RS]

In connection with Sirius B being the hairy, bestial Enkidu-figure, we see with interest that *hen* means 'to behave in a beast-like manner' and a *henti* is also specifically 'a beast-like person'. In addition to *henti* being a name for Osiris, who is the companion of Sirius, we find it describing 'a beast-like person' who is the archetypal companion in Sirius-related legends. And additionally, we find Hathor the cow-goddess, a form of Isis-Sirius, referred to as Hennu-Neferit. (Neferit simply means 'beautiful'.) But this word *hennu* with the double 'n' has the basic meaning of 'phallus' and has a phallic determinative hieroglyph, and therefore may not be related to the *hen* words with a single 'n'.

[**COMMENT:** And certainly "the golden pole with the golden cage on top" is reminiscent of a phallus. RS]

Pages 220-221

The amphibians must have a name, and the Dogon name for them of 'the Monitors' may be the best to consider using. 'Monitor' is more specific than 'Instructor', and 'Masters of the Water' is too long. There is no point using the euphemism the 'Annedoti', knowing that it means the 'Repulsive Ones'. A more generic and neutral term, I suppose, would be simply the 'Sirians'.

[COMMENT: As R.A. Boulay has pointed out in his introductory remarks to *Flying Serpents & Dragons*, these extraterrestrial "gods" are "shy". The "Lord" (Yahweh/Enlil) told Moses that if Moses looked upon His face, Moses would surely die. Die of what? Fright? Are these "saurian gods" really that "repulsive" when seen by humans? RS]

If we ever come into contact with them again, they will probably be called the 'Sirians' officially, and their civilization will be the 'Sirian civilization'. Their art will fall under the heading of 'Sirian culture' and their technology will be 'Sirian technology'. But what about their religion? There's a delicate point. It will be called the 'Sirian religion' and we will try to pretend it has nothing to do with us. But inevitably we will have to take into account that, whereas 'cultures' and 'technologies' can be localized, the greater problems of the nature of life itself and of an individual's relation to the universe — existential problems — are not localizable. There will in fact be no such ultimate thing as 'Sirian religion' except in the ethnographic sense. To speak of a 'Sirian' God will get us into deep waters. What do we mean when we speak of a 'Jewish' God or a 'Christian' God? There is no doubt that it is at the level of our deeper concerns — our religious and philosophical ones — that contact with an extraterrestrial civilization will make its deepest impact on us. And it is at this friable level of our preconceptions that we are most vulnerable. Here the foundations of our beliefs can crumble with the first shock wave. Here the entire edifice of our civilization can give way. Only by being prepared can we safeguard our own cultural integrity.

We must not dismiss speculations such as those we have just indulged ourselves in as idle, thinking that we will wait and see what turns up in a spaceship some day. If we are going to be coming into direct contact with amphibious extraterrestrials, we should try to get some thoughts together on their physical nature and requirements at the very least — if only to make them welcome. It is quite true, as Carl Sagan says: ' ... stories like the Oannes legend ... deserve much more critical studies than have been performed heretofore.' The critical studies should be institutionalized by the governments of the major powers, and made official programmes. The resources of the governments which pour into programmes to prevent their countries being overrun by military invasions, chemical warfare, nuclear blasts, should also pour into programmes to prevent our planet as a whole being overrun by a sudden extraterrestrial contact which gives little warning. No matter how much care is taken by any superior extraterrestrial civilization in dealing with us, it is really up to us to be ready for any contact. I would even venture that we may be under observation or surveillance at this very moment, with an extraterrestrial civilization based at the Sirius system monitoring our development to see when we will *ready ourselves* for their contacting us. In other words, we may very possibly be allowed to control the forthcoming contact ourselves. One wonders what any possible amphibious extraterrestrials living at Sirius would think roughly ten years later (speed of radio transmission at speed of light — across ten light years means a ten-year lag) upon receiving news from some automatic monitoring device which picked up a radio or television programme at Earth mentioning a book just published about amphibious extraterrestrials living at Sirius. Would they think that was their cue? If what I propose in this book really is true, then am I pulling a cosmic trigger?

Pages 226-227

In closing, I wish to make a final point of considerable importance. Let us assume that what I have proposed in this book really is true. Let us grant all the premises. Say that there really is an advanced civilization based at the Sirius system. No doubt we are under routine monitoring. No doubt they know by now roughly where we stand on the ladder of evolution. They have picked up our radio signals. They know we have been to the moon. Let us assume they wish us well. Let us assume even that they contact us someday when they think we are ready for it — or after we have discovered them by examining the Sirius system as I suggest and finding evidences of their existence.

Let us assume all this. Well, if that day comes — or if it doesn't and if some other day comes, some other civilization some day is known to us at some other star — there is one thing we must not forget. We must remember that no matter how grand and glorious *they* may be, they are still mortal beings in a universe which to them is still mysterious. They cannot and never will know all the answers. We may very well have a handful of answers that they have not. We may have some quirky skills which they cannot attain. We may have some peculiar native ingenuity which they lack, even if this is not obvious for centuries. There may be something about us that is so valuable that we are not just worthless primitives beside them. Let us never accept a view of ourselves as recipients of cosmic charity. We are men, and for all our faults, we have a few things about us which are worth some attention. We have had some remarkable characters in our history and we will have more. Whatever one's view of what lies beyond death — extinction, reincarnation, heaven and hell — the genetic stream goes on. There will be more men, and there will be great ones. We can rise to challenges. We have demonstrated courage throughout our history. Any superior civilizations may have even more superior civilizations behind them of whom they are curious. Let us not forget the principles of hierarchy, let us never blind ourselves to the possibility of a door behind the door behind the door. And if we ever find ourselves oppressed, let us be certain that there are others — somewhere — who would free us. The universe is finite but unbounded. There are between ten and a hundred million intelligent civilizations in our galaxy alone, in all likelihood. And there is always one more to contact than the one we have already contacted. We can afford to shop around in a shop the size of the universe.

[COMMENT: Well put indeed! And again I refer the reader to Temple's entire book for much more information about Earth's ancestral Nommo colonizers from the Sirius System. The following essay was sent to me by email, and I consider it to be most pertinent to this discussion of Temple's ideas. Footnotes, denoted by {braces}, are placed at the end. RS]

*

RIAP Bulletin (Volume 4, Number 4), October-December 1998

TRACKING THE ALIEN ASTROENGINEERS

An Essay By Vladimir V. Rubtsov
Chairman, Research Institute On Anomalous Phenomena (RIAP)
Kharkov, Ukraine

When studying the question of possible ancient visits of alien beings to the Earth, a researcher sometimes encounters data which cannot be interpreted as yet in a strictly scientific manner, but which, at the same time, are interesting enough to be regarded seriously and unbiasedly. Such data can be found, in particular, in the well-known "mythological astronomy" of the Dogon, an African people, living mainly in the West African Republic of Mali.

The Dogon believe that the Universe is "infinite, but measurable" and is filled with "spiral stellar worlds" (*yalu ulo*), one of which contains the Sun. This world may be seen in the sky as the Milky Way. The majority of heavenly bodies represent the "external" star system, whose influence on terrestrial life is, according to the Dogon, relatively small. There exists, however, also the "internal" system, which "participates directly in the life and development of men on the Earth". It includes Orion, Sirius, Pleiades and some other stars. These celestial bodies form the "support of the seat of the world". It is Sirius that occupies the main position among them, being called the "navel of the world".

Sirius is considered by the Dogon as a triple stellar system, consisting of the stars *Sigi tolo* (our Sirius A), *Po tolo* (Sirius B, a white dwarf) and *Emme ya tolo* (the hypothetical Sirius C, yet to be discovered). Close similarity between the characteristics of *Po tolo* and Sirius B (both bodies are white, small, very heavy, with fifty-year periods of revolution around the main star) stimulated a lively discussion on the pages of scientific — and not so scientific — journals about 20 years ago. Robert Temple, in his book *The Sirius Mystery* {1}, and Eric Guerrier in his *Essai sur La Cosmogonie des Dogon* {2} supposed that these data (as well as other astronomical information possessed by the Dogon) were brought to the Earth by cosmic visitors. However, their reasons could not break through the "armour of denial" of established science. The hypothesis of a recent adoption of this knowledgc from Europeans appeared convincing for most scientists.

It is natural that other components of the Dogon mythology, which have little in common with modern scientific knowledge, attracted even less scholarly interest. Yes, this is a real mythology, almost pure and not very simple. To analyze its content is not an easy task, and the results are not self-evident. Nonetheless, it is possible that we can derive from such an analysis some important information. Let us recall very briefly the main points of this mythology.

The supreme god Amma made the whole Universe within a grain of *po*, which is the Dogon name for *fonio*, the smallest kind of millet. This grain was located inside the "egg of the world"; it "spun and scattered the particles of matter in a sonorous and luminous motion", remaining, however, "inaudible and invisible" {3, p. 130}. Having opened this "egg", Amma let the spiral stellar worlds out, and it was thus that the Universe was realized. Then the god created the first living being — *Nommo anagonno.* This being is described either as a half-man, half-snake having flexible limbs, without any joints, red eyes and forked tongue, or just as a fish, namely a Silurus, sheat-fish, or cat-fish. This Nommo multiplied, and there appeared four Nommos: *Nommo die*, *Nommo titiyayne*, *O Nommo* and, at last, *Ogo*, a very harmful creature. As distinct from other Nommos, he is never represented as a fish. Instead of awaiting patiently the completion of the Amma's work, he hurriedly made an "ark" and rushed into space, wishing "to look at the world". Thus, he took disorder into the young world. After several voyages, Ogo landed on the Earth and turned into the pale fox or fennec, named Yurugu.

Made indignant by Ogo's escapades, Amma took everything he had created and put it back into the grain of *po*. To "purify" the Universe, he had to sacrifice one of the Nommos. After that "by whirling and ... acting as a spring, the *po* ... distributed all things in the Universe" {3, p. 423}. The empty shell of the grain became the star Po. In "the first year of the life of man on Earth" this star exploded, and its brightness decreased slowly during 240 years until it completely faded.

It is interesting that there is in the Dogon mythology another image of the Sirius system. According to it, the main star represents the Ogo's female twin, Yasigui, whom he chased with some dubious intentions. One of its satellites is Ogo himself, doomed to revolve eternally around his sister, remaining at a respectful distance from her.

Of course, this is only an outline of this very complicated genesis story. I am citing it here just as the basis for further considerations. Can this story be useful for paleovisitological studies?

Some time ago the present author suggested the idea of astroengineering interference by a cosmic supercivilization in the evolution of the Sirius system. This assumption was based on the Indo-European myth of the heavenly blacksmiths, who are fighting and chaining up the monstrous Dog, dangerous for the Universe, as well as on some astrophysical data from the history of Sirius (see: {4}).

It is known in astronomy that a white dwarf arises from a red giant as this loses its mass. This process is usually accompanied by a slow ejection of a planetary nebula which eventually dissipates into space. But sometimes the remaining core of the red giant can retain a mass exceeding the so-called Chandrasekhar limit (about 1.3 Sun masses). This leads inevitably to disastrous self-compression of the core and its explosion as a Supernova. As a result, powerful streams of matter and radiation are ejected into the surrounding space.

If such an event had ever happened in the Sirius system, at a small (on the cosmic scale) distance from the Solar system, it might have been fatal for the terrestrial biosphere. My idea was that some highly developed supercivilization could have tried to remove the excess of stellar matter from Sirius B, thus saving life and civilization on Earth.

Really, the only thing we know for sure about the evolution of the Sirius system is the fact that Sirius B was once a red giant whose mass exceeded that of Sirius A (that's why the former evolved more rapidly). The initial orbit of Sirius B was, most likely, circular; now it is a highly elongated ellipse. This suggests that the mass loss was accompanied by some considerable disturbances. Some part of the "lost" matter probably contaminated the atmosphere of Sirius A (see: {5}). But the real course of events is still very unclear. The situation will seem even more involved if we bear in mind the possible presence of the second satellite in this system, as is asserted by the Dogon and confirmed by recent astrophysical data (see: {6}).

It would be certainly very helpful to study thoroughly the Sirius system with modern astronomical equipment (say, by radio interferometers with a very long baseline). But it appears that relevant (and rather interesting) information can also be found in those vestiges of the great mythologies of Europe, Asia and Africa which have survived till now, however odd and strange they may appear to us. This information cannot be taken at face value, for the myth is a very special form of thinking and knowledge, much different from our modern mentality. We should carefully analyze and interpret mythological stories and characters to understand their profound sense and real significance. There are on this road many pitfalls and false turnings, but there may also be found some road-signs and important hints. Let us go through some of them.

It is well known that the most common (though not the only) name for Sirius in the ancient world was "The Dog" (with the variants: the wolf, the fox, the jackal). The ancient Egyptians called it, in particular, the Starry Dog and identified the star with Anubis, the jackal — or dog-headed god of the dead. The North American Indian Cherokee tribe believed that this Dog awaited the souls of the dead on the Milky Way; the Blackfeet Indians named the star "Dog-face". The oldest Hindu name for Sirius was *Sarama*, "one of the Twin Watch-Dogs of the Milky Way" {7, p. 119}. The Chinese knew this star as the Heavenly Wolf, and the Greeks as the Dog of Orion, or more specifically, as the dog Orthrus, a son of the monster Typhon. The Romans saw in it the Southern Cerberus, a watch-dog of their hell. As for the fennec Ogo, it is the smallest wild animal in the dog family (which hints probably at the small size of Sirius B).

What is more, Sirius represented not a decent house dog, but a terrible beast, monstrous and dangerous for everyone. It was related to death, hell and disaster. Orthrus' father Typhon was identified with the Egyptian evil god Seth (who, incidentally, was sometimes portrayed as a dog-headed creature) and was regarded as one of the monstrous adversaries of Zeus. The latter fought with Typhon and defeated him with much difficulty. Finally, Ogo himself is, as we know, a very harmful character in the Dogon mythology.

The worship of a dangerous dog was widespread in the ancient world, and this is rather strange: the dog was in fact the "first friend" of ancient man and played a very important part in his everyday life. Nonetheless, the fact remains: dogs (as well as wolves and jackals, which seems much more natural) were regarded as chthonian animals, guardians of the underworld. The "Inmost Story" of the Mongols contains a motif of monstrous metal dogs who feed on human flesh. The terrible dog Yarchuk, from Slavic mythology, had a wolf tooth in his mouth and two vipers under his lower lip. According to a Russian belief, a Solar eclipse happens when the heavenly wolf swallows the Sun (this idea was not unfamiliar to many other peoples).

The Ukrainians believed that Ursa Major was a team of horses with harness: "every night a black dog tries to bite through the harness, in order to destroy the world, but he does not achieve his disastrous aim; at dawn, when he runs to drink from a spring, the harness renews itself" {8, p. 168}. Another version of this story states that a dog was chained beside Ursa Minor; he tries in every way to gnaw through his iron chain, and when this happens, the world will perish. According to the famous ancient Greek philosopher Proclus, who lived in the 5th century A.D., "the fox star nibbles continuously at the thong of the yoke which holds together heaven and Earth"; the Germans added that "when the fox succeeds, the world will come to its end" {9, p. 385}. One can find some interesting details of this future event in the Nordic mythology. It has been called "Ragnaroek", and the wolf Fenrir, together with the great dog Garm, play leading parts in it. Having snapped his fetters (which, incidentally, were made of nothing), Fenrir will devour the Sun and the supreme god Odin.

[**COMMENT:** For additional information, see *Planet X Nibiru: Slow-Motion Doomsday*, Chapter 9 ("Yggdrasill & Ragnarok"). RS]

These fetters are of much importance for our subject. As was ascertained by the Russian philologist Dr. Vyacheslav Ivanov, the motif of the fight against the dragon in Slavic mythology grew out of an older motif of the hero-blacksmiths, chaining up a terrible dog. What is still more essential, "over the whole territory of Eurasia, this mythological complex is associated both with the Great Bear ..., with a star near it as a dog which is dangerous for the Universe, and also with blacksmiths ... " {10, p. 210}. One should remember that, although Sirius is far from this constellation in the firmament, it belongs to the same star-cluster.

Now, let us pay some attention to other Sirius names. There exists in mythology some kind of "principle of complementarity": you can describe a complex phenomenon, using a set of quite different, even incompatible, images. Thus, the first satellite of Sirius is at the same time an empty husk of a millet grain, and the Pale Fox himself. Just as much, Sirius may have been represented as the Dog, the Arrow, the Triangle, as well as in many other ways. This star was either the tip of the arrow (in Mesopotamia and Persia), or its target (in China, as well as in Ancient Egypt). The Chinese mythical emperor Huang-ti was both a smith and an archer; on an ancient picture he aims at the celestial jackal, located beside another star, which represent, probably, the A and B components of this system {9, ill. between pp. 216 and 217}. I would like to recall in this connection the hypothesis of the Russian scholar Dr. Igor Lissievich about possible paleovisits at the early stages of China's history. Huang-ti was the main character of these hypothetical events (see: {11}).

The Iranian mythology personified Sirius as Tishtrya, the divine archer (the corresponding character in the Vedic myths was Tishya). The name "Tishtrya" goes back to the Sanskrit term "three stars" and to an older Indo-European one of the same meaning. Some scholars prefer to see here a designation of the Belt of Orion, but it seems to be just an *ad hoc* conjecture. On the other hand, the name "three stars" is quite justified in terms of the Dogon concept of this stellar system. There is, by the way, a direct relationship between the word "Tishtrya" and the name of the hellish dog Cerberus.

Thus, there are in various parts of the world some traces of an ancient — and rather clear — concept of Sirius as a dangerous stellar system, consisting of three stars. Its transformation has been described, first, as the transition from Typhon (a fiery monster in rage, that is a red giant before its change into a Supernova) to Orphrus (a dangerous but suppressed beast, that is the core of the red giant in the process of its "calm" turning into a white dwarf). Second, the Dog is usually chained up by sacred blacksmiths, which can be interpreted as a description of astroengineering activity by a supercivilization. Nommos are also considered as heavenly blacksmiths, but they do not chain up the Fox; they simply circumcise him. This rather unexpected metaphor expresses very clearly the main point: it was necessary to remove the excess of stellar matter from Sirius B. The 240 years of increased brightness of the star looks like a slow explosion of this "cosmic bomb".

When did all this happen? Astrophysical data suggest that the lifetime of Sirius B as a white dwarf has been 30 to 100 million years. However, some classical authors, such as Ptolemy and Seneca, described Sirius as red, which is very different from its present white-bluish appearance. For instance, Seneca wrote: " ... The redness of the Dog star is deeper, that of Mars milder, that of Jupiter nothing at all." This enigma has been discussed by astronomers since the 18th century up to now, and it remains still unsolved. It is astrophysically very unlikely that Sirius B could have been a red giant as recently as 2,000 years ago; but we cannot rule out entirely the possibility of lasting astroengineering works in this system. In any case, attempts to explain the red color of Sirius by some atmospheric causes are not very convincing. There is some evidence that the epithet "red" was not unusual for Sirius in the past. Thus, Tistrya was called "aurusha", what can be translated either as "white" or as "red". In Egyptian hieroglyphic writing, Sirius was depicted as a red triangle with a small semicircle and a five-pointed star near it (see Ref. 12). The Babylonians referenced to the star as "shining like copper". Finally, the Dogon represent *Po tolo* by a red stone (let us note it is precisely *Po tolo*, not *Sigi tolo* or *Emme ya tolo*, which is represented in such a way!).

In a recent work {13} R. Ceragioli has made an attempt to solve the riddle of Sirius' redness in the context of classical philology: the color red was in antiquity a token of danger. The most typical cultural pattern for Sirius connected it with fire, fever, rage, bloodshed, heat and other perils; that is why it may have been called red even in spite of evidence. It is questionable, however, if Ptolemy and Seneca were so much devoted to the cultural tradition that they did not trust their own eyes and took a color of Sirius' scintillations for the intrinsic color of the star. It seems more appropriate to assume that they did in fact see Sirius as red, even though this can have been just a temporary reddening related to some physical (or astroengineering?) processes in this stellar system (*cf.* {14}).

What is even more important, the solution suggested by R. Ceragioli does not provide the answer to the main question: why the ancients attached so great "negative" importance to Sirius? Egyptian priests watched this star closely at its heliacal risings, believing that its bright and white color presaged abundance, and its redness betokened war. The inhabitants of the Greek island of Ceos, when expecting Sirius' rising "prayed for the north winds to cool the 'Dog's' heat, which in their myths had once threatened to burn the world" {13, p. 615}. All that fits well with the "astroengineering hypothesis", raising at the same time some doubts: was "the cosmic bomb" discharged completely? Let us remember that the myth of the Dogon tells us that the blacksmiths only chained up the Dog, but it does not mean they rendered it quite harmless.

Therefore, we can suppose that alien astroengineering activities inside the Sirius system were finished only recently (if at all). Yet, they could have started in a much earlier epoch, even a few millions of years ago. However strange this may sound, we have another evidence of a fantastically deep historical memory of the Dogon: they know quite well that the Lake Bosumtwi in Ghana was formed when a giant meteorite fell on the Earth {2, p. 1961}. According to the results of a special investigation, this infall happened not later than 1.3, or even 1.6, million years ago. It is rather doubtful that somebody, living then on our planet (it was the epoch of *Homo habilis* and maybe of the early *Pithecanthropus*), could have retained this information and conveyed it to the future *Homo sapiens*. This knowledge may also be of paleovisit origin. Of course, we should not understand the Dogon mention of the "first year of the life of man on Earth", when, as they believe, Sirius B exploded, too literally, but it would be a mistake to reject these data *a priori*.

Now, what can we conclude from all that has been said above? The astroengineering hypothesis seems to be worthy of further investigation. It can hardly be proved just on the basis of mythological studies, but such studies can lead us to a preliminary outline of those distant (in time, as well as in space) events. Mythology may be regarded as a special language, which has preserved for us fragmentary data from the dawn of the world. I mean here by the "world" not only the Earth, but rather all our region of the cosmos — which has been called by the Dogon the "internal system of the stars". Events, that once took place in various parts of this region, were "projected" onto the firmament with its visible luminaries, becoming subsequently the subjects of mythological stories. These stories have interacted and become partly confused, so that it is now almost impossible to go this way back and reach the initial point. It only remains to rely on human imagination as another instrument of knowledge. At the same time, we should be very careful when trying to prove our assumptions. Usually they are more temporary tools than faithful models of reality. Thus, the concept of paleovisits as arrivals of extraterrestrial starships whose crews taught our ancestors to the fundamentals of civilized life and science, may prove to be uncritical adoption from science fiction stories, whereas the real situation was much more complicated.

There can have been some events in the history of the "internal star system", which we can neither understand as yet, nor even assume. There are, for example, in the Dogon mythology some hints at the multidimensional structure of the Universe. Moreover, Nommos seem to be not a "simple" supercivilization whose origin is similar to that of our civilization, only the level of development being much higher, but rather an independent branch of evolution of cosmic intelligent beings, very different from such planetary offspring as we humans are. It is very important to go from questions to reliable facts and convincing answers, but it may be still more important to go from answers to new questions.

References:

1. R.K.G. Temple. *The Sirius Mystery*. London: Sidgwick & Jackson, 1976.
2. E. Guerrier. *Essai sur La Cosmogonie des Dogon: L 'Arche du Nommo.* Paris: Robert Laffont, 1975.
3. M. Griaule, G. Dieterlen. *The Pale Fox*. Chino Valley: Continuum Foundation, 1986.

4. V. Rubtsov. "Beyond the Sirius Lore". *Ancient Skies*, 1985, Vol. 12, No.4.
5. F. D'Antona. "The Binary System Sirius in the Context of Stellar Evolution". *Astronomy and Astrophysics*, 1982, Vol. 114, No.2.
6. V.N. Arskiy. "The Address of a Civilization?". *Zemlya i Vselennaya*, 1989, No.5 (in Russian).
7. R.H. Allen. *Star-Names and Their Meaning.* N.Y., 1899.
8. A.N. Afanasyev. *The Life Tree.* Moscow: Sovre-mennik, 1983 (in Russian).
9. G. de Santillana, H. von Dechend. *Hamlet's Mill.* Boston: D.R. Godine, 1983.
10. V.V. Ivanov. "The ancient Balkan and all-Indo-European text of the myth of the hero-killer of the Dog and some Eurasian parallels". *Slavyanskoye I Balkanskoye Yazykoznaniye.* Moscow: Nauka, 1977 (in Russian).
11. I.S. Lissevich. *Le vol interstellaire dans les légendes et les myths anciens.* J. Bergier, G.H. Gallet (Eds.) *Le Livre des Anciens Astronautes.* Paris: Albin Michel, 1977.
12. A. Stentzel. *Aegyptische Zeugnisse fuer die Farbe des Sirius im Altertum.* Astronomische Nachrichten, 1927, Bd. 231, Nr. 5542.
13. R. Ceragioli. "Behind the 'Red Sirius' Myth". *Sky and Telescope*, 1992, Vol. 83, No. 6.
14. F. D'Antona, I. Mazzitelli. "Constraints on the corona model for Sirius B". *Nature*, 1978, Vol. 275, No. 5682.

*

http://www.templarlodge.com/stargate.html#siriuslie

[COMMENT: This URL was no longer available at the time of the publication of this book. Thus, I could not obtain a date for the following material, which was probably written in 2001-2002. I am personally not aware of the particular details of these events, but it is my impression that this was written in an attempt to debunk Temple's assertions, for whatever motive. The very fact that it mentions "channeling" makes me suspicious of it, although I am including it here for the record, without comment. RS]

"The Sirius Lie"
Extract Of A Lecture For The Turn Of The Millennium
By Filip Coppins

Scientists learn that the Dogon do not possess secret knowledge about the star Sirius and its companions. What some consider to be the best evidence for extraterrestrial beings coming from Sirius is therefore dealt a devastating blow.

In 1976, two major books on extra-terrestrial visitation were published: Zecharia Sitchin's *The Twelfth Planet* and Robert Temple's *The Sirius Mystery*. Of the two, the latter became by far more famous and even attained the status of a semi-scientific work, as many were impressed with the scientific-looking train of logic of the book. Temple stated that the Dogon, a tribe in Africa, possessed extraordinary knowledge on the star system Sirius, the brightest star in the sky, the star which became the marker of an important ancient Egyptian calendar, the star which according to some is at the centre of beliefs held by the Freemasons, the star which according to some is where the forefathers of the human race might have come from.

Temple claimed that the Dogon possessed knowledge on Sirius B and Sirius C, companion stars to Sirius [A] that are, however, invisible to the naked eye. How did the Dogon know about their existence? Temple referred to legends of a mythical creature Oannes, who might have been an extraterrestrial being descending on Earth from the stars, to bring wisdom to our forefathers. In 1998, Temple republished the book with the subtitle "new scientific evidence of alien contact 5,000 years ago".

The book's glory came crashing down earlier this summer [2001, RS], when Lynn Picknett and Clive Prince published *The Stargate Conspiracy*. That book stated that Temple had been highly influenced in his thinking by his mentor, Arthur M. Young. Young was a fervent believer in "the Council of Nine", a group of channelled entities that claim they are the nine creator gods of ancient Egypt. "The Nine" are part of the UFO and New Age, and many claim to be in contact with them. "The Nine" also claim to be extraterrestrial beings, from the star Sirius. In 1952, Young was one of the nine people present during the "first contact" with the Council, where contact was initiated by Andrija Puharich, the man who brought the Israeli spoonbender and presumed psychic Uri Geller to America. It was Young who gave Temple in 1965 a French article on the secret star lore of the Dogon, an article written by Griaule and Dieterlen. In 1966, Temple, at the impressionable age of 21, became Secretary of Young's Foundation for the Study of Consciousness. In 1967, Temple began work on what would eventually become *The Sirius Mystery*. As Picknett and Prince have been able to show, Temple's arguments are often based on erroneous readings of encyclopaedic entries and misrepresentations of ancient Egyptian mythology. They conclude that Temple very much wanted to please his mentor. It is, however, a fact that the end-result is indeed a book that would have pleased Young and his beliefs in extraterrestrial beings from Sirius very much, whether or not this was the intention of Temple.

Though Temple's work is now therefore definitely challenged, the core of the mystery remained intact. At the centre of this enigma is the work of Marcel Griaule and Germaine Dieterlen, two French anthropologists, who wrote down the secret knowledge on "Sirius B" and "Sirius C" in their book *The Pale Fox*. But now, in another recent publication, *Ancient Mysteries* by Peter James and Nick Thorpe, this "mystery" is also uncloaked, as a hoax or a lie, perpetrated by Griaule.

To recapitulate, Griaule was initiated in the secret mysteries of the male Dogon, who allegedly told him the secrets of Sirius' invisible companions. Sirius ('*sigu tolo*' in their language) had two star companions. This was revealed in an article that was published by Griaule and Dieterlen in the French language in 1950.

In the 1930s, when their research occurred, Sirius B was known to have existed, even though it was only photographed in 1970. There was little if no possibility that the Dogon had learned this knowledge from Westerners that had visited them prior to Griaule and Dieterlen.

Griaule and Dieterlen published their findings on the Sirius companions without any reference or comment on how extra-ordinary the Dogon knowledge was. It would be others, particularly Temple in the sixties and seventies, who would zoom in on that aspect. To quote *Ancient Mysteries*: "While Temple, following Griaule, assumes that '*to polo*' is the invisible star Sirius B, the Dogon themselves, as reported by Griaule, say something quite different." To quote the Dogon: "When Digitaria ('*to polo*') is close to Sirius, the latter becomes brighter; when it is at its most distant from Sirius, Digitaria gives off a twinkling effect, suggesting several stars to the observer." James and Thorpe wonder — as anyone reading this should do — whether '*to polo*' is therefore an ordinary star near Sirius, not an invisible companion, as Griaule and Temple suggest.

The biggest challenge to Griaule, however, came from anthropologist Walter Van Beek. He points out that Griaule and Dieterlen stand alone in the world in their claims on the secrets of the Dogon. No other anthropologist supports their opinion — or claims.

In 1991, Van Beek led a team of anthropologists who declared that they could find absolutely no trace of the detailed Sirius lore reported by the French anthropologists. James and Thorpe understate the problem when they say that "this is very worrying".

Griaule had stated that about fifteen percent of the Dogon tribe knew about this secret knowledge, but Van Beek could, in a decade of research with the Dogon, find not a single trace of this knowledge. Van Beek was initially keen to find evidence for Griaule's claims, but had to admit that there may have been a major problem with Griaule's claims.

Even more worrying is Griaule's background. Though an anthropologist, Griaule was interested in astronomy, which he had studied in Paris. As James and Thorpe point out, he took star maps along with him on his field trips as a way of prompting his informants to divulge their knowledge of the stars. Griaule himself was aware of the discovery of Sirius B, and it is quite likely that he overinterpreted the Dogon responses to his questions. In the 1920s, before Griaule went to the Dogon, there were also unconfirmed sightings of Sirius C. Was Griaule told by his informants what he wanted to believe? It seems, alas, that the truth is even worse, at least for Griaule's reputation.

Van Beek actually spoke to the original informants of Griaule, who stated: "Though they do speak about '*sigu tolo*' [interpreted by Griaule as their name for Sirius], they disagree completely with each other as to which star is meant; for some, it is an invisible star that should rise to announce the '*sigu*' [festival]; for another it is Venus that through a different position appears as '*sigu tolo*'. All agree, however, that they learned about the star from Griaule."

So whatever knowledge they possessed, it was knowledge coming from Griaule, not knowledge native to the Dogon tribe. Van Beek also discovered that the Dogon are of course aware of the brightest star in the sky, which they do not, however, call '*sigu tolo*', as Griaule claimed, but '*dana tolo*'. To quote James and Thorpe: "As for Sirius B, only Griaule's informants had ever heard of it."

With this, the Dogon mystery comes to a crashing halt. *The Sirius Mystery* influenced more than twenty years of thinking about our possible ancestry from "forefathers" who have come from the stars. In 1996, Temple was quick to point out the new speculation in scientific circles on the possible existence of Sirius C, which made the claims by Griaule even more spectacular and accurate.

But Temple was apparently not aware of Van Beek's recent research. With this new research of both Van Beek and the authors of *Ancient Mysteries*, we uncover how Griaule himself was responsible for the creation of a modern myth, which, in retrospect, has created such an industry and almost religious belief that the scope and intensity can hardly be fathomed. Nigel Appleby, in his withdrawn publication *Hall of the Gods*, which was, according to Appleby himself, tremendously influenced by Temple's book, Appleby spoke about how Temple believed that present-day authorities were apparently unwilling to set aside the blinkers of orthodoxy or were unable to admit the validity of anything that lies outside their field or offer a challenge to its status quo. He further wondered whether there was also a modern arrogance that could not countenance the possible scientific superiority of earlier civilisations. It seems, alas, that Griaule, a scientist, wanted to give earlier civilisations more knowledge than they actually possessed. And various popular authors and readers have since been led into a modern mythology, the Age of the Dark Sirius Companion.

Chapter 5

The Sirius Connection

*

The Sirius Connection — Unlocking The Secrets Of Ancient Egypt
By Murry Hope (Element Books, Great Britain, 1996)

Chapter 5: Sirius — The Binary Star

Pages 79-81

Sirius, in the constellation of Canis Major, is one of the brightest stars in our night sky, and being only eight-and-a-half light years from the Earth also means it is one of our nearest neighbours. Around the middle of the last century the astronomer Bessel studied Sirius over a period of time and noted a perturbation in its movements, indicative of the presence of another body close enough to effect a gravitational pull of some considerable force. Yet, due to its extreme brightness, Bessel could find no trace of a mass large enough to affect a star of this size. Some years later, a very small white dwarf star was discovered circling Sirius, the orbital period of which was calculated to be around fifty Earth years. (Technically speaking, Sirius B does not orbit Sirius A, *per se*, rather both stars orbit a common centre of gravity.) This second star, lost in the glare of its companion, became known as Sirius B, sometimes called Digitaria, and has since been photographed.

Astronomers have now learned more about the nature of white dwarfs — stars that do not give out much light, but exert an enormous gravitational pull because of the extreme density of their atomic structure. A white dwarf is a star that has used up its lighter hydrogen and helium atomic fuel and has collapsed, which means that the remaining elements have become so densely packed that the nature of its substance hardly equates to matter as we know it. When atoms are compressed to such an extent the resulting mass becomes extremely heavy. [Robert K.G.] Temple tells us that a cubic foot of the surface of Sirius B would weigh 2000 tons, and a matchbox filled with star's core material would weight approximately fifty tons!

Some astronomers believe they have detected a third star in the Sirius system. A man named Fox claimed to have seen it in 1920; and again in 1926, 1928 and 1929 it was supposedly seen by Drs. van den Bos, Finsen and others at the Union Observatory. Then, suddenly, the elusive star disappeared! More recently, Irving W. Lindenblad, of the US Naval Observatory in Washington, probed the Sirius system. He failed to detect a third star, although he gleaned much additional information about Sirius B in the process.

In his controversial book, *The Sirius Mystery*, Temple postulated that beings from the Sirius system visited Earth many thousands of years ago and were partly (if not entirely) responsible for the leap from primitivity to the high standard of culture and civilisation achieved by the ancient Egyptians between the years 4500 BC and 3400 BC. His hypothesis was based in part on knowledge of the Sirius system possessed by the Dogons, an African people who live in Mali (the former French Sudan). These people he believes to be the direct descendants of those pre-dynastic Egyptians who could have witnessed the arrival here on Earth of aliens from the Sirius system, their information dating back to those early days when their ancestors received it directly from their galactic visitors. In what other way, he questions, could they have gained knowledge of the existence of the invisible Digitaria or the other astronomical details they apparently have possessed for centuries concerning this distant system? In the light of the information contained in the letter received from my Egyptian contacts, who, incidentally, had never heard of Temple or his book, his conjectures would appear to have a ring of authenticity about them. What a pity he did not come across the Tutsi data, it would certainly have helped him to tie up a few loose ends.

The whole of the Dogon religion and its accompanying rites are built around the concept of the Sirius system. The Dogons considered Digitaria to be of even greater significance than its larger and brighter companion, even though it was not visible to them. They also possessed knowledge of a third luminary which they called Sorghum-Female, *emme ya* the 'sun of women', the 'little sun' (which Temple has designated as Sirius C), which is accompanied by a satellite planet they call the 'star of women'. Sorghum-Female they believed to be larger than Digitaria and four times lighter in weight. To them it was the seat of the female souls of all living and future beings. This star of women is represented by an equidistant cross and outlined by three points, a male symbol of authority, surrounded by seven dots, or four (female) plus three (male) which are seen as the female and male soul.

Pages 84-85

The Egyptians and Dogons were by no means the only races to accord special powers to Sirius. Another African tribe, the Bambara, called it 'the star of foundation' *sigo dolo*, which is the same term used by the Dogon, while like the Dogon they also referred to the companion star (Sirius B or Digitaria) as *fin dolo*. Jointly the two stars were called *fã dolo fla* (the two stars of knowledge) because 'it represents in the sky the invisible body of Faro' conceived as a pair of twins — the implication being that the star is the seat of all learning. According to Bambara legend, Sirius was Mousso Koroni Koundye, whose twin, Pemba, maker of the earth, was a mythical woman whom he continually chased through space, but never managed to catch. This renders Mousso Koroni Koundye comparable to the Dogon Yasigui. Both of these female characters are said to have inaugurated circumcision and excision, as a result of which Sirius is known to both the Bambara and the Dogon as the 'star of circumcision'. The Bozo were equally familiar with the system. They called Sirius *sima kayne* (literally: Sitting Trouser), and its satellite *toñõ ñalema* (Eye Star). Can we possibly effect a connection here between this Eye Star and the fabled Eye of Ra which was passed between the Egyptian deities Hathor, Bast and Sekhmet?

[COMMENT: The God Yahweh commanded Moses to order that all Hebrew men be circumcised. This custom was passed on to the Falasha Jews of western Ethiopia; and it is still practiced today by all Abyssinian Christians, as well as Moslems, since Islam is an offshoot of Judaism. The custom is also found in other parts of Africa; but whether these particular tribes, such as the Bambara and Dogon, initiated circumcision independently of the Jews is unclear to me. And as we know, the Hebrew "Yahweh" is none other than Nibiruan Crown-Prince Enlil. Thus, the origin of the practice of circumcision also links Planet X Nibiru to the Sirius System. Regarding the "Eye Star" or "Eye of Horus", that is probably a symbolic remembrance of the "millwheel" appearance of Planet X as The Winged Disk, when seen from here on Earth. RS]

Pages 87-89

I find it fascinating that these simple African tribes should have sustained such a profound and accurate cosmic teaching for so many centuries, and one cannot help but view them, along with those mystically erudite Egyptian souls who were generous enough to trust me with the teachings of their inner traditions, as the worthy custodians of the Light and knowledge of the stars. Needless to say, orthodox astronomers tend to view all this with considerable scepticism, although they would appear to be at a loss to explain how the Dogons obtained this knowledge.

There would seem to be little difference between these accounts of Siriun landings and the information these gentle visitors bestowed upon their hosts. The fact that they left a sense of love and a yearning for their return may be taken as evidence of their kindness, generosity and cosmic maturity. The Dogon Ogo an the Egyptian Set have much in common, while Yourougou's eternal search for his anima re-echoes both the Isis-Osiris tale and that of the Gnostic Christos and Sophia.

[COMMENT: When the Dogon first had contact with the Sirian visitors, Sirius B was probably still a large, visible red star, part of a double-star system with white Sirius A. So if the Dogon asked these visitors whence they came, the visitors would have pointed at reddish Sirius B as their home. As Sirius B gradually shrank into neutron-dwarf invisibility, the Dogon priests would nevertheless have still had a memory of it. And yes — orthodox, establishment astronomers have absolutely no explanation at all as to how the Dogon obtained their accurate knowledge of Sirius B without the use of a high-powered telescope. RS]

Sirius A has always been associated with Isis, and Dogon information would seem to confirm the feminine nature of this beautiful blue-white star. Osiris could therefore be Sirius B. Could the Isis/Osiris legend be describing a series of actual events that took place in the Sirius system prior to the collapse of the Osirian or Sirius B star, and how does the third star feature in this saga?

[COMMENT: And connecting these dots is the purpose of my book, as was discussed in Chapter 1. RS]

Sirius C is sometimes linked with Nephthys, the hidden one, or her son Anubis, while [Robert K.G.] Temple theorises that the three Siriun stars might possibly be equated with the three Egyptian goddesses Isis, Anukis and Satis, and quotes Neugebnauer as stating specifically: 'The goddess Satis who, like her companion Anukis, is hardly to be taken as a separate constellation but rather as an associate of Sothis.' Anukis and Satis, the wives of Khnum, were frequently portrayed together with Sothis sailing in the same celestial barque. There is a theory popular in Europe that when Set usurped the Osirian throne and banished Isis and her baby son, Horus, the wilderness to which they were sent refers to a planet of harsh environment *outside of the Siriun system*, Earth, for instance, during an earlier stage of its development; and since, both metaphysically speaking and as embraced by the scientific concept of non-locality, all time is one, Isis is still with us, albeit unrecognised and rejected (the feminine principle in a male-orientated world!), endeavouring to raise her son, Horus (the knowledge, beauty and wisdom of the Aquarian Age?). It bears consideration.

General opinion tending to consider Sirius A as carrying feminine energies, Sirius B as masculine, and Sirius C or Sorghum-Female (if indeed the latter is a star and not a large planet), also in the feminine context, the inference would be that the feminine influence predominates in this system. Following through the idea that our solar system was seeded from Sirius, this would surely be a point in favour of the Great Mother or Creatrix principle rather than the masculine father-god so beloved of certain popular religions. I was interested to note Bauval and Gilbert's association of Osiris with Orion or Sah, the Sahu being considered the most elevated of the subtle 'bodies' which I chose to allot to Osiris (see *The Way of Cartouche*). In metaphysical terms, to become 'Osirified', indicated that the soul had attained to some exalted position in the spiritual hierarchy. My Egyptian contacts also informed me that according to their ancient beliefs not everyone has a *BA* or soul — the 'subtle vehicle' necessary to the ascent to the next stage of spiritual development. This has to be earned over many lifetimes. So until this has been achieved, we are bound to the confines of Earth. This equates with information of a paraphysical nature which I have recently covered in my book *Cosmic Connections*.

As Sirius was the most important star for the ancient Egyptians, the preceding four decans, which comprise the constellation of Orion also, as Bauval and Gilbert have suggested (see *The Orion Mystery*), assumed significance in their beliefs. The final portion of Orion rises above the horizon one 'hour' before Sirius, thus giving it the appearance of a kind of advanced guard. Sirius was known as Sept, *spd* or *spdt*, the 't' ending indicating the feminine, while the old name for Orion was Sah. Another interpretation is that the Isis/Osiris episode exemplifies a cosmological drama, with Isis (the Divine Mother) representing the Earth, being miraculously impregnated by the dead Osiris (Sirius seeding this solar system and our planet in particular) and giving birth to Horus (humankind) who is obliged to fight ill-health in youth (the learning of self-healing) before he is strong enough to overcome the evil of Set (his own lower nature, and other misplaced energies) and establish his father's kingdom on Earth (the spiritual maturity of *Homo sapiens*).

The now-spent Sirius B might well represent Egypt's Osiris, hence the association of that Neter with the dead, or those who have passed into another, more subtle dimension. When a great star collapses (dies), the guiding Essence (deva) behind its growth and development ascends to a higher plane or moves to a time-zone nearer to the Centre or Creative Force. On the other hand, the same principle could be applied to Ra, whose energies were also withdrawn, which would equate Sekhmet, or one of the lion goddesses, with Sirius A.

Pages 93-94

Until evidence of a more substantial nature is forthcoming one can only follow one's intuition in these matters, and my own tends to favour the fair people (Crystal People) and the leonids (Paschats), as the two main races of the Siriun star system, although I am given to understand that there were also some highly evolved forms of plant and amphibian life on the smaller planet believed to have orbited Sirius B which originally housed the leonids, so we may be talking about different planets and a different sun within the Siriun system (see my book *The Lion People*). But at this point we must needs take care, as the fine line between logical speculation and fantasy can easily and sometimes unwittingly become crossed. Besides, we may well be dealing with different time references from those normally accepted by modern science as referring to the past or future. For example, since the leonid legend dates back to a time *prior to the collapse of Sirius B*, which would count in millions of years according to our concept of linear time, these beings might, now, be operating from the subtle dimensions as far as we are concerned, or existing in a universe parallel to the one we view through our telescopes. If all this strains the credulity somewhat, consider Professor Fred Hoyle's Panspermia Theory, which conceives of alien micro-organisms distributed throughout interstellar space, and penetrating the Earth's atmosphere. Hoyle's answer as to whether life originated on Earth would be a resounding 'No'!

[COMMENT: And you, dear reader, yourself must be the ultimate judge here, if you can believe this or not. It is definitely a grand anomaly in our history. And it is certainly "ironic" (to put it mildly) that we radical Nibiruologists are to some extent in the same philosophical school as those fundamentalist evangelical Christians who believe in "intelligent design"! RS]

Page 96

Cosmic principles, whether they represent Sirius, Orion or any other part of the infinite universe do have a constancy which reaches beyond the boundaries of time and space. Whether we contact them in their capacity as givers of light, love and knowledge, or if, in invoking their energies, we trigger off the darker side of our own natures, will depend very much on us, like ever attracting like!

Chapter 6

Myth & Mystery

*

Myth & Mystery
An Introduction To The Pagan Religions Of The Biblical World
By Jack Finegan (Baker Books, Grand Rapids, Michigan, 1989)

Pages 48-51

The myth of Osiris was the most widespread of Egypt. In essential outline the story tells of a ruler slain by his brother, and of the ensuing struggle for the sovereignty between the ruler's son and the murderer — a story so conceivable in human life as perhaps to be accounted basically a legend, that is, to rest upon actual historical reminiscence. In the elaborated myth the slain ruler is Osiris and his brother, the murderer, is Seth. By a stratagem Osiris was encased in a coffin and cast into the Nile; in another view, his body was cut into many pieces and scattered over the land, and where the pieces fell, the land was fertile and green plants grew. Osiris, it is evident, is a god of vegetation and agriculture; Seth a god of the wilderness and of destruction.

Isis, in the myth, is the wife of Osiris. In grief and tribulation over the death of Osiris she searches for and finally recovers the body. (In the astronomical form of the myth the sorrowing Isis is the star Sirius and follows Orion, which is "the glorious soul [*ba*]" of Osiris.)

[COMMENT: Unlike other places, here we have a direct reference to the astronomical connection between Osiris/Isis and Orion/Sirius. Orion's Belt Stars point to Dog-Star Sirius. No wonder that they are intimately associated. How much more evidence do we need? And this summary of the legend is yet another "interpretation" that is consistent with those quoted elsewhere. RS]

When the body of Osiris is found, Isis prepares it for burial with the help of Anubis, the jackal-god of the western desert and the god of embalmment. Together Isis and her sister Nephthys (herself sister and wife of Seth) speak charms over the body of Osiris and he experiences resurrection. In union with him Isis becomes the mother of a son, Horus, while Osiris goes on into the beyond to be the god of the dead. The child Horus (Greek, Harsiese, "Horus, son of Isis"; also Harpokrates, "Horus the child", seen with a child's long lock of hair and the finger in the mouth) is raised in the safety of the papyrus swamps and and claims his father's throne. Then — in variant forms of the myth — Horus and Seth rule in the two parts of the land (Horus in Lower Egypt, Seth in Upper Egypt), or Horus rules the whole land while Seth is relegated to the wilderness. Thereafter every king of Egypt rules as Horus (or as both Horus and Seth, a representation that provides some inconsistencies in what follows) and, when he dies, becomes Osiris and rules in the underworld while his son, the new king, rules on earth as the new Horus.

Osiris is thus the dead king, and the myth has to do with the succession in the rulership of Egypt with its promise of stability in the land, but it is plain that the theme of resurrection is also prominently involved. Osiris is revived and becomes immortal. His immortality is also associated with the circumpolar stars that never set, and his resurrection is associated with such other natural events as the annual rising of the Nile and the death and life of vegetation. From this point of view, and when some of the prerogatives of the king gradually spread to his nobles and on to the common people, Osiris became not only the dead king and the ruler of the dead but also the prototype and savior of the common dead.

[COMMENT: *Certainly* the memory of Osiris is associated with the "circumpolar stars". When one fully grasps the concept of Hyperborea/ Yggdrasill, The Cosmic Tree, there is really no other explanation. If Planet X Nibiru originated in the Sirius System, was ejected as a result of the implosion of Sirius "B", thrown into our nearby system, captured by our Sun and now periodically stations itself above our North Pole, then the connection of Osiris to Hyperborea, The North Mountain, is not surprising. It simply becomes "a matter of perspective". Some of us "instinctively" understand this, and some of us don't. RS]

As such, the myth of Osiris was the basis of the Osirian mysteries, in which the death and life experiences of the god were presumably reenacted in dramatic form and, under the guidance of the priests, became also the experience of the devotee.

Many cities were centers of the Osirian cult, the most important being Busiris in the Delta (supposed birthplace of Osiris and site of the Djed pillar, his symbol) and Abydos in Upper Egypt (supposed place of his tomb; the present Umm el-Qaab).

[COMMENT: This "Djed Pillar" is merely The Cosmic Treetrunk, the electromagnetic tether-beam connecting our North Pole to Planet X Nibiru's South Pole (Bifrost, Rainbow Bridge To Asgard). These establishment mythographers just do not understand the "facts"! Any sort of "symbolic" Djed Pillar at Abydos, Egypt, would have been only an earthly representation, after the fact, of the historical Cosmic Pillar Hyperborea, which last disappeared around 687 BCE. RS]

As the cult spread throughout Egypt, other deities were incorporated into the framework of the myth — naturally gods of the dead such as Anubis and Khentamentiou (the wild dog of the dead in Abydos), the chthonian god Sokaris in Memphis, plus other local deities. At Heliopolis it was necessary to find accommodation with the doctrine of the world rule of Atum-Re, and this was done by establishing a genealogical relationship between the sun-god and the Osirian cycle.

The works known as the Coffin Texts and the "Book of the Dead" (represented both in texts and in tomb paintings) also express in detail the prevalent belief in and hope for life in the beyond. Here Osiris as the god of the dead is associated with strongly ethical ideas. In their resurrection the dead are pictured as standing in the presence of Osiris as their judge; to obtain blessed immortality their soul must be able to affirm moral worthiness by reciting a long list of sins that have not been committed and by reporting good deeds that have been performed, for example, "I have given bread to the hungry, water to the thirsty, clothing to the naked, and a ferry-boat to him who was marooned." Other works (originally of New Kingdom date, and also represented both in texts and in tomb paintings), the "Book of What is in the Underworld" (Amduat) and the "Book of the Gates", describe the sun-god Re as journeying in his barge during the twelve hours of the night through all the realms of the netherworld. For the blessed dead the promise is presumably implied of traveling through the night with Re, to then be brought with him out of the darkness into the light of the morning. For those condemned in the judgment, however, the Amduat pictures the most horrifying punishments.

[COMMENT: There are those like anthropologist and author "Acharya S" (*The Christ Conspiracy: The Greatest Story Ever Sold*) who have accumulated a lot of mythological information about various resurrectionist traditions. This Osiris resurrectionist legend could be compared to the resurrection of "The Jesus Christ" if it were not for the crucial fact that the image on the Shroud of Turin is that of Apollonius of Tyana. That, in and of itself, explains the reason that the "Mandaean False Messiah" (see below) did not die on the cross. As the *Quran* states, "Issa" (Jesus) was only made to appear as one who had been crucified and died. RS]

Pages 162-163

The residence of the chief [Greek] gods is on Mount Olympos (the highest peak in Greece, 9,600 feet), the place chosen by Zeus. Here the gods live in much the fashion of human beings, with the same virtues and vices (as some of the foregoing references well illustrated), but they are superior in power, and immortal, due to their divine food and their divine drink of ambrosia and nectar. As to exactly which deities dwell on Olympos there are some different reckonings; one relief, said to be of the early fifth century B.C. and to come from Tarentum in south Italy, shows twelve Olympians, each with distinguishing attributes, in the order: Apollo with lyre, Artemis with bow, Zeus with thunderbolt, Athena with her favorite bird the owl, Poseidon with trident, Hera with scepter (topped by pomegranate, symbol of fecundity), Hades with staff, Persephone with ears of corn, Ares wearing a helmet, Aphrodite holding a flower, Hermes in a cap and with his herald's staff (*caduceus*), and (probably) Demeter with a basket. Thus of the highest twelve, there are five children and one grandchild of Kronos and Rhea (Zeus, Poseidon, Hera, Hades, Persephone, Demeter), and six offspring of Zeus and several different wives (Apollo, Artemis, Athena, Ares, Aphrodite, Hermes).

[COMMENT: The Greek Mount Olympus is only "symbolic" of the true Hyperborean Mount Olympus. The twelve "Major Arcana" noted here obviously reflect the idea of the "Council Of Twelve" of Planet X Nibiru, although Finegan makes no reference to Zecharia Sitchin, whose list of the current "Council Of Twelve" includes the following (Greek names): Kronos, Rhea, Zeus, Maia, Poseidon, Libya, Hera, Hermes, Ares, Apollo, Artemis and Aphrodite. Thus, Finegan's list is slightly different than Sitchin's. RS]

Powerful as the Olympian deities are, they are not always able to accomplish their will, nor are they exempt from inevitable results that come from their actions. In fact they themselves, as well as human beings, are in the last analysis subject to Fate (*moiksa*), the ultimate determiner of destiny. Here too, however, this concept was expressed mythologically. Fate is a goddess bearing the several names of Ananke (necessity), Tyche (fortune), Heimarmene (allotted portion), Dike (justice), and Nemesis (retribution). Again there are three Fates, and they are pictured as weavers: Lachesis (the measurer), Clotho (the spinner), and Atropos (the cutter of the thread of life).

Pages 262-263

[COMMENT: The following information from Finegan's book, as well as Appendix B, which also references the Mandaeans, is included here for the benefit of those who have previously read my *Apollonius Of Tyana & The Shroud Of Turin*, since it is not my intention to write another book about Apollonius. Those readers who are not interested in the subject of Apollonius can immediately jump ahead to Chapter 7, and readers who may indeed be interested can return to this material later if they so desire. The "Mandaean Religion", which later evolved into Manichaean Christianity, was found in southeastern Iraq, Kuwait and southwestern Iran, in that area where the Tigris and Euphrates Rivers converge in Iraq and not far downstream merge with the Karun River flowing out of Iran. The "Mandaic Language" was an eastern dialect of Aramaic. Mandaean literature spans the period from about 250-1530 CE. In their earliest times they were identified with the "Nasoraeans", a word that is derived from the Aramaic verb *nasar* which means "to keep secret". RS]

Other [Mandaean] texts are known as "secret scrolls", being accessible only to the Mandaean priests. These include a rite of marriage called "The Marriage Ceremony of the Great Shishlam", and a rite for the consecration or "coronation" of priests called "The Coronation of the Great Shishlam". As the titles indicate, the ceremony is in each case that of Shishlam Rba.

The name *Shishlam* means "perfected perfection", and is the Mandaean name for the Divine Man, who personified perfected and perfect humanity. His title *rba* means "great", and is also the word for a teacher, master, and initiator. Thus in the rites here described the Great Shishlam is regarded as the personification or prototype of the bridegroom and of the priest in their respective cases.

A very lengthy scroll known as "The Thousand and Twelve Questions" contains secret ritual and moral instructions, which are never imparted to the laity and are only given to the young priest at his initiation into the priesthood. The seven texts that compose the scroll deal with the mass of the soul, the ceremonies of marriage, burial, baptism, and the like. Two other "secret scrolls", short commentaries that have to do chiefly with the ceremony for the dead, are called "The Great First World" and "The Lesser First World". Both are illustrated with relatively crude drawings of holy trees and plants, probably pictures of spiritual worlds and beings.

[COMMENT: These drawings of holy trees and spiritual beings undoubtedly refer symbolically to The Cosmic Tree and the élite nobility of Planet X Nibiru. RS]

Another category of works is that of the so-called *Diwans*, a term that usually means a miscellaneous collection from various authors. The *Diwans* are said to number twenty-four, all in the form of scrolls, some with illustrations. The *Diwan Abatur* or "Progress Through the Purgatories" is preserved in a scroll about twenty feet long and is named for the light-being (Abatur) who weighs the soul in the beyond. Along with the text the illustrations, in an almost cubist and evidently conventionalized art style, depict the purgatories and their demons, the moon ship, the sun ship, the ship that ferries the souls of the righteous from the earth to the world of Abatur, the scales for the weighing of the soul, the heavenly baptizers in the world of Abatur who baptize the souls in the heavenly Jordan, and so on.

The "Diwan of the Baptism of Hibil Ziwa" describes a baptism (*masbuta*) that a Mandaean priest must undergo if he has suffered pollution by infringement, even accidental or involuntary, of ritual rules. As the prototype of such happening, it seems that the light-being Hibil Ziwa ("Light Giver") was sent into the seven worlds of darkness to perform certain tasks and incurred such pollution that he could not return to the world of light until he underwent a ritual meal (*zidqa brika*, "blessed oblation") and a baptism, and these happenings are related and pictured in the work in question.

Hibil Ziwa also figures in the "Book of the Zodiac" (*Sfar Malwasha*), which he is said to have given to Adam Paghra (the physical Adam) so that Adam might be able to foresee coming events. The work is indeed consulted by Mandaean priests for all sorts of astronomical knowledge in respect of the naming of a child, the choice of auspicious days, and the like.

[COMMENT: A discussion of the origin and significance of baptism is not within the purview of this book, except for Finegan's brief remarks which follow. Suffice it to say that the practice of baptism existed *before* the advent of Biblical Christianity and did not originate as a direct result of it. The practice became popular, however, with Christianity in the Levant, and from there it was later incorporated into the rituals of Manichaean Christianity. RS]

Page 265

Therefore the Sabeans in the Quran are probably also to be identified as the Mandaeans or as including the Mandaeans, thus indicating that Mandaean written texts must already have been brought together in some collected form so as to constitute a "book" and to authenticate the Mandaeans as "people of the book" by the time of Muhammad and the time when the Arabs brought Islam to Mesopotamia in the seventh century A.D. Very possibly it was in order to substantiate further their position in this very regard that the Mandaean *Ginza* was also named the "Book of Adam", and their "Book of the Kings" was also named for John (the Baptist) — both Adam and John the Baptist being prominent in the Quran. ...

It has also been shown that some of the poems in the Coptic "Psalms of Thomas", ascribed to Thomas, one of the first disciples of Mani (A.D. 216-277), are virtual adaptations of some of the Mandaean poetry (especially in the *masiqta*-hymns); thus again it is indicated that Mandaean texts existed in written form already in the third century A.D.

Pages 269-274

When four hundred years went by in Jerusalem, Jesus, son of Mary (*Mariam*) and head of the Christians, was born; he created a church for himself and gathered a community to himself (410.31ff.).

[COMMENT: These peculiar parenthetical citations refer to Mandaean manuscripts. RS]

Jesus Christ was baptized by John (*Johana*) and became wise through the wisdom of John, but then distorted the words of John, altered the baptism in the Jordan, and preached sacrilege and deceit in the world (51.14-17). His name is Immanuel (*Amunel*) and he called himself Jesus the Savior, but he is a false Messiah (29.19: 47.16-19). He was the son of Ruha (248.8-9) and is one of the seven angels who lead people astray: (1) the Sun, (2) Venus, (3) "Christ, the falsifier, who falsifies the original doctrine", (4) the Moon, (5) Saturn, (6) Jupiter, and (7) Mars (46.29-35). *Ginza* tells its readers to not fear Christ the Roman (*i.e.*, Byzantine, presumably so-called in reflection of the hostility between the Byzantine church and the Mandaeans), who alters the genuine revelation, nor to allow the Planets, Ruha, and Christ to exercise lordship over them (49.14-15; 437:12-13). One of the purgatories is that of Jesus Christ, in which are found those who denied the life and confessed Christ (186.40ff.). On the day when heaven and earth end, Ruha, Christ, the sun, the moon, and the planets, and those who have made confession to them, will fall to the king of darkness and to the final fire (203.11ff.; 255.22ff.). Thus the Mandaeans, in spite of some knowledge of biblical data, express as violent opposition to Christianity as to Judaism.

[COMMENT: The Mandaeans were expressing "violent opposition" to the teachings of the false Messiah, Jesus Barabbas, a topic which is discussed at length in my *Apollonius Of Tyana & The Shroud Of Turin.* The revolutionary bandit Jesus Barabbas, cousin of John The Baptist, was the man who was released from prison by Pontius Pilate in exchange for the crucifixion of Apollonius of Tyana. Regarding Jesus Barabbas, the Roman historian Celsus wrote in *Logos Alethes* in 180 CE: "It was Jesus himself who fabricated the story that he had been born of a virgin. In fact, however, his mother was a poor country woman who earned her living by spinning. She had been driven out by her carpenter-husband when she was convicted of adultery with a soldier named Panthera. She then wandered about and secretly gave birth to Jesus. Later, because he was poor, he hired himself out in Egypt where he became adept in magical powers. Puffed up by these, he claimed for himself the title of God." RS]

In addition to the mention of John the Baptist in the *Ginza* 51.12-51.14, there is another section in the *Right Ginza* (190.22-186.23) about John, in which we are told of the baptism of *Manda d-Hira* by John and of John's death and ascent to the world of light. Also there is a long section about Jahja/Johana in the book that bears his name ("Book of John" 75-123). In all, however, except for John's baptismal work in the Jordan River (including the baptism of Jesus), and a few details such as the names (Zakhria and Enishbai) and advanced age of his parents, there is very little contact between the Mandaean tradition and the New Testament narratives concerning John; rather John appears fully as a prophet of the Mandaean religion (so also in the *Haran Gawaita*).

After a final reference to Jesus, the head of the Christians, in the eighteenth tractate of the *Right Ginza*, the text continues (411.6ff.) with further respect to world history with a listing of early Iranian kings including the Kayanians Luhrasp and Gushtasp (Vishtaspa), of others including Solomon, son of David, and of Sandar (Alexander) the Greek. Somewhat thereafter we come to the name of King Ardban, and are therefore evidently in the time of the Parthian (Arsacid) dynasty (*c.* 250 B.C. - A.D. 226), for there were no less than five kings of that dynasty named Artabanus, with dates as follows: I (211-191 B.C.), II (128-123 B.C.), III (A.D. 12-38), IV (A.D. 80-81), and V (A.D. 212-226).

[COMMENT: King Artabanus III and his son Crown-Prince Bardanes were in power when Apollonius of Tyana and Damis of Ninevah visited Babylon in 24 CE. RS]

This Ardban, who is the only one named in the *Ginza*, must be Artabanus V, for after attributing to him a reign of fourteen years, it is said that "after him were Persian kings", with Ardshir at their head. The latter were the Sasanians or Neo-Persians, with Ardashir I (A.D. 226-241) as the founder of the Sasanian Empire. Then, after the "Persian kings", come the "Arabian kings", and of them it is said that they will reign for seventy-one years (414.18). Seventy-one years after the first Arab forces entered Persia (A.D. 632) and the last Sasanian king was killed (Yazdegerd III, d. 651) would reach only into the early eighth century. Since the Arabs actually ruled much longer, this tractate must have been compiled sometime around the middle of the seventh century (407).

MANDAEAN HISTORY

The broken text of the *Haran Gawaita*, which alone in the Mandaean literature professes to give historical information about the early history of the Mandaeans (whom it calls Nasoraeans), begins in the middle of a sentence with reference to an anonymous "him" whose identity has been lost: "And Haran Gawaita receiveth him and that city in which there were Nasoraeans, because there was no road for the Jewish rulers. Over them was King Ardban. And sixty thousand Nasoraeans abandoned the Sign of the Seven and entered the Median hills, a place where we were free from domination by all other races."

In the continuation of the text we are told that the Nasoraeans loved the Lord, that is, Adonai, until Mary, "a daughter of Moses", became the mother of the "False Messiah", who dwelt on Mount Sinai with his brother, and gathered to himself a people called Christians (3-4). Jahja/Johana (John the Baptist) was born too, and he was a "high healer whose medicine was Water of Life" and an "envoy of the High King of Light" (5-7). Sixty years after his death the Jews persecuted the Nasoraeans (7-8). Apparently in divine reprisal for this hostile action, Jerusalem was destroyed and made "like heaps of ruins"; also the Jews in Baghdad (Babylonia) where four hundred of their rulers had reigned for eight hundred years, were brought to an end (9). In succession to the descendants of King Ardban (the Parthian rulers, the Arsacid dynasty), the Hardabaean dynasty (evidently the Sasanians, since their rule followed that of the Parthians) ruled for 360 years (14-15). "Then the Son of Slaughter, the Arab [Muhammad and his successors]," took the sovereignty (15). After four hundred years of Arab rule, the False Messiah will return and perform miracles by fraud and sorcery during a reign of six thousand years (19-20). Thereafter, for some unexplained reason, "fifty thousand years will pass in calm ease ... and ill-will shall be removed from the minds of all peoples, nations, and tongues" (20). But finally there will be decadence and the end of the world will come (21).

Returning to the opening sentences of the text, the Nasoraeans are described as abandoning "the Sign of the Seven [planets]". This is surely a reference to Jerusalem, for we have already seen that in the *Right Ginza* it was the seven planets with Ruha their mother who built Jerusalem at the command of Adonai (410.8-11), and Christ "the falsifier" is named as one of the seven planets whose evil influence is warned against (46.24-35), while in another text the Seven are brought into connection with the law in Jerusalem.

[COMMENT: In the movie *The Passion Of The Christ* the Aramaic word "Adonai" is used for "God" or "Heavenly Father". For a full discussion of "Nasoreans" or Nazarenes, see Chapter 11 of *Apollonius Of Tyana & The Shroud Of Turin*. RS]

Also the statement in the *Haran Gawaita* (3) that the Nasoraeans "loved the Lord, that is, Adonai", until the birth of Jesus and the rise of Christianity, suggests that originally they had some relationship to the Jewish environment, although, as we have seen, they are later violently opposed to Judaism. The dating also seems to be very confused, for Mary, the mother of Jesus, is called a daughter of Moses, and Jesus is associated with Mount Sinai.

[**COMMENT:** This is most peculiar because Chilean Hermeticist Darío Salas Sommer, using the pseudonym of John Baines, wrote in *The Stellar Man* that "Jesus" (conceived by Hermes) was sent to Earth to rid it of "Archon Y" who was the "Yahweh" affiliated with Moses and who deceived Moses into helping him attain a foothold of power over the Earth. This Hebrew "Yahweh" is, of course, none other than Nibiruan Crown-Prince Enlil. RS]

Abandoning Jerusalem, then, we are told that the Nasoraeans were received in Haran Gawaita and entered the Median hills. Haran Gawaita means "the inner Haran", and is possibly a reference to Haran in northwestern Mesopotamia. The Median hills (*Tura d-Madai*) must be the mountainous regions of southwestern Iran, ancient Media. In the *Haran Gawaita* (10) it is said that the Median mountains are called Haran Gawaita. Therefore perhaps it was the whole territory from northwestern Mesopotamia to southwestern Iran into which the Nasoraeans moved.

The ruler who was "over them", presumably meaning the one under whom they had a favorable reception and received protection, was King Ardban. As we have already seen, there is only one Ardban mentioned in the *Ginza* and this is most probably the Parthian king Artabanus V; as thus apparently the most well-known king of that name in Mandaean tradition, it is probably he who is also meant here. The empire of the Parthians (*c.* 250 B.C. - A.D. 226) included Babylonia, and from the second century B.C. their major capital was Ctesiphon on the left bank of the the Tigris (opposite Seleucia). With the defeat and death of Artabanus V in A.D. 226 and the occupation of the palace of Ctesiphon by the victorious Ardashir, the Parthians were succeeded by the Sasanians. If the identification of King Ardban with Artabanus V is correct, the main movement of the Nasoraeans from the west into Parthian territories in the east was in the early third century A.D.

[COMMENT: And the early third century corresponds to the rise of Manichaean Christianity in the same region. The Manichaeans revered Apollonius of Tyana, not Jesus Barabbas, as the true Messiah. Finegan devotes his final 50-page chapter to the Manichaean Religion, only a small portion of which is included here. RS]

That in the course of time the Nasoraeans/Mandaeans lived both in the west and in the east is also suggested by other general references in their literature. As to the west, there is much information given about the Palestinian region, the Jews, and Judaism. For example, the *Jordna* (*i.e.*, the Jordan River) is of great importance and is the name for all "living" (*i.e.*, flowing) water, in which it is requisite that the numerous Mandaean baptisms and ablutions be performed. This fact is most convincingly explained by the supposition that their original baptismal practice was in the Jordan River in Palestine, near where they lived, and that when they were compelled to leave their homeland they went on calling each flowing water which they used in this way the "Jordan".

As to the east, there are many reflections of life in Mesopotamia and Iran. Linguistically, some eighty loanwords from the Akkadian and some 125 loanwords from the Iranian languages have been counted in the Mandaic language. Geographically, the Euphrates and the Tigris rivers are frequently mentioned (*e.g.*, *Right Ginza* 61.25; 414.7; *etc.*). The emphasis upon the seven planets may reflect a Babylonian background, while the contrast of light and darkness is much like the same contrast in Persia.

Continuing in the *Haran Gawaita* and in respect to the life of the Nasoraeans in the east, we gather that their situation was relatively favorable under the Parthians. In this time in Baghdad (Babylonia), it is reported, the Nasoraeans "multiplied and became many", and they had as many as 400 temples (*masknas*; 10). Under the Sasanians, however, the number of their temples shrank to 170 (14). This corresponds with the known fact that the Sasanians (Neo-Persians) were strong Zoroastrians and suppressed other faiths: the inscription of their high priest Kartir at Naqsh-i Rustam, for example, tells how he persecuted Jews, Buddhist ascetics, Brahmans, Nasoraeans, and Christians. As for "the Son of Slaughter, the Arab", when he "emerged" (as the *Haran Gawaita* puts it), "he drew the sword and converted people to himself by the sword". By then only sixty Nasoraean sanctuaries remained in Babylonia (12.15).

On the other side we have seen that when the *Right Ginza* speaks harshly against Christ it may call him a "Roman" (49.14; see also 51.26, 28; 52.5, 8, 22; 416.19), doubtless meaning a Byzantine and reflecting the hostility of the Byzantine church. Thus in the long run it was probably pressure both from the Byzantine world on the one hand and from Islam on the other that pushed the Mandaeans more and more into the swampy regions of southern Mesopotamia where their chief centers were found in modern times.

BAPTISM

In the Mandaean cult baptism is of central importance, and was obviously that which earned for the Mandaeans the name of "those who wash themselves" (*mughtasilah*). It was requisite that this washing be done in "living" (*i.e.*, flowing) water, which was called the Jordan even when it was no longer the actual river of that name in Palestine. It was also believed that these waters were descended from the heavenly Jordan, or Jordans (plural), which exist in the worlds of light. "All the world attests that the living water comes forth from beneath the throne of God" (*Right Ginza* 281.21-22). There is also a "Light-Euphrates" (*prash ziua*), which is the heavenly prototype of the earthly Euphrates River, the latter in many ways taking the place of the original Jordan for the Mesopotamian Mandaeans, and the two rivers (Jordan and Euphrates) being mingled in their view. "Let us ascend to the mouth of the Light-Euphrates, to the bank of the great Jordan of Life, and to the great edifice of our Father" (*Left Ginza* 425.35-36). The water of Mandaean baptism has, therefore, both the natural sense of cleansing and making pure, and also the sense of communicating life because, in the Mandaean view, it comes from the world of heavenly life.

[COMMENT: This idea that "living water comes forth from beneath the throne of God" is reminiscent of the Velikovskian "Saturn Theory". Before his death in November 1979, Dr. Immanuel Velikovsky suggested in an article published by *KRONOS Journal* that the Planet Saturn used to be tethered to our North Pole by a "watery" electromagnetic beam; and that when Saturn detethered and moved into its current orbit, a great flood of water descended down upon the Earth, from, as it were, "beneath the throne of God". This "Saturn Theory" has been promulgated and popularized by the Kronia Group, in particular David Talbott, who produced a "Saturn Theory" video titled *Remembering The End Of The World.* Of course, the object that was tethered to our North Pole was not the Planet Saturn, but the Planet X Nibiru. See Chapter 14. RS]

Pages 294-296

MANICHAEAN RELIGION

In relation to Buddhism it has already been noted that the death of Mani is described in the Parthian text M 5569 by a word equivalent to *parinirvana*, and the same Parthian term (or cognate forms) is used again for the death of Mani (in M 5), for the death of Buddha (in M 42), and for the death of Jesus (in M 104). There is also (in M8171) mention of "the district of the transmigration of souls". In Buddhist terminology the Sanskrit *parinirvana* signifies entry into the state of final release from the cycles of birth and rebirth, and the concepts of rebirth and of ultimate release must therefore have been a part of Manichaean belief. It is not unlikely that Mani became familiar with such terminology and concepts when he was in India (as al-Biruni says), and used the same in his own preaching.

In relation to Jesus and Christianity, both al-Nadim and al-Biruni report that Mani declared himself to be the *paksaxletoz* (promised by Jesus in John 14:16). Not a few Manichaean texts also reflect acquaintance with and use of Christian literature, much of which acquaintance and use may go back to Mani himself. In his own *Shabuhragan*, as known in Middle Persian fragments (M 473, M 475, M 477, M 482), Mani employs and cites at length the account of the last judgment from Matthew 25:31-46. In a Parthian text (M 4570) about the trial of Jesus before the Jewish authorities there are quotations from Matthew 26-27, Mark 14, and Luke 22, which probably come from Tatian's *Diatessaron*. In another text (M 18), which tells of the trial of Jesus before Pilate and of the women at the tomb, there is use not only of the canonical Gospels (probably from a harmony like that of Tatian) but also of the apocryphal Gospel of Peter. There is also demonstrable use of other apocryphal gospels, including the Gospel According to Philip and the Gospel According to Thomas. For example, Logion 5 in the Gospel According to Thomas reads: "Know what is before your face, and what is hidden from you will be revealed to you." This saying is incorporated in *Kephalaia* in an extended disquisition on the sun (chap. 65). The sun, it is remarked at this point, is high over all, corresponding to the mystery of the Father of Greatness. "[See], my beloved," the text continues, "I have instructed you in two mysteries, the mystery of the night and the mystery of the day. ... Now the mystery of the Light and the mystery of the Darkness are daily revealed in the creation."

The sun, the text goes on to explain, represents the mystery of the Light, as it comes daily into the world and reveals the wonders of its free gifts (*xaksisma*); the fearful night, on the other hand, corresponds to the mystery of the Darkness, and reveals itself in evil deeds. In their error (*plaen*), however, the sects do not understand this (and here comes the quotation from the Gospel According to Thomas): "They do not distinguish the mystery of the Light and that of the Darkness. Concerning this mystery, which to the sects is hidden, the Savior [*Sowteks*] gave an intimation to his disciples: 'Know what is before your face, and what is hidden from you will be revealed to you.'"

[COMMENT: The New Testament Thomas, twin-brother of "Jesus" and author of the gnostic Gospel of Thomas, was actually Damis of Ninevah, lifelong companion and "twin" of Apollonius of Tyana. RS]

There are also two direct quotations and one indirect reference that, according to al-Biruni in his *India*, Mani attributed to Jesus, but are otherwise unknown. In the case of the second of these quotations al-Biruni says the passage was in Mani's "Book of Mysteries"; perhaps the other two passages were in the same book. The three passages read as follows:

"The apostles asked Jesus about the life of inanimate nature, whereupon he said: 'If that which is inanimate is separated from the living element which is commingled with it, and appears alone by itself, it is again inanimate and is not capable of living, while the living element which has left it, retaining its vital energy unimpaired, never dies.'"

[COMMENT: Compare that to these words of Apollonius: "There is no death of anything save in appearance. That which passes over from essence to nature seems to be birth, and what passes over from nature to essence seems to be death. Nothing really is originated, and nothing ever perishes; but only now comes into sight and now vanishes. It appears by reason of the density of matter, and disappears by reason of the tenuity of essence. But it is always the same, differing only in motion and condition." RS]

"Since the apostles knew that the souls are immortal, and that in their migrations they array themselves in every form, that they are shaped in every animal, and are cast in the mould of every figure, they asked the Messiah what would be the end of those souls which did not receive the truth nor learn the origin of their existence. Whereupon he said. 'Any weak soul which has not received all that belongs to her of truth perishes without any rest or bliss.'"

"The other religious bodies blame us because we worship sun and moon, and represent them as an image. But they do not know their real natures; they do not know that sun and moon are our path, the door whence we march forth into the world of our existence [into heaven], as this has been declared by Jesus."

It becomes evident, however, in the Manichaean literature that although Mani and the Manichaeans refer to many events in the life of Jesus, especially those connected with his suffering and death, they hold a docetic view, as is also familiar in Christian Gnosticism. In Manichaean terminology the true Son of God is "Jesus the Splendor"; as a divine being he cannot really participate in a world that is material and evil; he can only manifest an apparent corporeality in the form of Jesus, the son of Mary. It was only Mary's son who suffered and died, not the Son of God. Along this line of thought a Parthian fragment (M 24) states that the secret truth of Christ is that he changed his form and his appearance; and a Middle Persian fragment (M 25 I) holds the Christians guilty of blasphemy because they call upon the son of Mary (*bar Mariam*) as the Son of the Lord (Adonai).

Thus it is to the divine figure, to Jesus the Splendor (or the Glorious), that many Manichaean hymns are addressed, for example, the Middle Persian M 28 II, which calls upon Jesus as the physician who heals spiritual sicknesses: "We will open the mouth to call upon thee, and make the tongue ready for praise. We call upon thee, [who are] entire life. We praise thee, Jesus the Splendor, the New Kingdom. Thou art, thou art ... a healer, the most beloved Son. ... Come for salvation ... Helper of the tender and conqueror of the attacker! Freer for those who are bound and Healer of the wounded! Awakener of the sleeping and Arouser of the drowsy! We will fill our eyes with praise and open our mouth for invocation. And we will bring reverence to thy Greatness, to thee, Jesus the Splendor."

[COMMENT: And "Jesus the Splendor" was Apollonius of Tyana, not the Jesus Barabbas described by Celsus. RS]

Chapter 7

Fingerprints Of The Gods

*

Fingerprints Of The Gods
By Graham Hancock (Three Rivers Press, New York, 1995)

INTRODUCTORY COMMENT
By Rob Solàrion

Graham Hancock travelled the world, collecting information. Ultimately, his purpose was to prove that "gods" or "ancient astronauts" played a part in our early global history, as evidenced by such monuments as The Great Pyramid and Macchu Pichu. So I picked up Hancock's book and looked through the index for various keywords. Finding lots of references, I decided to "speed read" the entire 578-page book, searching for passages that might be pertinent to me. He cited a lot of books that I have already read, such as *Hamlet's Mill, Maps Of The Ancient Sea Kings, New Larousse Encyclopedia Of Mythology*, Gerald Hawkins, Robert Bauval, Robert K.G. Temple, Peter Tompkins, Ignatius Donnelly, Piazzi Smyth, Sir E.A. Wallis Budge and others. Hancock includes formidable footnotes and bibliography. It was easy for me to "speed read" his book because I was already familiar with most of the information. And I suddenly realized what a wonderful "reference book" it is, what a great store of basic "source material". In that respect, Graham Hancock deserves admiration.

His great failure is that ultimately he comes to no original conclusions about what he is presenting, or at best he arrives at spurious conclusions based upon other people's previous postulations. For example, he cites *Hamlet's Mill* by Giorgio de Santillana and Hertha von Dechend throughout his book, and even devotes one whole section to an examination of their ideas. But he buys right into their false conclusion that this "Cosmic Millwheel" refers only to the Zodiac and the Sun's Precession through it over a period of 25,920 years. And when he mentions Herodotus' critical passage that in Ancient Egypt twice in the past the Sun rose where it now sets and set where it now rises, he falls back on *Hamlet's Mill*, concluding that the Egyptians had observed the Zodiac for tens of thousands of years; and that what they really meant was that, for example, at today's Vernal Equinox if the Sun is aligned with Pisces, then 12,960 years ago, it was aligned with Virgo, opposite of Pisces, and 12,960 years before that it was again aligned with Pisces. Thus, the Egyptians were also referring to Precession of the Equinoxes, with the Sun's changing its alignment with respect to the background Zodiac, not to the Sun's actually moving from one directional position to another because of a Polar Axis Shift. Hancock allowed the improper conclusion of de Santillana and von Dechend to influence his own interpretation of this most important passage about the rising and setting of the Sun in Egypt.

Although he cites Zecharia Sitchin a couple of times in footnotes, as well as *Earth In Upheaval* by Dr. Immanuel Velikovsky in several footnotes and *Worlds In Collision* in one, never does Graham Hancock mention Planet X Nibiru. Sometimes he will end a chapter with a question like "What could have been the reason for this?" If one ignores the existence of Planet X Nibiru, then one simply cannot find the truth. And as far as I am concerned, from my personal view of these ancient wonders and periodic cataclysms, if you ignore the physical Cosmic Tree Yggdrasill, then you also totally miss the point. The Cosmic Tree was *not* some sort of arcane reference to Precession of the Equinoxes. It was real. It was Hyperborea, "land beyond the mountains where the North Wind rises"!

Graham Hancock's book would be an excellent reference addition to anyone's library, but don't expect to find any answers to ultimate questions about the "secrets" of the Universe.

*

Pages 204-206

A MONSTER CHASED THE SUN

There is one ancient culture that perhaps preserves more vivid memories in its myths than any other; that of the so-called Teutonic tribes of Germany and Scandinavia, a culture best remembered through the songs of the Norse scalds and sages. The stories those songs retell have their roots in a past which may be much older than scholars imagine and which combine familiar images with strange symbolic devices and allegorical language to recall a cataclysm of awesome magnitude:

In a distant forest in the east an aged giantess brought the world a whole brood of young wolves whose father was Fenrir. One of these monsters chased the sun to take possession of it. The chase was for long in vain, but each season the wolf grew in strength, and at last he reached the sun. Its bright rays were one by one extinguished. It took on a blood red hue, then entirely disappeared.

[COMMENT: "Fenrir" is also known as "Fenris Wolf". Ignatius Donnelly discussed the Fenris Wolf in *Ragnarok.* See Chapter 9 of *Slow-Motion Doomsday.* The "Fenris Wolf" battled the "Midgard Serpent" just prior to the onset of the cataclysm. This legend certainly refers to the fact that when Earth passes through the "serpentine tail" of Planet X Nibiru, the sky will turn red before total Nightfall sets in for several days. See also *The Wayward Sun* by Rand & Rose Flam-Ath (*The Velikovskian* Journal, 1997), available at my website. RS]

Thereafter the world was enveloped in hideous winter. Snow-storms descended from all points of the horizon. War broke out all over the earth. Brother slew brother, children no longer respected the ties of blood. It was a world when men were no better than wolves, eager to destroy each other. Soon the world was going to sink into the abyss of nothingness.

Meanwhile, the wolf Fenrir, whom the gods had long ago so carefully chained up, broke his bonds at last and escaped. He shook himself and the world trembled. The ash tree Yggdrasil [envisaged as the axis of the earth — GH] was shaken from its roots to its topmost branches. Mountains crumbled or split from top to bottom, and the dwarfs who had their subterranean dwellings in them sought desperately and in vain for entrances so long familiar but now disappeared.

[COMMENT: This is describing the idea that during this period of chaos and darkness, suddenly Yggdrasill springs forth from the North Pole. When Planet X Nibiru (or its gigantic mothership) hurls down its "thundering" electromagnetic tether-beam from its South Pole to our North Pole, our world would be "shaken from its roots" because our Polar Axis would shift. RS]

Abandoned by the gods, men were driven from their hearths and the human race was swept from the surface of the earth. The earth itself was beginning to lose its shape. Already the stars were coming adrift from the sky and falling into the gaping void. They were like swallows, weary from too long a voyage, who drop and sink into the waves.

[COMMENT: This is such a curious passage. As Rand & Rose Flam-Ath ask in bewilderment at the end of *The Wayward Sun*, "But why did the sky fall?" I have thought about this question many times, but I can find no answer. If in two different mythologies (Ute Indians and Scandinavians) on two different continents (North America and Europe) both describe some "physical event" using identical terminology, then obviously whatever happened seemed to resemble the idea that "the sky is falling". (Such an idea carries over even today in the nursery rhyme about Chicken Little.) After Earth enters the "serpentine tail" and Nightfall occurs, do all of the cascading fiery boulders from Planet X resemble the stars falling from the sky, or the sky itself falling? The only way we shall be able to understand what the "falling sky" indicates is by watching the arrival sequence again with our own, more modern eyes. I'm sure that we'll know at once what they meant. RS]

The giant Surt set the entire earth on fire; the universe was no longer more than an immense furnace. Flames spurted from fissures in the rocks; everywhere there was the hissing of steam. All living things, all plant life, were blotted out. Only the naked soil remained, but like the sky itself the earth was no more than cracks and crevasses.

And now all the rivers, all the seas, rose and overflowed. From every side waves lashed against waves. They swelled and boiled slowly over all things. The earth sank beneath the sea. ...

Yet not all men perished in the great catastrophe. Enclosed in the wood itself of the ash tree Yggdrasil — which the devouring flames of the universal conflagration had been unable to consume — the ancestors of a future race of men had escaped death. In this asylum they had found that their only nourishment had been the morning dew.

[COMMENT: Morning Dew. Madhu. Manna. Ambrosia. See Chapter 1 of *Slow-Motion Doomsday.* This idea of morning-dew nourishment following the cataclysm is universal on our planet. RS]

Thus it was from that wreckage of the ancient world a new world was born. Slowly the earth emerged from the waves. Mountains rose again and from them streamed cataracts of singing waters.

[COMMENT: Hancock was quoting the *New Larousse Encyclopedia Of Mythology* here. RS]

The new world this Teutonic myth announces is our own. Needless to say, like the Fifth Sun of the Aztecs and the Maya, it was created long ago and is new no longer. Can it be a coincidence that one of the many Central American flood myths about the fourth epoch, 4 *Atl* ('water'), does not install the Noah couple in an ark but places them instead in a great tree just like Yggdrasil? '4 *Atl* was ended by the floods. The mountains disappeared. ... Two persons survived because they were ordered by one of the gods to bore a hole in the trunk of a very large tree and to crawl inside when the skies fell. The pair entered and survived. Their offspring repopulated the world.'

[COMMENT: Actually two comments. First, notice how this Mayan "myth" resembles the survival of Osiris in the tamarisk tree of Phoenicia, a tree which also grew to great size. Second, if the "sky was falling", perhaps the "gods" saved various people by keeping them under the protection of Hyperborea, The Cosmic Tree. However people might have survived, the point is that some people *do* survive in all parts of the world. But we simply do not have sufficient ancient scientific records to be able to predict what would happen today. RS]

Isn't it odd that the same symbolic language keeps cropping up in ancient traditions from so many widely scattered regions of the world? How can this be explained? Are we talking about some vast, subconscious wave of intercultural telepathy, or could elements of these remarkable universal myths have been engineered, long ages ago, by clever and purposeful people? Which of these improbable propositions is the more likely to be true? Or are there other possible explanations for the enigma of the myths?

We shall return to these questions in due course. Meanwhile, what are we to conclude about the apocalyptic visions of fire and ice, floods, volcanism and earthquakes, which the myths contain? They have about them a haunting and familiar realism. Could this be because they speak to us of a past we suspect to be our own but can neither remember clearly nor forget completely?

[COMMENT: Dr. Immanuel Velikovsky referred to this same idea as "Mankind's Collective Amnesia". See *Worlds In Collision*, for example. RS]

Pages 253-255

OPENERS OF THE WAY

What is remarkable about all this is the way that the mill (which continues to serve as an allegory for cosmic processes) stubbornly keeps on resurfacing, all over the world, even where the context has been jumbled or lost. Indeed in [Giorgio de] Santillana and [Hertha] von Dechend's argument [in *Hamlet's Mill*], it doesn't really matter if the context is lost. 'The particular merit of mythical terminology,' they say, 'is that is can be used as a vehicle for handing down solid knowledge independently from the degree of insight of the people who do the actual telling of stories, fables, *etc.*' What matters, in other words, is that certain central imagery should survive and continue to be passed on in retellings, however far these may drift from the original storyline.

An example of such drift (coupled with the retention of essential imagery and information) is found among the Cherokees, whose name for the Milky Way (our own galaxy) is 'Where The Dog Ran'. In ancient times, according to Cherokee tradition, the 'people in the South had a corn mill', from which meal was stolen again and again. In due course the owners discovered the thief, a dog, who 'ran off howling to his home in the North, with the meal dropping from his mouth as he ran, and leaving behind a white trail where now we see the Milky Way, which the Cherokee call to this day ... "Where The Dog Ran".'

[**COMMENT:** Assuming, as I do, that "myth" reflects primitive "science", this is a new one for me. Basically, I have no idea what it means. Sirius is the Dog-Star, but why did this "Dog" run off "howling to his home in the North, with the meal dropping from his mouth"? One could imagine all sorts of things. RS]

In Central America, one of the many myths concerning Quetzalcoatl depicts him playing a key role in the regeneration of mankind after the all-destroying flood that ended the Fourth Sun. Together with his dog-headed companion Xolotl, he descends into the underworld to retrieve the skeletons of the people killed by the deluge. This he succeeds in doing, after tricking Miclantechuhtli, the god of death, and the bones are brought to a place called Tamoanchan. There, like corn, they are milled into a fine meal on a grindstone. Upon this ground meal the gods then release blood, thus creating the flesh of the current age of men.

Santillana and von Dechend do not think that the presence of a canine character in both the above variants of the myth of the cosmic mill is likely to be accidental. They point out that Kullervo, the Finnish Hamlet, is also accompanied by 'the black dog Musti'. Likewise, after his return to his estates in Ithaca, Odysseus is first recognized by his faithful dog, and as anyone who has been to Sunday school will remember, Samson is associated with foxes (300 of them, to be precise), which are members of the dog family. In the Danish version of the Amleth/Hamlet saga, 'Amleth went on and a wolf crossed his path amid the thicket.' Last but not least an alternative recension of the Kullervo story from Finland has the hero (rather weirdly) being 'sent to Estonia to bark under the fence, he barked one year. ... '

Santillana and von Dechend are confident that all this 'doggishness' is purposive: another piece of the ancient code, as yet unbroken, persistently tapping out its message from place to place. They list these and many other canine symbols among a series of 'morphological markers' which they have identified as likely to suggest the presence, in ancient myths, of scientific information concerning precession of the equinoxes. These markers may have had meanings of their own or been intended simply to alert the target audience that a piece of hard data was coming up in the story being told. Beguilingly, sometimes they may also have been designed to serve as 'openers of the way' — conduits to enable initiates to follow the trail of scientific information from one myth to another.

Thus, even though none of the familiar mills and whirlpools is in sight, we should perhaps sit up and pay attention when we learn that Orion, the great hunter of Greek myth, was the owner of a dog. When Orion tried to ravish the virgin goddess Artemis, she produced a scorpion from the earth which killed him and the dog. Orion was transported to the skies where he became the constellation that bears his name today; his dog was transformed into Sirius, the Dog Star.

[COMMENT: And as we all well know, the belt stars of Orion point to Sirius. RS]

Precisely the same identification of Sirius was made by the ancient Egyptians, who linked the Orion constellation specifically to their god Osiris. It is in Ancient Egypt too that the character of the faithful celestial dog achieves its fullest and most explicit mythical elaboration in the form of Upuaut, a jackal-headed deity whose name means 'Opener of the Ways'. If we follow this way opener to Egypt, turn our eyes to the constellation of Orion, and enter the potent myth of Osiris, we find ourselves enveloped in a net of familiar symbols.

The reader will recall that the myth presents Osiris as the victim of a plot. The conspirators initially dispose of him by sealing him in a box and casting him adrift on the waters of the Nile. In this respect does he not resemble Utnapishtim, and Noah and Coxcoxtli and all the other deluge heroes in their arks (or boxes, or chests) riding out the waters of the flood?

Another familiar element is the classic precessional image of the world-tree and/or roof-pillar (in this case combined). The myth tells us how Osiris, still sealed inside his coffer, is carried out into the sea and washed up at Byblos. The waves lay him to rest among the branches of a tamarisk tree, which rapidly grows to a magnificent size, enclosing the coffer within its trunk. The king of the country, who much admires the tamarisk tree, cuts it down and fashions the part which contains Osiris into a roof pillar for his palace. Later Isis, the wife of Osiris, removes her husband's body from the pillar and takes it back to Egypt to undergo rebirth.

The Osiris myth also includes certain key numbers. Whether by accident or by design, these numbers give access to a 'science' of precession, as we shall see in the next chapter.

Page 262

In the Hebrew Cabala there are 72 angels through whom the Sephiroth (divine powers) may be approached, or invoked, by those who know their names and numbers. Rosicrucian tradition speaks of cycles of 108 years (72 plus 36) according to which the secret brotherhood makes its influence felt. Similarly the number 72 and its permutations and subdivisions are of great significance to the Chinese secret societies known as Triads. An ancient ritual requires that each candidate for initiation pay a fee including '360 cash for "making clothes", 108 cash "for the purse", 72 cash for instruction, and 36 cash for decapitating the "traitorous subject".' The 'cash' (the old universal brass coin of China with a square hole in the centre) is of course no longer in circulation but the *numbers* passed down in the ritual since times immemorial have survived. Thus in modern Singapore, candidates for Triad membership pay an entrance fee which is calculated according to their financial circumstances but which must always consist of multiples of $1.80, $3.60, $7.20, $10.80 (and thus, $18, $36, $72, $108, or $360, $720, $1,080, and so on.

[COMMENT: As noted earlier, this idea of 72 "angels" or Hermetic Archons, Greek Titans, Solomonic Hebrew Lemegeton Spirits or the miscellaneous "72 Names Of God" or "72 Goetian Spirits" refers to the ruling Council, their Consorts and the "Parliament" of Planet X Nibiru. RS]

Pages 372-377

THE TRAIN OF THE SUN AND THE DWELLER IN SIRIUS

Of course the ability to recognize and define precessional world ages in myth implies that the ancient Egyptians possessed better observational astronomy and a more sophisticated understanding of the mechanics of the solar system than any ancient people have hitherto been credited with. There is no doubt that knowledge of this calibre, if it existed at all, would have been highly regarded by the Ancient Egyptians, who would have transmitted it from generation to generation in a secretive manner. Indeed, it would have ranked among the highest arcana entrusted to the keeping of the priestly elite at Heliopolis and would have been passed on, in the main, through an oral and initiatory tradition. If, by chance, it had found its way into the Pyramid Texts, is it not likely that its form would have been veiled by metaphors and allegories?

[COMMENT: Not necessarily. What you see may be what you get. What these "allegories" may actually refer to are the "shars" (orbits) of Planet X Nibiru, *not* Precession of the Equinoxes! RS]

I walked slowly across the dusty floor of the tomb chamber of Unas, noting the heavy stillness in the air, casting my eyes over the faded blue and gold inscriptions. Expressed in coded language several millennia before Copernicus and Galileo, some of the passages inscribed on these walls seemed to offer clues to the true heliocentric nature of the solar system.

In one, for example, Ra, the Sun God, was depicted as seated upon an iron throne encircled by lesser gods who moved around him constantly and who were said to be 'in his train'. Likewise, in another passage, the deceased Pharaoh was urged to 'stand at the head of the two halves of the sky and weigh the words of the gods, the aged ones, who revolve around Ra'.

[COMMENT: The "head of the two halves of the sky" refers to Planet X Nibiru (Asgard) atop the electromagnetic "Cosmic Treetrunk" (Bifrost), which seems to divide the sky into western and eastern halves. RS]

If the 'aged ones' and the 'encircling gods' revolving around Ra should prove to be parts of a terminology referring to the planets of our solar system, the original authors of the Pyramid Texts must have enjoyed access to some remarkably advanced astronomical data. They must have known that the earth and the planets revolved around the sun rather than vice versa. The problem this raises is that neither the Ancient Egyptians at any stage in their history, nor even their successors the Greeks, or for that matter the Europeans until the Renaissance, are supposed to have possessed cosmological data of anything approaching this quality. How, therefore, can its presence be explained in compositions which date back to the dawn of Egyptian civilization?

[COMMENT: The "encircling gods" are the various satellites that dangle from Planet X, also tethered electromagnetically. See the cover design of *Slow-Motion Doomsday*. Also see the diagram from the so-called "Voynich Manuscript". Both can be seen at my website. In his video *Remembering The End Of The World*, David Talbott uses computer animation to correlate the "orbits" of these "encircling gods" with various ancient symbols. It is fascinating to watch it. See Chapter 14 and Appendix A. RS]

Another (and perhaps related) mystery concerns the star Sirius, which the Egyptians identified with Isis, the sister and consort of Osiris and the mother of Horus. In a passage addressed to Osiris himself, the Pyramid Texts state: 'Thy sister Isis cometh unto thee rejoicing in her love for thee. Thou settest her upon thee, thy issue entereth into her, and she becometh great with child like the star Sept [Sirius, the Dog Star — GH]. Horus-Sept cometh forth from thee in the form of Horus, dweller in Sept.'

Many interpretations of this passage are, of course, possible. What intrigued me, however, was the clear implication that Sirius was to be regarded as a *dual entity* in some way comparable to a woman 'great with child'. Moreover, after the birth (or coming forth) of that child, the text makes a special point of reminding us that Horus remained a 'dweller in Sept', presumably suggesting that he stayed close to his mother.

[**COMMENT:** One might conclude that when red-giant Sirius "B" (Digitaria) imploded and ejected Planet X Nibiru into our neighboring Solar System, this was remembered as "Isis" giving birth to "Horus", *i.e.*, Planet X's being "born" into our Solar System. RS]

Sirius is an unusual star. A sparkling point of light particularly prominent in the winter months in the night skies of the northern hemisphere, it consists of a *binary* star system, *i.e.*, it is in fact, as the Pyramid Texts suggest, a 'dual entity'. The major component, Sirius-A, is what we see. Sirius-B, on the other hand — the dwarf-star which revolves around Sirius-A — is absolutely invisible to the naked eye. Its existence did not become known to Western science until 1862, when US astronomer Alvin Clark spotted it through one of the largest and most advanced telescopes of the day. How could the scribes who wrote the Pyramid Texts possibly have obtained the information that Sirius was two stars in one?

In *The Sirius Mystery*, an important book published in 1976, I knew that the American author Robert Temple had offered some extraordinary answers to this question. His study focused on the traditional beliefs of the Dogon tribe of West Africa — beliefs in which the binary character of Sirius was explicitly described and in which the correct figure of fifty years was given for the period of the orbit of Sirius-B around Sirius-A. Temple argued cogently that this high quality technical information had been passed down to the Dogon from *the Ancient Egyptians* through a process of cultural diffusion, and that it was to the Ancient Egyptians that we should look for an answer to the Sirius mystery. He also concluded that the ancient

Egyptians must have received the information from intelligent beings from the region of Sirius.

Like Temple, I had begun to suspect that the more advanced and sophisticated elements of Egyptian science made sense only if they were understood as parts of an inheritance. Unlike Temple, I saw no urgent reason to attribute that inheritance to extra-terrestrials. To my mind the anomalous star knowledge the Heliopolitan priests had apparently possessed was more plausibly explained as the legacy of a lost human civilization which, against the current of history, had achieved a high level of technological advancement in remote antiquity. It seemed to me that the building of an instrument capable of detecting Sirius-B might not have been beyond the ingenuity of the unknown explorers and scientists who originated the remarkable maps of the prehistoric world discussed in Part I.

[COMMENT: Hancock is referring to the Piri Reis Maps and other ancient maps described in the book *Maps Of The Ancient Sea Kings* by Charles H. Hapgood (New York, 1966). RS]

Nor would it have daunted the unknown astronomers and measurers of time who bequeathed to the Ancient Maya a calendar of amazing complexity, a data-base about the movements of the heavenly bodies which could have been the product of thousands of years of accurately recorded observations, and a facility with very large numbers that seemed more appropriate to the needs of a complex technological society than to those of a 'primitive' Central American kingdom.

MILLIONS OF YEARS AND THE MOVEMENTS OF THE STARS

Very large numbers also appeared in the Pyramid Texts, in the symbolic 'boat of millions of years', for example, in which the Sun God was said to navigate the dark and airless wastes of interstellar space. Thoth, the god of wisdom ('he who reckons in heaven, the counter of the stars, the measurer of the earth') was specifically empowered to grant a life of millions of years to the deceased pharaoh. Osiris, 'king of eternity, lord of everlasting', was described as traversing millions of years in his life. And figures like 'tens of millions of years' (as well as the more mind-boggling 'one million of millions of years') occurred often enough to suggest that some elements at least of Ancient Egyptian culture must have evolved for the convenience of scientifically minded people with more than passing insight into the immensity of time.

Such a people would, of course, have required an excellent calendar — one that would have facilitated complex and accurate calculations. It was therefore not surprising to learn that the Ancient Egyptians, like the Maya, had possessed such a calendar and that their understanding of its workings seemed to have declined, rather than improved, as the ages went by. It was tempting to see this as the gradual erosion of a corpus of knowledge inherited an extremely long time ago, an impression supported by the Ancient Egyptians themselves, who made no secret of their belief that their calendar was a legacy which they had received 'from the gods'.

[COMMENT: This "Boat of Millions of Years" is also mentioned by Zecharia Sitchin in *The Stairway To Heaven*. See Chapter 9. If "the Sun God was said to navigate the dark and airless wastes of interstellar space" over "millions of years", might not this refer to the fact that Planet X Nibiru, home of "immortal gods", moves regularly *shar* after *shar* between our system and the Sirius System, or between our Inner Solar System and the distant boundary of the Oort Cloud? See Chapter 11 herein, as well as Chapter 12 of *Slow-Motion Doomsday*, but we can only speculate at this point in time. RS]

We consider the possible identity of these gods in more detail in the following chapters. Whoever they were, they must have spent a great deal of their time observing the stars, and they had accumulated a fund of advanced and specialized knowledge concerning the star Sirius in particular. Further evidence for this came in the form of the most useful calendrical gift which the gods supposedly gave to the Egyptians: the *Sothic* (or Sirian) cycle.

The Sothic cycle was based on what is referred to in technical jargon as 'the periodic return of the heliacal rising of Sirius', which is the first appearance of this star after a seasonal absence, rising at dawn just ahead of the sun in the eastern portion of the sky. In the case of Sirius the interval between one such rising and the next amounts to *exactly* 365.25 days — a mathematically harmonious figure, uncomplicated by further decimal points, which is just twelve minutes longer than the duration of the solar year.

[COMMENT: For an analysis of the Egyptian Sothis Period, see Chapter 17 ("Introduction To Galactic Mathematics"). This idea of a heliacal, or morning, rising of Sirius might also imply that when Planet X Nibiru first becomes visible to the naked eye at Crossover, it appears as a morning star. RS]

The curious thing about Sirius is that out of an estimated 2000 stars in the heavens visible to the naked eye it is the only one to rise heliacally at this precise and nicely rounded interval of 365 and a quarter days — a unique product of its 'proper motion' (the speed of its own movement through space) combined with the effects of precession of the equinoxes. Moreover, it is known that the day of the heliacal rising of Sirius — New Year's Day in the Ancient Egyptian calendar — was traditionally calculated at Heliopolis, where the Pyramid Texts were compiled, and announced ahead of time to all the other major temples up and down the Nile.

I remembered that Sirius was referred to directly in the pyramid Texts by 'her name of the New Year'. Together with other relevant utterances (*e.g.*, 669), this confirmed that the Sothic calendar was *at least* as old as the Texts themselves, and their origins stretched back into the mists of distant antiquity. The great enigma, therefore, is this: in such an early period, who could have possessed the necessary knowhow to observe and take note of the coincidence of the period of 365.25 days with the heliacal rising of Sirius — a coincidence described by the French mathematician R.A. Schwaller de Lubicz as 'an entirely exceptional celestial phenomenon'?

"We cannot but admire the greatness of a science capable of discovering such a coincidence. The double star of Sirius was chosen because it was the only star that moves the needed distance and in the right direction against the background of the other stars. This fact, known four thousand years before our time and forgotten until our day, obviously demands an extraordinary and prolonged observation of the sky."

It was such a legacy — built out of long centuries of precise observational astronomy and scientific record-keeping — that Egypt seems to have benefited from at the beginning of the historical period and that was expressed in the Pyramid Texts.

In this, too, there lies a mystery ...

[**COMMENT:** There may be a "mystery" connected with this period of 669 years, which I have never seen before. But there is no longer a "mystery" about the Egyptian Sothis Period, at least as far as I am concerned. See Chapter 17. RS]

Pages 384-385

Long before Diodorus [Siculus, first century BCE, RS], Egypt was visited by another and more illustrious Greek historian: the great Herodotus, who lived in the fifth century BC. He too, it seems, consorted with priests and he too managed to tune in to traditions that spoke of the presence of a high civilization in the Nile Valley at some unspecified date in remote antiquity. Herodotus outlines these traditions of an immense prehistoric period of Egyptian civilization in Book II of his *History*. In the same document he also hands on to us, without comment, a peculiar nugget of information which had originated with the priests of Heliopolis:

"During this time, they said, there were four occasions when the sun rose out of his wonted place — twice rising where he now sets, and twice setting where he now rises."

What is this all about?

[COMMENT: Hancock then falls back on the Precession of the Equinoxes to explain this "mystery", trying, I suppose, to follow along with de Santillana and von Dechend in *Hamlet's Mill*. But this has nothing to do with Precession of the Equinoxes. When the Polar Axis shifts at the beginning of every *shar*, of course the rising and the setting of the Sun in Egypt and elsewhere would be different. RS]

Pages 391-395

OSIRIS AND THE LORDS OF ETERNITY

Occasionally referred to in the texts as a *neb tem*, or 'universal master', Osiris is depicted as human but also superhuman, suffering but at the same time commanding. Moreover, he expresses his essential dualism by ruling in heaven (as the constellation of Orion) and on earth as a king among men. Like Viracocha in the Andes and Quetzalcoatl in Central America, his ways are subtle and mysterious. Like them, he is exceptionally tall and always depicted wearing the curved beard of divinity. And like them too, although he has supernatural powers at his disposal, he avoids the use of force whenever possible.

[COMMENT: Technically speaking, the names Osiris, Viracocha and Quetzalcoatl do not refer to the same "god". The Egyptian Osiris is the Nibiruan Enlil (Incan Catequil/Pillan and Mayan Tlaloc); the Incan Viracocha is the Nibiruan Enki (Egyptian Seth and Mayan Itzamna); and the Mayan Quetzalcoatl is the Nibiruan Nannar (Egyptian Thoth and Incan Pihuechenyl). These "gods" can range in height to about 20 feet (6-7 meters), and many of the Saurian males have beards. The reason that human males have beards, unlike the other primates, is that this genetic characteristic came from Planet X Nibiru. See my essay on "Nibiruan Physiology" which is available at my website. For more on "Names Of The Gods", see Chapter 12. It is also of note that when Apollonius of Tyana and his scribe Damis of Ninevah were travelling to India in 24-25 BCE, they encountered caves of "dragons", some of which were bearded. RS]

We saw in Chapter Sixteen that Quetzalcoatl, the god-king of the Mexicans, was believed to have departed from Central America by sea, sailing away on a raft of serpents. It is therefore hard to avoid a sense of *déjà vu* when we read in the *Egyptian Book of the Dead* that the abode of Osiris also 'rested on water' and had walls made of 'living serpents'. At the very least, the convergence of symbolism linking these two gods and two farflung regions is striking.

There are other obvious parallels as well.

The central details of the story of Osiris have been recounted in earlier chapters and we need not go over them again. The reader will not have forgotten that this god — once again like Quetzalcoatl and Viracocha — was remembered principally as a benefactor of mankind, as a bringer of enlightenment and as a great civilizing leader. He was credited, for example, with having abolished cannibalism and was said to have introduced the Egyptians to agriculture — in particular to the cultivation of wheat and barley — and to have taught them the art of fashioning agricultural implements. Since he had an especial liking for fine wines (the myths do not say where he acquired this taste), he made a point of 'teaching mankind the culture of the vine, as well as the way to harvest the grape and to store the wine ... ' In addition to the gifts of good living he brought to his subjects, Osiris helped to wean them 'from their miserable and barbarous manners' by providing them with a code of laws and inaugurating the cult of the gods in Egypt.

When he had set everything in order, he handed over the control of the kingdom to Isis, quit Egypt for many years, and roamed about the world with the sole intention, Diodorus Siculus was told, 'of visiting all the inhabited earth and teaching the race of men how to cultivate the vine and sow wheat and barley; for he supposed that if he made men give up their savagery and adopt a gentle manner of life he would receive immortal honours because of the magnitude of his benefactions ... '

Osiris travelled first to Ethiopia, where he taught tillage and husbandry to the primitive hunter-gatherers he encountered. He also undertook a number of large-scale engineering and hydraulics works: 'He built canals, with flood gates and regulators ... he raised the river banks and took precautions to prevent the Nile from overflowing ... Later he made his way to Arabia and thence to India, where he established many cities. Moving on to Thrace he killed a barbarian king for refusing to adopt his system of government. This was out of character; in general, Osiris was remembered by the Egyptians for having forced no man to carry out his instructions, but by means of gentle persuasion and an appeal to their reason he succeeded in inducing them to practise what he preached. Many of his wise counsels were imparted to his listeners in hymns and songs, which were sung to the accompaniment of instruments of music.'

Once again the parallels with Quetzalcoatl and Viracocha are hard to avoid. During a time of darkness and chaos — quite possibly linked to a flood — a bearded god, or man, materializes in Egypt (or Bolivia, or Mexico). He is equipped with a wealth of practical and scientific skills, of the kind associated with mature and highly developed civilizations, which he uses unselfishly for the benefit of humanity. He is instinctively gentle but capable of great firmness when necessary. He is motivated by a strong sense of purpose and, after establishing his headquarters at Heliopolis (or Tiahuanaco, or Teotihuacan), he sets forth with a select band of companions to impose order and to reinstate the lost balance of the world.

[**COMMENT:** Hancock is simply confusing the various qualities of three different "gods", all of whom had a hand in "civilizing" primitive humans. But did it work? Are we truly "civilized" today? One wonders sometimes. To quote Darío Salas from *The Stellar Man*, "Although the world has progressed from a barbaric state to civilization in the course of history, the 'savage sapiens' is today basically as primitive as in the distant past, covered only with layers of cultural and educational varnish." RS]

Leaving aside for the present the issue of whether we are dealing here with gods or men, with figments of the primitive imagination or with flesh-and-blood beings, the fact remains that the myths *always* speak of a *company* of civilizers: Viracocha has his 'companions', as have both Quetzalcoatl and Osiris. Sometimes there are fierce internal conflicts within these groups, and perhaps struggles for power: the battles between Seth and Horus, and between Tezcatlipoca and Quetzalcoatl are obvious examples. Moreover, whether the mythical events unfold in Central America, or in the Andes, or in Egypt, the upshot is also always pretty much the same: the civilizer is eventually plotted against and either driven out or killed.

The myths say that Quetzalcoatl and Viracocha never came back (although, as we have seen, their return to the Americas was expected at the time of the Spanish conquest). Osiris, on the other hand, did come back. Although he was murdered by Set soon after the completion of his worldwide mission to make men 'give up their savagery', he won eternal life through his resurrection in the constellation of Orion as the all-powerful god of the dead. Thereafter, judging souls and providing an immortal example of responsible and benevolent kingship, he dominated the religion (and the culture) of ancient Egypt for the entire span of its known history.

SERENE STABILITY

Who can guess what the civilizations of the Andes and of Mexico might have achieved if they too had benefited from such powerful symbolic continuity. In this respect, however, Egypt is unique. Indeed, although the Pyramid Texts and other archaic sources recognize a period of disruption and attempted usurpation by Set (and his seventy-two 'precessional' conspirators), they also depict the transition to the reigns of Horus, Thoth and the later divine pharaohs as being relatively smooth and inevitable.

[COMMENT: These were *not* "precessional" conspirators! They were merely the other "Archons" of the Nibiruan élite. RS]

This transition was mimicked, through thousands of years, by the mortal kings of Egypt. From the beginning to the end, they saw themselves as the lineal descendants and living representatives of Horus, son of Osiris. As generation succeeded generation, it was supposed that each deceased pharaoh was reborn in the sky as 'an Osiris' and that each successor to the throne became a 'Horus'.

This simple, refined, and stable scheme *was already fully evolved and in place at the beginning of the First Dynasty* — around 3100 BC. Scholars accept this; the majority also accept that what we are dealing with here is a highly developed and sophisticated religion. Strangely, very few Egyptologists or archaeologists have questioned where and when this religion took shape.

Is it not to defy logic to suppose that well-rounded social and metaphysical ideas like those of the Osiris cult sprung up fully formed in 3100 BC, or that they could have taken such perfect shape in the 300 years which Egyptologists sometimes grudgingly allow for them to have done so? There must have been a far longer period of development than that, spread over several thousands rather than several hundreds of years. Moreover, as we have seen, every surviving record in which the Ancient Egyptians speak directly about their past asserts that their civilization was a legacy of 'the gods' who were 'the first to hold sway in Egypt'.

The records are not internally consistent: some attribute much greater antiquity to the civilization of Egypt than others. All, however, clearly and firmly direct our attention to an epoch far, far in the past — anything from 8000 to almost 40,000 years before the foundation of the First Dynasty. Archaeologists insist that no material artefacts have ever been found in Egypt to suggest that an evolved civilization existed at such early dates, but this is not strictly true. As we saw in Part VI, a handful of objects and structures exist which have not yet been conclusively dated by any scientific means.

The ancient city of Abydos conceals one of the most extraordinary of these undatable enigmas ...

Pages 438-441

NAVIGATORS IN THE BOAT OF MILLIONS OF YEARS

Ta-Neteru was thought to have had a definite earthly location a very long way south of Ancient Egypt — seas and oceans away — farther even than the spice country of Punt (which probably lay along East Africa's Somali coast). To confuse matters, however, Punt was also spoken sometimes as the 'Divine Land', or 'God's Land', and was the source of the sweet-smelling frankincense and myrrh especially favoured by the gods. ...

[COMMENT: Hancock was obviously not familiar with, or simply ignored, the identification of the Land of Punt with Hebrew Palestine, as proven by Dr. Immanuel Velikovsky. When Queen Hatshepsut Makeda

Saba, the Queen of Sheba, journeyed to the Land of Punt, she paid a state visit to King Solomon in Jerusalem. See *Ages In Chaos* by Dr. Velikovsky, as well as my chronological discussion in Chapter 18. This Land of Ta-Neteru is obviously identical with "Tuat, City Of God" which was also south of Egypt and was ruled by Osiris. Here were located the fabled "Elysian Fields" of Osiris. See *Slow-Motion Doomsday*, Chapter 7 ("Heaven & Hell"). RS]

Was it from a land such as this, superbly irrigated and scientifically farmed, that the agriculture bringer Osiris, whose title was '**PRESIDENT OF THE LAND OF THE SOUTH**' [bold caps added by me — RS], had voyaged to Egypt at the dawn of the First Time? And was it from a land such as this, accessible only by boat, that ibis-masked Thoth had also made his way, crossing seas and oceans to deliver the priceless gifts of astronomy and earth-measurement to the primitive inhabitants of the prehistoric Nile Valley?

Whatever the truth behind the tradition, Thoth was remembered and revered by the Ancient Egyptians as the inventor of mathematics, astronomy and engineering. 'It was his will and power,' according to Wallis Budge, 'that were believed to keep the forces of heaven and earth in equilibrium. It was his great skill in celestial mathematics which made proper use of the laws upon which the foundation and maintenance of the universe rested.' Thoth was also credited with teaching the ancestral Egyptians the skills of geometry and land-surveying, medicine and botany. He was believed to have been the inventor 'of figures, of the letters of the alphabet, and of the arts of reading and writing.' He was the 'Great Lord of Magic' who could move objects with the power of his voice, 'the author of every work on every branch of knowledge, both human and divine'.

[COMMENT: The Egyptian Thoth was, of course, identical with the Greek Hermes, founder of the Hermetic Tradition and supposed author of "The Emerald Tablet Of Thoth", as it made its way from ancient times to Alexander The Great to Apollonius Tyanaeus to Roman Emperor Hadrian. Where this "Emerald Tablet" is today is anybody's guess, but I would venture to speculate that it is stored unrecognized in the basement of some musty European provincial or monastic museum, probably in France or Italy. For more information about this "Emerald Table", see my book *Apollonius Of Tyana & The Shroud Of Turin*, Chapter 24 ("Apollonius Historical Bibliography"). Furthermore, the doctrinal school founded at

Crotona, Italy, by Pythagoras, of which Apollonius was a member, taught "Hermetic secrets" of mathematics, medicine and magic. RS]

It was to the teachings of Thoth — which they guarded jealously in their temples and claimed to have been handed down from generation to generation in the form of forty-two books of instruction — that the Ancient Egyptians ascribed their world-renowned wisdom and knowledge of the skies. This knowledge was spoken of almost in awe, by the classical commentators who visited Egypt from the fifth century BC onwards.

Herodotus, the earliest of these travellers, noted: 'The Egyptians were the first to discover the solar year, and to portion out its course into twelve parts. ... It was observation of the course of the stars which led them to adopt this division. ... '

[COMMENT: All of those people who lived during 1588-687 BCE would have been "eyewitnesses" to Hyperborea, The Cosmic Tree. Then, suddenly, it was no longer there. A century later, it would have passed into legend with a new generation of Earthlings who'd never seen it with their own eyes. It goes without saying, I think, that religious cults all over the world would have preserved a "mythological tradition" of the "Millennium Of The Gods" that existed in their past. When Greek historian Herodotus visited Egypt around 450 BCE, what the priests would have told him was what they remembered from only a short 300 years ago. If the Solar Year had been *exactly* 360 days *before* Planet X Nibiru departed, and then stabilized at the "Sothic" Year of 365.25 days during the period from 762-687 BCE, of course the "priests" (scientists!) in Egypt and elsewhere would have had to devise a method to measure the length of the now-modified year because their old system of 360 days was no longer valid. It wouldn't have taken them too many years to figure this out, and it is my opinion that this modification is what is being referred to here, as well as in so many other reference sources. See also *Worlds In Collision* by Dr. Velikovsky. RS]

Plato (fourth century BC) reported that the Egyptians had observed the stars 'for ten thousand years'. And later, in the first century BC Diodorus Siculus left this more detailed account: 'The positions and arrangements of the stars as well as their motions have always been the subject of careful observation among the Egyptians. ... From ancient times to this day they have preserved the records concerning each of these stars over an incredible number of years. ... '

Why should the Ancient Egyptians have cultivated an almost obsessional interest in the long-term observation of the stars, and why in particular should they have kept records of their movements 'over an incredible number of years'? Such detailed observations would not have been necessary if their only interest, as a number of scholars have seriously suggested, had been agricultural (the need to predict the seasons, which any country-born person can do). There must have been some other purpose.

[**COMMENT:** Of course. They were trying to predict the regular arrivals, tetherings and departures of Planet X Nibiru. Most people today can't even predict it, let alone think seriously about it. Most people are completely clueless about Planet X Nibiru. If we can't "predict it" (at least publicly!) with all our modern scientific knowledge, then how can we expect that the Ancient Egyptians could have done it? The only difference between now and then, I suppose, is that the Ancient Egyptians seem to have taken the whole idea a lot more seriously over a longer period of time than we humans seem to have done in later centuries. But once again, the clock of the *shar* is ticking, as one day passes, comes another! RS]

Moreover, how did the Ancient Egyptians get started on astronomy in the first place? It is not an obvious hobby for a valley-dwelling landlocked people to develop on their own initiative. Perhaps we should take more seriously the explanation they themselves offer: that their ancestors were taught the study of the stars by a god. We might also pay closer attention to the many unmistakably maritime references in the Pyramid Texts. And there could be important new inferences to draw from ancient Egyptian religious art in which the gods are shown travelling in beautiful, high-prowed, streamlined boats, built to the same advanced ocean-going specifications as the pyramid boats at Giza and the mysterious fleet moored in the desert sands at Abydos.

[**COMMENT:** The reader is advised to go to my website and download the rare (PDF) manuscript by Miss E. Valentia Straiton titled *The Celestial Ship Of The North.* RS]

Landlocked people do not as a rule become astronomers; seafaring people do. Is it not possible that the maritime iconography of the Ancient Egyptians, the design of their ships, and also their splendid obsession with observing the stars, could have been part of an inheritance passed on to their ancestors by an unidentified seafaring, *navigating* race, in remote prehistory? It is really only such an archaic race, such a forgotten maritime civilization, that could have left its fingerprints behind in the form of maps which accurately depict the world as it looked before the end of the last Ice Age. It is really only such a civilization, steering its course by the stars 'for ten thousand years' that could have observed and accurately timed the phenomenon of equinoctial precession with the exactitude attested in the ancient myths. And, although hypothetical, it is only such a civilization that could have measured the earth with sufficient precision to have arrived at the dimensions scaled down in the Great Pyramid.

[**COMMENT:** See *Slow-Motion Doomsday*, Chapter 2 ("The Polar Pivotal Axis"). A north-south longitude and an east-west latitude drawn through The Great Pyramid of Egypt divide the Planet Earth into equal volumes of land and water. This "pivotal" place on the Earth denotes the northern extremity of the "Polar Pivotal Axis". There is another idea, mentioned in *Slow-Motion Doomsday*, about a theory that The Great Pyramids were placed in that exact spot, in much the same way as we affix "wheel balances" to the tires of our automobiles, to "modify" the effects of a Polar Axis Shift, originating in that "pivotal" place on Earth! So many mysteries! So many "fingerprints of the gods"! RS]

THE SIGNATURE OF A DISTANT DATE

It was almost midnight by the time that we reached Giza. We checked into the Siag, a hotel with an excellent pyramid view, and sat out on our balcony as the three stars of Orion's belt tracked slowly across the southern heavens.

It was the disposition of these three stars, as archaeo-astronomer Robert Bauval had recently demonstrated, that served as the celestial template for the site-plan of the three Giza pyramids. This, in itself, was a remarkable discovery, suggesting a far higher level of observational astronomy, and of surveying and setting-out skills, than scholars had attributed to the Ancient Egyptians. Even more remarkable, however — and the reason that I had arranged to meet him at Giza the next morning — was Bauval's contention that the pattern traced out on the ground (in almost fifteen million tons of perfectly dressed stone) matched exactly the pattern in the sky during the epoch of 10,450 BC. If Bauval was correct, the pyramids had been devised, using the changes precession effects in the positions of the stars, as the permanent architectural signature of the eleventh millennium BC.

[COMMENT: It is the contention of Frenchman Robert Bauval that the three primary pyramids of Giza are symmetrically and identically aligned with each other so as to "map" the three stars in Orion's Belt. If you overlay (to scale) Orion's Belt on top of the Giza Complex, there is no doubt about such a match. Whether this was intentional is anybody's guess; but given all the other evidence, the similarities of everything else, it is difficult to disregard this identification as a mere "coincidence". RS]

Pages 482-484

MEMORIES OF THE POLAR DAWN

Our ancestors may have preserved in their most ancient traditions memories of a displacement [of the Polar Axis, RS]. We saw some of these memories in Part IV: cataclysm myths that appear to be eyewitness accounts of the series of geological disasters which accompanied the end of the last Ice-Age in the northern hemisphere. There are other myths too, which may have come down to us from that epoch between 15,000 and 10,000 BC. Among these are several which speak of lands of the gods and of former paradises, all of which are described as being in the south (for example, the Ta-Neteru of the Egyptians) and many of which seem to have experienced polar conditions.

The great Indian epic, *Mahabaratha*, speaks of Mount Meru, the land of the gods: 'At Meru the sun and moon go round from left to right every day, and so do all the stars. ... The mountain by its lustre, so overcomes the darkness of night, that the night can hardly be distinguished from the day. ... The day and night are together equal to a year to the residents of the place. ... '

[COMMENT: See *Slow-Motion Doomsday*, Chapter 6 ("The Night Sun"). RS]

Similarly, as the reader will recall from Chapter Twenty-five, Airyana Vaejo, the mythical paradise and former homeland of the Avestic Aryans of Iran, seems to have been rendered uninhabitable by the sudden onset of glaciation. In later years it was spoken of as a place in which: 'the stars, the moon and the sun are only once a year seen to rise and set, and a year seems only as a day.'

In the *Surya Siddhanta*, an ancient Indian text, we read, 'The gods behold the sun, after it has once arisen, for half a year.' The seventh Mandala of the *Rigveda* contains a number of 'Dawn' hymns. One of these (VII, 76) says that the dawn has raised its banner on the horizon with its usual splendour and reports in Verse 3 that a period of *several days* elapsed between the first appearance of the dawn and the rising of the sun that followed it. Another passage states, 'many were the days between the first beams of the dawn and actual sunrise'.

Are these eyewitness accounts of polar conditions?

Although we can never be sure, it may be relevant that in Indian tradition the Vedas are believed to be revealed texts, passed down from the time of the gods. It may also be relevant that in describing the processes of transmission, all the traditions refer to the *pralayas* (cataclysms) which occasionally overtake the world and claim that in each of these the written scriptures are physically destroyed. After each destruction, however, certain *Rishis* or 'wise men' survive who 'repromulgate, at the beginning of the new age, the knowledge inherited by them as a sacred trust from their forefathers in the preceding age. ... Each *manvantara* or age thus has a Veda of its own which differs only in expression and not in sense from the antediluvian Veda.'

[COMMENT: Again let me emphasize that Graham Hancock's *Fingerprints Of The Gods* would be an excellent reference book addition to any "metaphysical" library. RS]

Chapter 8

Gods Of The New Millennium

*

Gods Of The New Millennium
By Alan F. Alford (Eridu Books, United Kingdom, 1996)

[COMMENT: This is an expensive, large, thick, hardcover book with a glossy dust-jacket, and it is of very high quality, much better quality than Graham Hancock's book. And like Hancock's, it also contains formidable footnotes and bibliography. RS]

Pages 108-109

THE SIRIUS SECRET

In 1976, an American scholar, with interests in astronomy and ancient civilisations, published an astonishing book. In *The Sirius Mystery*, Robert Temple produced overwhelmingly detailed evidence that an African tribe, known as the Dogon, possessed an extraordinary knowledge of the Sirius star system.

Robert Temple began his studies following an earlier report by two French anthropologists, Marcel Griaule and Germaine Dieterlen, who claimed to have found knowledge of Sirius in four Sudanese tribes. The French scientists had focused their investigations on a people known as the Dogon, who lived in Mali, West Africa. Between 1946 and 1950, they gathered information from four Dogon priests concerning their sacred religious traditions. These traditions were apparently based on a myth which had been passed down orally from one generation to another.

Every sixty years, the Dogon practised a ceremony known as *Sigui*, which re-enacted the re-creation of the world by the god Amma, the crushing of the primitive Ogo-man, and the subsequent granting of civilisation by Amma's son Nommo. The day of the gods' arrival was known to the Dogon as the "day of the fish", and the gods themselves were regarded as amphibious beings.

According to Dogon tradition, these gods had come from a planet orbiting Sirius B, one of three stars in the Sirius star system. The Dogon accurately described the 50-year orbit of Sirius B around Sirius A. This is quite amazing, because Sirius B is a "white dwarf", the tiniest form of visible star in the universe. As such, it is invisible to the naked eye, and barely visible even with a good telescope. If the tale was a myth, why did the Dogon not worship Sirius A, the so-called "dog star", which is one of the brightest stars in the sky?

Robert Temple provides incontrovertible evidence that the Dogon knew of the existence of the invisible Sirius B. How could they have known? Some cynics have attributed this knowledge to visiting missionaries, but as Temple points out, these missionaries arrived more than a hundred years before Sirius B was photographed for the first time in 1970.

Nor was the Dogon knowledge of astronomy restricted to Sirius. Robert Temple also demonstrates that they knew of the Earth's rotation on its axis and its 365-day solar orbit, which they split into a calendar of 12 months. As for the Moon, the Dogon knew that it was dry and dead. And among their other remarkable knowledge (allegedly) is the existence of Saturn's ring and Jupiter's four largest satellites. Where did all this knowledge come from? Temple concludes his research as follows: "The result, in 1974, seven years later, is that I have been able to show that the information which the Dogon possess is really more than five thousand years old and was possessed by the ancient Egyptians in the *pre-dynastic times* before 3200 BC [emphasis added by AA]."

Pages 179-180

GEOMETRY OF THE GODS

The permanent disabling of the Great Pyramid led to the immediate need for a new beacon site to guide the incoming *shems* ("sky-chambers"). Baalbek had served its purpose following the Flood, but the gods were now planning something far more sophisticated.

[COMMENT: This "permanent disabling of the Great Pyramid" refers to the removal of the golden capstone, which according to Alford quoting Zecharia Sitchin occurred during "The Pyramid Wars", discussed in Sitchin's *The Wars Of Gods And Men*. If I had read this before in Sitchin, I'd forgotten it, because it has always been my opinion that the golden capstone was removed at a much later time. On the other hand, "The Pyramid Wars" may have been fought in the early part of the current *shar* which began in 1588 BCE, centuries after Sitchin dated it, with the capstone's removal occurring before 762 BCE. This is simply something that we shall never know with certainty, unless our "gods" deign to reveal all this now-hidden "official history" to us. But that is a major digression and could be the subject of a whole book, in and of itself. Finally, for those who may not know, *shem* is interpreted by Sitchin to mean "spaceship". For more information, see Chapter 11 ("Brief History Of Planet X Nibiru"). RS]

Whilst work was in progress, Baalbek continued as the central focus and a new beacon was established at Heliopolis, just 16 miles north-east of Giza. The Heliopolis beacon was located in a position where it could continue to be used after the completion of the new space facilities, but in the meantime it was used to point the way to Baalbek, and this necessitated another equidistant beacon site to be temporarily set up on the eastern coast of the Sinai peninsula.

[COMMENT: Years ago I visited Baalbek. It is quite an impressive sight, and one wonders how "they" flattened out the top of that rock mountain as they did. The Temple of Jupiter, ruins of which still remain, was probably added much later in time, after the "Nibiruan Airport" had long been abandoned in favor of the new aerospace facilities at the Nazca Plain in Peru. RS]

It is no coincidence that Heliopolis was once the most sacred city of Egypt, where its earliest kings were consecrated. This small city was the site of the enigmatic "benben" stone and the site from which the legendary phoenix rose from the ashes. As with the Sumerian culture, the powerful Egyptian priesthood at Heliopolis also safeguarded the scientific knowledge bestowed by the gods, along with the records of the divine succession which stemmed from Ra.

The turbulent history of northern Egypt has left little remaining at Heliopolis today, other than a single obelisk of red granite, 170 feet high and weighing 350 tons. It is generally believed that this obelisk, attributed to Senuseret I in the early second millennium BC, replaced an earlier construction.

[COMMENT: Rather as an aside, let me interject something here. The State Capitol of Texas was built well over 100 years ago. It is a massive red granite edifice which was deliberately constructed to be taller than the Capitol in Washington. (Everything's bigger in Texas, you know?) There are records of where these early Texans got the granite, but nobody is certain about exactly what sort of equipment was used to build the Capitol. And absolutely nobody knows how their statue of Lady Liberty got hoisted to the top of the dome in those pre-technological days. When they removed Lady Liberty a few years ago to clean the statue, they had to use a construction helicopter. So how did the early Texans place that statue on top of the Capitol **without** a helicopter? There were newspaper articles about this in Texas. So we should never underestimate the ingenuity of "ancient people" to construct such "mysterious" monuments and buildings. RS]

The Greek-given name Heliopolis meant "City of the Sun", a reference to the Sun god, Shamash.

[COMMENT: Shamash is Utu-Shamash, Greek Sun-God Prince Apollo, who was the commander of the Sinai Spaceport. See Chapter 13 ("Vengeful Birth Of Lord Hellespontiacus"). RS]

In so naming it, the Greeks recognised its original link with the other city of Heliopolis, also known as Baalbek. The original name of the Egyptian Heliopolis was Annu — a clear reference to the Sumerian AN, representing both "Heaven" and Anu, the heavenly father of the gods. Several writers have noted that Annu meant "Pillar City" and its hieroglyphic sign indeed resembled a high sloping tower (Figure 26a), sometimes surmounted by a *mu* or sky-chamber. The original function of "Pillar City" may also shed light on the mysterious djed symbol which is often associated with Heliopolis. The Egyptologists usually refer to this strange object (shown in Figure 26b) as "the backbone of Osiris", a meaningless expression of contrived symbolism. In fact, the djed symbol looks rather like a tower or lighthouse, and it was often depicted in pairs, sometimes in the mysterious Duat, flanking the Gateway to Heaven. Did there once exist a second djed pillar with a similar function? The second, temporary, flight path would suggest that such a site must have once existed in the Sinai peninsula. It was almost certainly for this reason that the Pyramid Texts referred to the Heliopolitan gods as the "Lords of the Dual Shrines".

[COMMENT: This "backbone of Osiris" was **not** a "meaningless expression of contrived symbolism". Far from it. It was the electromagnetic tether-beam that connected Earth's North Pole to Planet X Nibiru's South Pole. It was "Bifrost, Rainbow Bridge To Asgard". Alford's Figure 26a depicts a foundation structure with a diamond-shaped object on top. See the cover of *Slow-Motion Doomsday* for my own artistic rendition of this "djed pillar". And Ancient Egypt is not the only country in the world where we find records of a "Cosmic Pillar" or "Cosmic Tree" or "Cosmic Mountain". Clearly all of these references point to Yggdrasill and Hyperborea. There is no other logical explanation. As for Alford's speculation regarding a second "djed pillar" in the Duat (or "Tuat, City Of God" — see Chapter 7 of *Slow-Motion Doomsday*), I feel that this is incorrect, because there was only one Cosmic Tree, not two. RS]

Pages 217-219

GIZA'S CHRONOLOGY

As promised, I will now pass a few comments on Robert Bauval and Adrian Gilbert's claim that the "airshafts" [of the Great Pyramid, RS] were aligned to certain stars, and thus fulfilled a symbolic purpose. Using Rudolf Gantenbrink's measurements of the slope of the shafts, they found that *c*.2450 BC the southern shaft of the Queen's Chamber had been aligned with Sirius, the northern shaft of the King's Chamber had been aligned with Alpha Draconis and the southern shaft of the King's Chamber had been aligned with the lowest star in Orion's Belt.

The first point to note is that the date 2450 BC has no particular significance, since Khufu's reign is generally accepted to be 2550 BC.

[COMMENT: For the record, I could debate these dates, but it would be a digression from the matter at hand. See Chapter 18. RS]

More importantly, however, it must be recognised that 2450 BC is a convenient average adopted by Bauval and Gilbert; their precise results showed three different dates: 2400 BC for the alignment of the Queen's Chamber and 2425 and 2475 BC for the alignments of the King's Chamber. This is not very convincing. First, the lower Queen's Chamber must have been constructed first and should therefore align at the earlier date and not at the later date! Secondly, the King's Chamber shafts, which should have been constructed simultaneously, aligned at two different dates, with an unexplained 50 years difference!

One of Bauval and Gilbert's findings which does interest me, however, is the exact mirror image of the three Giza pyramids with Orion's Belt at 10,450 BC, both in relative position and size. Several writers have claimed that the Pyramid has alignments to stars, but their claims are unconvincing, since the Earth's wobble causes the positions of stars to move, and there are bound to be random alignments from time to time. Bauval and Gilbert's discovery is in a different league, because what they found was a mirror image rather than a chance alignment. The fact that they found that mirror image to be exact at 10,450 BC is all the more convincing because they did not go looking for it (in contrast to a deliberate search for alignments around the time of Khufu). Indeed, they have not been able to offer a satisfactory explanation of this remote date.

According to Khufu, the alleged builder of the Pyramid, its owner was the goddess Isis, whom he referred to as the "Mistress of the Pyramid". The evidence concerning Isis is in the form of an inscription on a stele, found in the 1850s in the temple, or "House", of Isis near to the Great Pyramid. It has been translated as follows:

Eternal life to Horus Mezdau.
To King of Upper and Lower Egypt,
Khufu, eternal life is given!
He founded the House of Isis,
Mistress of the Pyramid,
Beside the House of the Sphinx.

[COMMENT: This was quite a surprising linguistic connection to run across at random, so to speak. I didn't expect to find this. A "chief god" of the Hindus, and also perhaps the Persians, was "Ahura Mazda". The linguistic radical *aHuRa* is similar to the radical *HoRuS*, and *MaZDa* is identical with *MeZDau*! And for your "trivial pursuit" amusement, the Mazda automobile is named after this "god"! My opinion is that "Mazda" is the equivalent of Nibiruan ex-Emperor Alalu who was deposed in a *coup d'état* by his son, current Emperor Anu (an event remembered in Greek mythology as "The Battle Of The 72 Titans"), after which Emperor Alalu and Empress Lilitu were exiled to permanent oblivion on "Tiamat", probably, if they are still alive, to a gilded cavern underneath "Grand Teton" Mountain near American Yellowstone National Park, as recounted by William Bramley in *The Gods Of Eden*. See also Chapter 11. RS]

The stele on which this inscription was found came to be known as the "Inventory Stele", and is today exhibited in the Cairo Museum. Ironically, the experts have proclaimed it a forgery, because it contradicts the evidence from [British Lord] Vyse's fraudulent inscription just over a decade earlier! Whilst it is true that the writing style on the stele may indicate that it was produced some time after Khufu, it is perfectly possible that it was a copy of an earlier original. Scholars have been unable to suggest why such a forgery would have been made 4,000 years ago in a manner which praised Khufu and yet attributed the Pyramid to the ancient goddess, Isis.

Whilst not constituting a chronological proof, the Inventory Stele has all the hallmarks of authenticity, since Manetho's history of Egypt dates the rule of Isis and Osiris in Egypt to around 10,000 BC. Furthermore, it corroborates Bauval and Gilbert's dating of the Giza pyramids to 10,450 BC.

As we have seen in chapter 7, the Flood occurred *c.*11,000 BC, shortly before the above dates for Isis and the Giza pyramids. All of this evidence corresponds to the ancient texts, as interpreted by Zecharia Sitchin, which link the pyramids to a flight path built by the gods as soon as practical following the Flood.

[COMMENT: As I have postulated in *Slow-Motion Doomsday*, the dates of "shars" prior to 1588 BCE were 5188, 8788 and 12,388 BCE. These dates of 10,450 and 11,000 BCE undoubtedly refer to either the arrival or the departure of Planet X Nibiru between 12,388 and 11,488 BCE, four "shars" ago. RS]

Chapter 9

The Stairway To Heaven

*

The Stairway To Heaven
By Zecharia Sitchin (Avon Books, New York, 1983)

INTRODUCTORY COMMENT
By Rob Solàrion

The following information, transcribed almost verbatim, comes from pages 47-71 of *The Stairway To Heaven* by Zecharia Sitchin. This is Volume 2 of *THE EARTH CHRONICLES* series, which I first read in its entirety in the summer and fall of 1993. Mr. Sitchin always includes numerous graphics to accompany his text. For those, you should consult his printed edition.

Chapter IV ("The Stairway To Heaven") is presented in a rather "factual" manner, whereas in Chapter V ("The Gods Who Came To Planet Earth") Mr. Sitchin tends to philosophize about these facts from a modern perspective. Mr. Sitchin wrote in *THE EARTH CHRONICLES* that there was a "spaceport" in the Sinai and an "airport" at Baalbek, Phoenicia/ Lebanon. Thus, by inference, when the Pharaoh went "eastwards", he went to the Sinai Spaceport with all its elaborate security precautions, just like Cape Canaveral today, I am sure. Mr. Sitchin is attempting to bolster his theory that there was a "spaceport" in the Sinai, which Nibiruan Prince Utu/Apollo sabotaged during The Pyramid Wars. I have no disagreement with that. However, as I shall note in my comments below, I think that this "Stairway To Heaven" was located in the far-north, possibly accessible from "east of Egypt" (as from central Asia) by underground tunnels and the like. But the actual "Stairway" itself was the electromagnetic tether-beam connecting our North Pole to Planet X Nibiru's South Pole. The "Ascender" described below is the same "animal" as the "tomcat" of Siberian legend, a "tomcat" that climbed up and down the "North Pole". This "tomcat" was remembered as a "squirrel" named "Ratatosk" that climbed up and down the Yggdrasill Tree in the Nordic legends, and as a "monkey" in the Central American legends. Even in modern Alaska there are Indian "totem poles" on which are carved some sort of animal shown climbing up or down, or perched on top. See Illustrations 23-26 of my book *Planet X Nibiru: Slow-Motion Doomsday.*

"There is a mill which grinds by itself, swings of itself, and scatters the dust a hundred versts away. And there is a golden pole with a golden cage on top which is also the Nail of the North. And there is a very wise tomcat which climbs up and down this pole. When he climbs down, he sings songs; and when he climbs up, he tells tales." Legend of the Ostyaks of the Siberian Irtysh River Valley

This "animal" was a space "elevator" or "conveyance" of some sort that "ferried" gods and men up the "treetrunk" from North Polar Regions (Scandinavia, Siberia, Alaska) to Planet X, or to its "mothership", tethered about 60,000 miles (about 100,000 kilometers) above our North Pole. When this "configuration" was still in place over 2,700 years ago, it is said in the Egyptian legend that it required a journey of 8 days to travel up this "ladder" or "pillar" to the "Celestial Disk" or "Heaven", from northern Siberia, for example. That seems a bit long to me, but then I do not really understand what sort of "conveyance" the Pyramid Texts were referring to in the first place. Neither does anyone else.

As for the northern celestial millwheel that "scatters the dust a hundred versts away", in Norse mythology there is reference to a sacred "Vigrid Plain" which is associated with their "Ragnarok" legends. It is stated: "The sons of Muspel direct their course to the plain which is called Vigrid. Thither repair also the Fenris-Wolf and the Midgard-Serpent." And: "To this place have also come Loke and Hrym, and with him all the Frost Giants. In Loke's company are all the friends of Hel. The sons of Muspel have then their efficient bands alone by themselves. The plain Vigrid is one hundred rasts on each side." The "Vigrid Plain" is the same size as the area over which the millwheel scatters the dust, and it is significant that in two completely different mythologies there are references to a northern, polar area that is 100 "rasts" or "versts" wide. The words "rast" and "verst" are identical, like the Spanish word "amigo" and the Italian "amico". When "coincidental" details turn up in two or more cultures, one knows that one is on the trail of "hidden secrets"!

The facts presented in Chapter IV obviously indicate that someone, such as Pharaoh Pepi I (Apopi I), actually travelled up the "Rainbow Bridge Bifrost" to "Asgard"; then returned and lived to tell about it. One presumes that after The Cosmic Tree finally "dissolved" in 687 BCE, it was remembered only as a "mythical" place for the Pharaohs' "eternal life"; but there is no doubt that when you consider such legends in our contemporary technological context, as Mr. Sitchin does at the beginning of Chapter V, this is clearly a description of a journey from our planet to theirs, or to their gigantic tethered/anchored mothership.

Some of us researchers may have minor disagreements with Mr. Sitchin over this detail or that, but we must be grateful for his most voluminous writings; for without them, we would all still be in the dark as to what force or object causes these "periodic" cataclysms on our Planet "Tiamat"!

*

CHAPTER IV, THE STAIRWAY TO HEAVEN

Let us imagine ourselves in the Pharaoh's magnificent funerary temple. Having mummified and prepared the Pharaoh for his Journey, the *Shem* priests now intone the gods to open for the king a path and a gateway. The divine messenger has arrived on the other side of the false door, ready to take the Pharaoh through the stone wall and launch him on his journey.

[**COMMENT:** The term *Shem* is interpreted by Sitchin elsewhere as being a word for "spaceship" or "rocketship". RS]

Emerging through the false door on the eastern side of his tomb, the Pharaoh was instructed to set his course eastward. Lest he misunderstand, he was explicitly warned against going west: "Those who go thither, they return not!" His goal was the *Duat*, in the "Land of the Mountain Gods". He was to enter there "The Great House of Two ... the House of Fire";

where, during a "night of computing years", he shall be transformed into a Divine Being and ascend "to the east side of Heaven".

[COMMENT: In my book *Planet X Nibiru: Slow-Motion Doomsday*, in Chapter 7 ("Heaven & Hell"), I discuss "Tuat, City Of God". The word *Tuat* is identical with the word *Duat*. There I postulated that *Tuat* was that part of the Planet Earth in the Southern Hemisphere, where The Cosmic Tree was not visible. It was a land of gloom and terror in the "partially lighted regions" where the "solid darkness" began — hence, Hell. See also Illustrations 3-8 of that book. Here, however, Sitchin presents the idea that the *Duat* was in the Northern Hemisphere. This seems logical when you read through what follows. This matter deserves further investigation. However, the "Land Of The Mountain Gods" certainly refers to The Sacred Mountain, Mount Olympus, Mount Zion, Mount Meru, Hyperborea, Yggdrasill, Asgard. "The Great House Of Two" is probably a reference to the fact that Planet X Nibiru is ruled by two: Emperor Anu & Empress Antu. The "City Of An" mentioned below is certainly a reference to Anu. Furthermore, if one were looking upwards at Planet X above our North Pole, one would be looking at its Southern Hemisphere. Perhaps the "landing pad" of the "City Of Anu" is located on the eastern side of their Planet, or this idea simply indicated the cardinal direction on their Planet where the "tomcat" arrived before "landing" on top of the "North Pole", in the "Golden Cage" or the turning millwheel, that is, the rotating Planet X. We shall not know this for sure until after 2012, when we can once again see this Sacred World Tree with our own eyes. And finally, I am at a loss to explain this reference, repeated later, to a "Night Of Computing Years". Our years, or their "shars", or both? Is this mysterious idea the Planet X equivalent of our own "atomic clock"? But they are coming back, and they'll be "computing years" again, we can be certain of that! RS]

The first obstacle in the Pharaoh's course was the Lake of Reeds — a long body of marshy waters made up of a series of adjoining lakes. Symbolically, he had the blessing of his guardian god to cross the lake by parting its waters; physically, the crossing was possible because the lake was served by the Divine Ferryman, who ferried the gods across in a boat made by Khnum, the Divine Craftsman. But the Ferryman was stationed on the far side of the lake, and the Pharaoh had a hard time convincing him that he was entitled to be fetched and ferried over.

The Ferryman questioned the Pharaoh about his origins. Was he the son of a god or goddess? Was he listed in the "Register of the Two Great

Gods"? The Pharaoh explained his claims to being of "divine seed", and gave assurances of his righteousness. In some cases it worked. In other instances the Pharaoh had to appeal to Ra or to Thoth to get him across; in which instances, the boat and its oars or rudder came alive with uncanny forces; the ferryboat began to move by itself, the steering-oar grasped by the king directed itself. All, in short, became self-propelled. One way or another, the Pharaoh managed to cross the lake and be on his way toward "The Two That Bring Closer the Heavens":

He descends into the boat, like Ra,
On the shores of the Winding Watercourse.
The king rows in his *Hanbu*-boat;
He takes the helm toward the
Plain of "The Two That Bring Closer the Heavens",
In the land beginning from the Lake of Reeds.

[**COMMENT:** The exact location of this "Lake Of Reeds" is unknown. However, one gets the distinct impression when reading this, that access to the *Duat* was highly restricted; even getting to its location in the first place was impossible because of all the "security checks" along the way. Nibiruan Duke Dumuzi/Adonis once worked for King (now Duke) Nergal/Hades as a "ferryman" of Adamu slaves to the "Underworld", probably a reference to the gold mines that Nergal supervised in southern Africa, in the fiery land of the "partially lighted regions" — Hell. Duke Dumuzi was murdered by Baron Marduk during The Pyramid Wars, when the evil Baron Marduk and his treacherous consort Baroness Sarpanit stole the powerful MEs from Emperor Anu and Empress Antu and declared themselves the new Emperor and Empress. Eventually they were defeated, and the MEs were returned to Anu and Antu. This period of war probably preceded the Egyptian legend; thus, a new "ferryman" might have replaced Duke Dumuzi. But this reference could also be to a completely different type of "ferryman". See Chapter 11 for additional details about Nibiruan history. RS]

The Lake of Reeds was situated at the eastern end of the domain of Horus. Beyond lay the territories of his adversary Seth, the "lands of Asia". As would be expected on such a sensitive boundary, the king discovers the lake's eastern shore is patrolled by four "Crossing guards, the wearers of side locks". The way these guards wore their hair was truly their most conspicuous feature. "Black as coal," it was "arranged in curls on their foreheads, at their temples and at the back of their heads, with braids in the center of their heads."

[COMMENT: Clearly by now the king has arrived in Central Asia, but this hairstyle made me think of Orthodox Hasidic Jews. This is "mysterious", to say the least. In northern Siberia, just south of the Arctic Ocean there are high mountain peaks. There are also myriad rivers that empty into the Arctic Ocean, and one presumes that there are numerous lakes as well. The Rhipaean Mountains which led to Hyperborea were thought to have been located around the Altai Mountains of central Siberia and western Mongolia. See Chapter 8 ("Hyperborea & The Rhipaean Mountains") of *Slow-Motion Doomsday.* The *Duat* could have been located north of the Altais. This is highly inhospitable, middle-of-nowhere territory. Even today, it would be a daunting task to try to hike with only a backpack from, say, Ulan Bator, Mongolia, to Khatanga, Siberia. RS]

Combining diplomacy with firmness, the king again proclaimed his divine origins, claiming he was summoned by "my father Ra". One Pharaoh is reported to have used threats: "Delay my crossing, and I will pluck out your locks as lotus flowers are plucked in the lotus pond!" Another had some of the gods come to his assistance. One way or another, the Pharaoh managed to proceed.

The king has now left the lands of Horus. The eastward place which he seeks to reach — though under the aegis of Ra — is "in the region of Seth". His goal is a mountainous area, the Mountains of the East. His course is set toward a pass between two mountains, "the two mountains which stand in awe of Seth." But first he has to traverse an arid and barren area, a kind of no-god's land between the domains of Horus and Seth. Just as the pace and urgency of the Utterances increase, for the king is getting closer to the Hidden Place where the Doors of Heaven are located, he is challenged again by guards. "Where goest thou?" they demand to know.

The king's sponsors answer for him: "The king goes to Heaven, to possess life and joy; that the king may see his father, that the king may see Ra." As the guards contemplate the request, the king himself pleads with them: "Open the frontier ... incline its barrier ... let me pass as the gods pass through!"

Having come from Egypt, from the domain of Horus, the king and his sponsors recognize the need for prudence. Many Utterances and verses are employed to present the king as neutral in the feud between the gods. The king is introduced both as "born of Horus, he at whose name the Earth quakes", and as "conceived by Seth, he at whose name Heaven trembles". The king stresses not only his affinity to Ra, but declares that he proceeds "in the service of Ra"; producing thereby a *laissez-passer* from higher authority. With shrewd evenhandedness, the texts point out to the two gods their own self-interest in the king's continued journey, for Ra would surely appreciate their aid to one who comes in his service.

[COMMENT: It is unclear to whom this "Ra" refers. Sitchin equates Duke Dumuzi with the Egyptian "Aten-Ra" and Baron Marduk with the Egyptian "Amon-Ra". Horus (Prince Ishkur/Ares) was the son of Osiris (Crown-Prince Enlil/Zeus), who was dismembered by his brother Seth (Prince Enki/Poseidon). Isis (Queen Ninkhursag/Hera) collected the pieces of Osiris and reassembled them. This episode resulted in the "feud between the gods" mentioned above. At any rate, this particular Pharaoh was allowed to proceed. RS]

Finally, the guards of the Land of Seth let the king proceed toward a mountain pass. The king's sponsors make sure that he realizes the import of the moment:

Thou are now on the way to the high places
In the land of Seth.
In the land of Seth
Thou will be set on the high places,
On that high Tree of the Eastern Sky
On which the gods sit.

The king has arrived at the *Duat.*

The *Duat* was conceived as a completely enclosed Circle of the Gods, at the head-point of which there was an opening to the skies (symbolized by the goddess Nut) through which the Imperishable Star (symbolized by the Celestial Disk) could be reached. Other sources suggested in reality a more oblong or oval valley, enclosed by mountains. A river which divided into many streams flowed through this land, but it was hardly navigable and most of the time Ra's barge had to be towed, or moved by its own power as a "boat of earth", as a sled.

[COMMENT: The "Imperishable Star" and "Celestial Disk" refer to Planet X Nibiru tethered and floating like a cosmic jewel above our North Pole. The "Goddess Nut" is a reference to the Egyptian Goddess Neith, who is the equivalent of Empress Antu, the Greek Rhea (wife of Kronos, or Emperor Anu). She, of course, resides and rules on Planet X Nibiru, Mount Olympus — "on that high Tree of the Eastern Sky on which the gods sit" — yet another reference to The Cosmic Tree. RS]

The *Duat* was divided into twelve divisions, variably described as fields, plains, walled circles, caverns or halls, beginning above ground and continuing underground. It took the departed king twelve hours to journey through this enchanted and awesome realm; this he could achieve, because Ra had put at his disposal his magical barge or sled, in which the king traveled aided and protected by his sponsoring gods.

[COMMENT: The crucial "key" for entrance by a "visitor" into the *Duat* was the presence of these so-called "sponsoring gods". Certainly no one who was not "sponsored" could enter, just as some modern tourist who is not "sponsored" cannot visit the classified areas of Cape Canaveral. The Elysian Fields, or Champs Elysées, were supposedly part of the *Duat.* RS]

There were seven gaps or passes in the mountains that enclosed the *Duat*, and two of them were in the mountains on the east side of Egypt (*i.e.*, in the mountains on the west of the *Duat*), which were called "The Horizon" or "The Horn" of "The Hidden Place". The pass through which Ra had traveled was 220 *atru* (some twenty-seven miles) long, and followed the course of a stream: the stream, however, ran dry and Ra's barge had to be towed. The pass was guarded and had fortifications "whose doors were strong".

[COMMENT: The "Hidden Place" was the "airport" where this "tomcat" conveyance climbed up and down the North Nail, North Pole, to the "Golden Cage" on top. Just as a fast-moving freight train whistle sounds different when the train is approaching than when the train is departing, so also this "tomcat" sounded different: when it descended, it sounded as if it were "singing" (high pitch?); when it ascended, it sounded as if it were "talking" (low pitch?). RS]

The Pharaoh, as some papyri indicate, took the course leading through the second, shorter pass (only some fifteen miles long). The papyrus drawings show him upon the barge or sled of Ra, passing between two mountain peaks on each of which there is stationed a company of twelve guardian gods. The texts describe a "Lake of Boiling Waters" nearby — waters which, despite their fiery nature, are cool to the touch. A fire burns below the ground. The place has a strong bituminous or "natron" stench which drives away the birds. Yet not too far away, there is depicted an oasis with shrubs or low trees around it.

Once across the pass, the king encounters other companies of gods. "Come in peace," they say. He has arrived at the second division.

It is called, after the stream that runs through it, *Ur-nes* (a name which some scholars equate with *Uranus*, the Greek god of the skies). Measuring some fifteen by thirty-nine miles, it is inhabited by people with long hair, who eat the flesh of their asses and depend on the gods for water and sustenance, for the place is arid and the streams are mostly dry. Even Ra's barge turns here into a "boat of earth". It is a domain associated with the Moon god, and with Hathor, the Goddess of Turquoise.

[**COMMENT:** Hathor is the Egyptian equivalent of Nibiruan Princess-Royal Inanna, the Greek Aphrodite and the Hindu Lakshmi. She was the passionate lover of Duke Adonis Dumuzi at the time of his murder during The Pyramid Wars. See Chapter 13 regarding Lord Hellespontiacus. RS]

Aided by the gods, the king passes safely through the second division and in the Third Hour arrives at *Net-Asar*, "The Stream of Osiris". Similar in size to the second division, this third division is inhabited by "The Fighters". It is there that the four gods, who are in charge of the four cardinal points of the compass, are stationed.

The pictorial depictions which accompanied the hieroglyphic texts surprisingly showed the Stream of Osiris as meandering its way from an agricultural area, through a chain of mountains, to where the stream divided into tributaries. There, watched over by the legendary Phoenix birds, the *Stairway to Heaven* was situated; there, the Celestial Boat of Ra was depicted as sitting atop a mountain, or rising heavenward upon streams of fire.

Here, the pace of prayers and Utterances increases again. The king invokes the "magical protectors", that "this man of Earth may enter the *Neter-*

Khert" unmolested. The king is nearing the heart of the *Duat*; he is near the *Amen-Ta*, the "Hidden Place".

It was there that Osiris himself had risen to the Eternal Afterlife. It was there that the "Two That Bring Closer the Heaven" stood out "yonder against the sky", as two magical trees. The king offers a prayer to Osiris (the Chapter's title in the Book of the Dead is "Chapter of Making His *Name* in the *Neter-Khert* Granted"):

May be given to me my *Name*
In the Great House of Two;
May in the House of Fire
My Name be granted.
In the night of computing years,
And of telling the months,
May I be a Divine Being,
May I sit at the east side of Heaven.
Let the god advance me from behind;
Everlasting is his *Name*.

The king is within sight of the "Mountain of Light".

He has reached the STAIRWAY TO HEAVEN.

The Pyramid Texts said of the place that it was "the stairway in order to reach the heights". Its stairs were described as "the stairs to the sky, which are laid out for the king, that he may ascend thereon to the heavens". The hieroglyphic pictograph for the Stairway to Heaven was sometimes a single stairway (which was also cast in gold and worn as a charm), or more often a double stairway, as a step pyramid. This Stairway to Heaven was constructed by the gods of the city of An — the location of the principal temple of Ra — so that they, the gods, could be "united with the Above".

The king's goal is the Celestial Ladder, an Ascender which would actually carry him aloft. But to reach it in the House of Fire, the Great House of Two, he must enter the *Amen-Ta*, the Hidden Land of Seker, God of the Wilderness.

It is a domain described as a fortified circle. It is the subterranean Land of Darkness, reachable by entering into a mountain and going down spiraling hidden paths protected by secret doors. It is the fourth division of the *Duat* which the king must now enter; but the mountain entrance is protected by two walls and the passage between them is swept by flames and manned by guarding gods.

When Ra himself had arrived at this entrance to the Hidden Place, "he performed the designs" — followed the procedures — "of the gods who are therein by means of his voice, without seeing them." But can the king's voice alone achieve for him admission? The texts remind the challenger that only "he who knoweth the plan of the hidden shaftways which are in the Land of Seker" shall have the ability to journey through the Place of Underground Passages and eat the bread of the gods.

Once again the king offers his credentials. "I am the Bull, a son of the ancestors of Osiris," he announces. Then the gods who sponsor him, pronounce in his behalf the crucial words for admission:

Admittance is not refused thee
At the gate of the *Duat*;
The folding doors of the Mountain of Light
Are opened to thee;
The bolts open to thee of themselves.
Thou treadest the Hall of the Two Truths;
The god who is in it greets thee.

The right formula or password having thus been pronounced, a god named Sa uttered a command; at his word, the flames ceased, the guards withdrew, the doors opened automatically, and the Pharaoh was admitted into the subterranean world.

"The mouth of the earth opens for thee, the eastern door of heaven is open for thee," the gods of the *Duat* announce to the king. He is reassured that though he enters the mouth of the earth, it is indeed the Gateway to Heaven, the coveted eastern door.

[**COMMENT:** It was at this point that the king could enter the "underground installations" which led to the port of departure. RS]

The journey in the fourth and following Hours leads the king through caverns and tunnels where gods of diverse functions are sometimes seen, sometimes only heard. There are underground canals, on which gods move about in soundless barques. There are eerie lights, phosphorous waters, torches that light the way. Mystified and terrified, the king moves on, toward "the pillars that reach the Heaven".

[COMMENT: Nowadays our space facilities are generally above ground and totally manmade. It could be that the Nibiruans simply, culturally, liked to incorporate dramatic vistas created by Nature into their architectural designs; hence, the peculiarity of the descriptions of some of these "chambers" or "levels" in the *Duat.* RS]

The gods seen along the way are mostly organized in groups of twelve, and bear such epithets as "Gods of the Mountain", "Gods of the Mountain of the Hidden Land", or "The Holders of the Time of Life in the Hidden Land". The drawings that accompanied some of the ancient texts provide identification of these gods through the different scepters held by them, their particular headgear, or by depicting their animal attributes — hawk-headed, jackal-headed, lion-headed. Serpents also make an appearance, representing subterranean guards or servants of the gods in the Hidden Land.

[COMMENT: Here Sitchin makes reference to serpents, but he certainly does not think of these Nibiruans as being "saurian" or "reptilian". I can only refer the reader to my website where you can find the original edition of *Flying Serpents & Dragons* by R.A. Boulay, with which I am in almost total agreement, although Sitchin would vehemently disagree. RS]

The texts and the ancient illustrations suggest that the king has entered a circular underground complex, within which a vast tunnel first spirals down and then up. The depictions, presented in a cross-section fashion, show a gradually sloping tunnel some forty feet high, with a smooth ceiling and a smooth floor, both made of some solid material two to three feet thick. The tunnel is partitioned into three levels, and the king moves within the middle level or corridor. The upper and lower levels are occupied by gods, serpents and structures of diverse functions.

The king's sled, pulled by four gods, begins its journey by gliding silently along the middle corridor; only a beam emitted from the vehicle's bow lights the way. But soon the passage is blocked by a sharply slanting partition, and the king must get off and continue on foot.

The partition, as the cross-section depictions show, is one wall of a shaft that cuts across the three tunnel levels (which slope at about 15°) at a sharper angle of some 40°. It apparently begins above the tunnel, perhaps at ground level or somewhere higher within the mountain; it seems to end as it reaches the floor of the lowest, third level. It is called *Re-Stau*, "The Path of the Hidden Doors; and at the first and second levels, it is indeed provided with chambers that look like air-locks. These chambers enable Seker and other "hidden gods" to pass through, though "the door has no leaves". The king, who has left his sled, mysteriously passes through this slanting wall simply by virtue of the command of some god, whose voice had activated the air-lock. He is greeted on the other side by representatives of Horus and Thoth, and is passed along from god to god.

On his way down, the king sees "faceless gods" — gods whose faces cannot be seen. Offended or simply curious, he pleads with them:

Uncover your faces,
Take off your head coverings,
When ye meet me;
For, behold, I [too] am a mighty god
Come to be among you.

But they do not heed his plea to show their faces; and the texts explain that even they, "these hidden beings, neither see or look upon" their own chief, the god Seker "when he is in this form himself, when he is inside his abode in the earth".

Spiraling his way down, the king passes through a door and finds himself on the third, lowest level. He enters an antechamber which bears the emblem of the Celestial Disk, and is greeted by the god who is "The Messenger of Heaven" and a goddess who wears the feathered emblem of Shu, "He who rested the sky upon the Stairway to Heaven." As called for by the formula in the Book of the Dead, the king proclaims:

Hail, two children of Shu!
Hail, children of the Place Of The Horizon ...
May I ascend?
May I journey forth like Osiris?

The answer must be positive, for the king is admitted by them, through a massive door, into the shafts which only the hidden gods use.

In the Fifth Hour, the Pharaoh reaches the deepest subterranean parts which are the secret ways of Seker. Following shafts that incline up, over and down, the Pharaoh cannot see Seker; but the cross-section drawings depict the god as a hawk-headed person, standing upon a serpent and holding two wings within a completely enclosed oval structure deep underground, guarded by two sphinxes. Though the king cannot see this chamber, he hears coming from it "a mighty noise, like that heard in the heights of the heavens when they are disturbed by a storm". From the sealed chamber there flows a subterranean pool whose "waters are like fire". Chamber and pool alike are in turn enclosed by a bunkerlike door on the right side. As further protection, a mound of soil is piled up atop the sealed chamber. The mound is topped by a goddess, whose head only is seen, protruding into the descending corridor. A beetle symbol (meaning "to roll, to come into being") connects the head of the goddess with a conical chamber or object in the uppermost corridor; two birds are perched upon it.

The text and symbols inform us that, though Seker was hidden, his presence could be made known even in the darkness, because of a glowing "through the head and eyes of the great god, whose flesh radiates forth light". The triple arrangement — goddess, beetle (*Kheper*) and conical object or chamber — apparently served to enable the hidden god to be informed of what goes on outside his hermetically sealed chamber. The hieroglyphic text adjoining the beetle symbol states: "Behold Kheper who, immediately the (boat?) is towed to the top of this circle, connects himself with the ways of the *Duat*. When this god standeth on the head of the goddess, he speaks words to Seker every day."

[COMMENT: Think about it. This is quite peculiar language here: "When the god standeth on the head of the goddess, he speaks words to Seker every day." Does this mean that when the "tomcat" conveyance is climbing to Planet X, over a period of 8 days, it sounds like it is "telling tales" to someone, in this case Seker? Or is there an unknowable alternative explanation? RS]

The passage by the Pharaoh over the hidden chamber of Seker and by the setup through which Seker was informed of such passage, was deemed a crucial phase in his progress. The Egyptians were not the only ones in antiquity who believed that each departed person faced a moment of judgment, a spot where their deeds or hearts would be weighed and evaluated and their soul or Double either condemned to the Fiery Waters of Hell, or blessed to enjoy the cool and lifegiving waters of paradise. By ancient accounts, here was such a Moment of Truth for the Pharaoh.

Speaking for the Lord of the *Duat*, the goddess whose head only was seen announced to the Pharaoh the favorable decision: "Come in peace to the *Duat* ... advance in thy boat on the road which is in the earth." Naming herself Ament (the female Hidden One), she added: "Ament calls to thee, so that thou mayest go forward in the sky, as the Great One who is in the Horizon."

Passing the test, not dying a second time, the king was born again. The way now led by a row of gods whose task it was to punish the condemned; but the king proceeds unharmed. He rejoins his boat or sled; it is accompanied by a procession of gods, one of them holds the emblem of the Tree of Life.

The king has been found worthy of Afterlife.

[**COMMENT:** In other words, the king has passed through all the necessary "utterances", these "secret passwords", these "security codes", to allow him to board the "tomcat" up the electromagnetic "Golden Pole" to the tethered Celestial Winged Disk, Planet X Nibiru, or its monstrous "mothership". RS]

Leaving the zone of Seker, the king enters the sixth division, associated with Osiris. (In versions of the Book of the Gates, it was in this Sixth Hour that Osiris judged the departed.) Jackal-headed gods "Who Open the Ways" invite the king to take a refreshing dip in the subterranean pool or Lake of Life, as the Great God himself had done when he passed here before. Other gods, "humming as bees", reside in cubicles who doors fly open by themselves as the king moves by. As he progresses, the epithets of the gods assume more technical aspects. There are twelve gods "who hold the rope in the *Duat*", and the twelve "who hold the measuring cord".

[**COMMENT:** Obviously these Saurian Nibiruans had such an advanced technology that this ancient Egyptian priest didn't have the vocabulary necessary to describe it properly. One wonders what are meant by "the rope in the *Duat*" and "the measuring cord". RS]

The sixth division is occupied by a series of chambers set close together. A curving path is called "The Secret Path of the Hidden Place". The king's boat is towed by gods clad in leopard skins, just as the *Shem* priests who performed the Opening of the Mouth ceremonies were clad.

Is the king nearing the Opening or Mouth of the Mountain? In the Book of the Dead, the chapters indeed now bear such titles as "The chapter of sniffing the air and of getting power". His vehicle is now "endowed with magical powers ... he journeyeth where there is no stream and where there are none to tow him; he performeth this by words of power" which proceed from the mouth of a god.

As the king passes through a guarded gate into the seventh division, the gods and the surroundings lose their "underworld" aspects and begin to assume celestial affiliations. The king encounters the falcon-headed god *Heru-Her-Khent*, whose hieroglyphic name included the stairway symbol and who wore on his head the Celestial Disk emblem. His task is "to send the star-gods on their way and to make the constellation-goddesses go on their way". These were a group of twelve gods and twelve goddesses who were depicted with star emblems. The incantations to them were addressed to "the starry gods" —

Who are divine in flesh,
Whose magical powers have come into being ...
Who are united into your stars, who rise up for Ra ...
Let your stars guide his two hands
So that he may journey to the Hidden Place in peace.

[**COMMENT:** If these "underground surroundings" began to take on more "celestial affiliations", then this must indicate that the king had entered a more artificial or manmade environment than the one he'd just passed through. He was approaching the actual "launch site". One wonders, in passing, how an ancient Egyptian scribe would portray Cape Canaveral today, being surrounded by water, exotic vegetation and wildlife, including all sorts of reptiles, aquatic and otherwise. A difference between then and now would be that today, the people running Cape Canaveral would be "sapien humanoids" like the scribe himself, so the element of "godlike mystery" would be removed from the modern scenario. RS]

In this division, there are also present two companies of gods associated with the *Ben-ben*, the mysterious object of Ra that was kept at his temple in the city of An (Heliopolis). They "are those who possess the mystery", guarding it inside the *Het-Benben* (The Ben-ben House); and eight who guard outside but also "enter unto the Hidden Object". Here there are also nine objects, set up in a row, representing the symbol *Shem* which hieroglyphically meant "Follower".

The king has indeed arrived in parts of the *Duat* associated with An, after whom Heliopolis was named. In the Ninth Hour, he sees the resting place of the twelve "Divine Rowers of the Boat of Ra", they who operate Ra's celestial "Boat of Millions of Years". In the Tenth Hour, passing through a gate, the king enters a place astir with activity. The task of the gods there is to provide Flame and Fire to the boat of Ra. One of the gods is called "Captain of the gods of the boat". Two others are those "Who order the course of the stars". They and other gods are depicted with one, two or three star symbols, as though showing some rank associated with the heavens.

[COMMENT: This "Boat Of Millions Of Years" may indeed be a literal rendition of words used by the ancient scribe. Perhaps this spaceship or "conveyance" is indeed millions of Earth Years old. A million Earth Years would be only about 277 "Shars"! Today we call our vehicles "space shuttles" and "space stations", but we could just as easily have called them "boats" instead of "shuttles" or "stations". RS]

Passing from the tenth to the eleventh division, the affinity to the heavens rapidly increases. Gods bear the Celestial Disk and star emblems. There are eight goddesses with star emblems "who have come from the abode of Ra". The king sees the "Star Lady" and the "Star Lord", and gods whose task it is to provide "power for emerging" from the *Duat*, "to make the Object of Ra advance to the Hidden House in the Upper Heavens".

[COMMENT: The "Hidden House In The Upper Heavens" undoubtedly refers to the "palace" of Emperor Anu and Empress Antu on Planet X Nibiru, atop the "Golden Pole" to Mount Olympus. RS]

In this place there are also gods and goddesses whose task it is to equip the king for a celestial trip "over the sky". Together with some gods he is made to enter a "serpent" inside which he is to "shed the skin" and emerge "in the form of a rejuvenated Ra". Some of the terms here employed in the texts are still not understood, but the process is clearly explained: the king, having entered dressed as he came, emerges as a falcon, "equipped as a god"; the king "lays down on the ground the *Mshdt*-garment"; he puts on his back the "Mark-garment"; he "takes his divine *Shuh*-vestment" and he puts on "the collar of beloved Horus" which is like "a collar on the neck of Ra". Having done all that, "the king has established himself there as a god, like them". And he tells the god who is with him: "If thou goest to Heaven, so will the king go to Heaven."

[COMMENT: Obviously, as Mr. Sitchin notes below, the king had gone into a "dressing room" of some sort, took off his royal garments and donned a "spacesuit" before boarding the "tomcat" and climbing the Golden Pole. The fact that there is a reference to a serpent's shedding the skin, rather than, say, to a sheep's being shorn of its wool, further indicates the "saurian" or "reptilian" nature of these technologically advanced super-beings, these "custodial gods" as William Bramley refers to them in his thrilling book *The Gods Of Eden*. Reptiles can easily hibernate; it is only a matter of ambient air temperature. They would be much more efficient space travellers than Earth mammals, who would have to transport ultra-elaborate life-support systems. RS]

The illustrations in the ancient texts depict here a group of gods dressed in unusual garb, like tightly fitting overalls adorned with circular collar bands.

[COMMENT: It sounds like a spacesuit to me! RS]

They are led or directed by a god with the emblem of the Celestial Disk upon his head who stands with outstretched arms between the wings of a serpent with four human legs. Against a starry background, the god and the serpent face another serpent which, though wingless, clearly flies as it carries aloft a seated Osiris.

Having been properly equipped, the king is led to an opening in the center of a semi-circular wall. He passes the hidden door. Now he moves within a tunnel which is "1300 cubits long" called "Dawn at the End". He reaches a vestibule; the emblems of the Winged Disk are seen everywhere. He encounters goddesses "who shed light upon the road of Ra" and a magical scepter representing "Seth, the Watcher".

The gods explain to the awed king:

This cavern is the broad hall of Osiris
Wherein the wind is brought;
The north wind, refreshing,
Will raise thee, O king, as Osiris.

[COMMENT: Here is a reference to the "North Wind". Obviously some sort of propulsion mechanism is indicated, and it reinforces the idea that "Hyperborea" (literally, from the Greek, "beyond the north") is indeed a "land beyond the mountains where the North Wind rises"! RS]

It is now the twelfth division, the final Hour of the king's subterranean journey. It is "the uttermost limit of the thick darkness". The point which he has reached is named "Mountain of the Ascent of Ra". The king looks up and is startled; the celestial boat of Ra looms in front of his eyes, in all its awesome majesty.

He has reached an object which is called "The Ascender to the Sky". Some texts suggest that Ra himself prepared the Ascender for the king, "that the king may ascend upon it to the heavens"; other texts say that the Ascender was made or set up by several other gods. It is "the Ascender which had carried Seth" heavenward. Osiris could not reach the Firmament of Heaven except by means of such an Ascender; thus the king too requires it in order to be translated, as Osiris, to eternal life.

The Ascender or Divine Ladder was not a common ladder. It was bound together by copper cables; "its sinews (like those) of the Bull of Heaven". The "uprights at its sides" were covered over tightly with a kind of "skin"; its rungs were "*Shesha*-hewn (meaning unknown); and "a great support (was) placed under it by He Who binds".

[COMMENT: This reference to the "Bull Of Heaven", as will be explained by Sitchin below, indicates the actual "conveyance" that "ferried" people from our North Polar Regions to Planet X Nibiru. This is the fourth animal metaphor for whatever climbed up and down the "Golden Pole" or "Totem Pole": tomcat, squirrel, monkey, or bull. One presumes that close-up this "Ascender" seemed large like a bull, but when viewed from afar, it seemed more like a smaller animal. It is quite interesting, and I am sure that Mr. Sitchin did not make this connection, that this "ladder" was bound together by copper cables. In the totally unrelated Finnish creation epic *Kalevala* we find the following passages for comparison: "Now rejoiced the Crone of Podja, And conveyed the bulky Sampo, To the rocky hills of Pohja, And within the Mount of Copper, And behind nine locks secured it. There it struck its roots around it, Fathoms nine in depth that measured, One in Mother Earth deep-rooted, In the strand the next was planted, In the nearest mount the third one. ... Again commenced his songs of magic, For the last time sang them loudly, Sang himself a boat of copper, With a copper deck provided. In the stern himself he seated, Sailing o'er the sparkling billows, Still he sang as he was sailing: 'May the time pass quickly o'er us, One day passes, comes another, And again shall I be needed. Men will look for me and miss me, To construct another Sampo, And another harp to make me Make another Moon for gleaming, And another Sun for shining. When my Sun and Moon are absent, In the air no joy remaineth.'" The Finnish "Sampo" is the same "object" as the Siberian "Golden Pole" with the "Golden Cage" on top, the Egyptian *Ded* or *Djed* Pillar(s), the Nordic Yggdrasill Tree, and the ancient Korean Crown with Cosmic Tree pictured by Illustration 32 of *Slow-Motion Doomsday.* See also Chapter 9 ("Yggdrasill & Ragnarok") of that book. RS]

Illustrations to the Book of the Dead showed such a Divine Ladder — sometimes with the *Ankh* ("Life") sign symbolically reaching toward the Celestial Disk in the heavens — in the shape of a high tower with a superstructure. In stylized form, the tower by itself was written hieroglyphically ("Ded") and meant "Everlastingness". It was a symbol most closely associated with Osiris, for a pair of such pillars commemorate the two objects which stood in the Land of Seker and made possible the ascent of Osiris heavenward.

[COMMENT: The *Ankh* would be an excellent symbol to represent The Cosmic Tree. Compare my modernistic illustration on the cover of *Slow-Motion Doomsday* with the *Ankh* symbol. RS]

A long Utterance in the Pyramid Texts is both a hymn to the Ascender — the "Divine Ladder" — and a prayer for its granting to the king Pepi:

Greetings to thee, divine Ascender;
Greetings to thee, Ascender of Seth.
Stand thou upright, Ascender of god;
Stand upright, Ascender of Seth;
Stand upright, Ascender of Horus
Whereby Osiris came forth into Heaven ...
Lord of the Ascender ...
To whom shalt thou give the Ladder of god?
To whom shalt thou give the Ladder of Seth,
That Pepi may ascend to Heaven on it,
To do service as a courtier of Ra?
Let also the Ladder of god be given to Pepi,
Let the Ladder of Seth be given to Pepi
That Pepi may ascend to Heaven on it.

[COMMENT: The Pharaoh "Pepi I" was "Apopi I", one of the Assyrian Hyksos conquerors, or an Egyptian puppet ruler, in that Egyptian "Dark-Age Interregnum" between the end of the Middle Kingdom and the start of the New Kingdom. Dr. Immanuel Velikovsky's synchronization of the Israelite Exodus from Egypt with the end of the Middle Kingdom is the "cornerstone" or "signature" event mathematically, chronologically. According to the Velikovskian Historical Reconstruction, Apopi I ruled Egypt from about 1482 until about 1440 BCE. This was well after the establishment of The Cosmic Tree a century earlier. Apopi I was the contemporary of Israelite Judges Ehud and Shamgar and Chaldean Queen Bilatat. Let me state in passing here that Dr. Velikovsky will ultimately be more remembered for his historical reconstruction in the *Ages In Chaos* series than for his accompanying scientific writings. Without his historical reconstruction, it would have been literally impossible for me or anyone else to have determined the mathematics for this current *shar*, which is set to end in 2012 CE. See Chapter 18 for details. RS]

The Ascender was operated by four falcon-men, "Children of Horus" the Falcon-god, who were "the sailors of the boat of Ra". They were "four youths", who were "Children of the Sky". It is they "who come from the eastern side of the sky ... who prepare the two floats for the king, that the king may thereby go to the horizon, to Ra". It is they who "join together" — assemble, prepare — the Ascender for the king. "They bring the Ascender ... they set up the Ascender ... they raise up the Ascender for the king ... that he might ascend to Heaven on it."

The king offers a prayer:

May my "Name" to me be given
In the Great House of Two;
May my "Name" be called
In the House of Fire,
In the Night of Computing Years.

Some illustrations show the king being granted a *Ded* — "Everlastingness". Blessed by Isis and Nephtys, he is led by a falcon-god to a rocket-like *Ded*, equipped with fins.

The king's prayer to be given Everlastingness, a "Name", a Divine Ladder, has been granted. He is about to begin his actual ascent to the Heavens.

Though he requires only one Divine Ladder for himself, not one but two Ascenders are raised together. Both the "Eye of Ra" and the "Eye of Horus" are prepared and put into position, one on the "wing of Thoth" and the other on the "wing of Seth". To the puzzled king, the gods explain that the second boat is for the "son of Aten", a god descended of the Winged Disk — perhaps the god to whom the king had spoken in the "equipping chamber":

The Eye of Horus is mounted
Upon the wing of Seth.
The cables are tied,
The boats are assembled,
That the son of the Aten
Be not without a boat.
The king is with the son of Aten;
He is not without a boat.

"Equipped as a god", the king is assisted by two goddesses "who seize his cables" to step into the Eye of Horus. The term "Eye" (of Horus, of Ra), which has gradually replaced the term Ascender or Ladder, now is being increasingly displaced by the term "boat". The "Eye" or "boat" into which the king steps is 770 cubits (*circa* 1000 feet) long. A god who is in charge of the boat sits at its bow. He is instructed: "Take this king with thee in the cabin of thy boat."

As the king "steps down into the perch" — a term denoting an elevated resting place, especially of birds — he can see the face of the god who is in the cabin. "for the face of the god is open". The king "takes a seat in the divine boat" between two gods; the seat is called "Truth which makes alive". Two "horns" protrude from the king's head (or helmet): "he attaches to himself that which went forth from the head of Horus." He is plugged-in for action.

The texts dealing with the Journey to the Afterlife by King Pepi I describe the moment: "Pepi is arrayed in the apparel of Horus, and in the dress of Thoth; Isis is before him and Nephtys is behind him; Ap-uat who is Opener of the Ways hath opened a way unto him; Shu the Sky Bearer hath lifted him up; the gods of An make him ascend the Stairway and set him before the Firmament of the Heaven; Nut the sky goddess extends her hand to him."

[COMMENT: Nephtys, or Nephthys, is Princess Ninti, the Greek Libya, the wife of Prince Enki, the Greek Poseidon, the Egyptian Seth. This reference to "Ap-uat" certainly indicates Empress Antu, also known as Apas, Apsu, Apo and Ops. It is unclear to me whether "Shu" refers to Emperor Anu, consort of Antu, or to the wicked deposed Emperor Alalu, who is now in pathetic exile on Planet Earth, probably in a gilded cavern in the Grand Teton Mountains at Yellowstone National Park, Wyoming. But that is just my "opinion" for what it's worth. Whether all these "bigwig dignitaries" were actually present at the launch site is questionable, but certainly "sponsors" or representatives from "The Council Of Twelve" were present there. RS]

The magical moment has arrived; there are only two more doors to be opened, and the king — as Ra and Osiris had done before — will emerge triumphantly from the *Duat* and his boat will float on the Celestial Waters. The king says a silent prayer: "O Lofty one ... thou Door of Heaven: the king has come to thee; cause this door to be opened for him." The "two *Ded* pillars are standing" upright, motionless.

And suddenly "the double doors of heaven are open!"

The texts break out in ecstatic pronouncements:

The Door to Heaven is open!
The Door of Earth is open!
The aperture of the celestial windows is open!
The Stairway to Heaven is open;
The Steps of Light are revealed ...
The double Doors to Heaven are open;
The double doors of *Khebhu* are open
For Horus of the east, at daybreak.

Ape-gods symbolizing the waning moon ("Daybreak") begin to pronounce magical "words of power which will cause splendor to issue from the Eye of Horus". The "radiance" — reported earlier as the hallmark of the twin-peaked Mountain of Light — intensifies:

The sky-god
Has strengthened the Radiance for the king
That the king may lift himself to Heaven
Like the Eye of Ra.

The king is in this Eye of Horus,
Where the command of the gods is heard.

[COMMENT: "Ape-gods"? Fascinating. RS]

The "Eye of Horus" begins to change hues: first it is blue, then it is red. There are excitement and much activity all around:

The red Eye of Horus is furious in wrath,
Its might no one can withstand.
His messengers hurry, his runner hastens.
They announce to him who lifts up his arm
In the East: "let this one pass."
Let the god command the fathers, the gods:
"Be silent ... lay your hands upon your mouth ...
Stand at the doorway of the horizon,
Open the double doors (of heaven)."

The silence is broken; now there are sound and fury, roaring and quaking:

The Heaven speaks, the Earth quakes;
The Earth trembles;
The two districts of the gods shout;
The ground is come apart ...
When the king ascends to Heaven
When he ferries over the vault (to Heaven) ...

The Earth laughs, the Sky smiles
When the king ascends to Heaven.
Heaven shouts in joy for him;
The Earth quakes for him.
The roaring tempest drives him,
It roars like Seth.
The guardians of Heaven's parts
Open the doors of Heaven for him.

Then "the two mountains divide", and there is a lift-off into a cloudy sky of dawn from which the stars of night are gone:

The sky is overcast,
The stars are darkened.
The bows are agitated,
The bones of Earth quake.

Amid the agitation, quaking and thundering, the "Bull of Heaven" ("whose belly is full of magic") rises from the "Isle of Flame". Then the agitation ceases; and the king is aloft — "dawning as a falcon":

They see the king dawning as a falcon, as a god;
To live with his fathers,
To feed with his mothers ...
The king is a Bull of Heaven ...
Whose belly is full of magic
From the Isle of Flame.

Utterance 422 speaks eloquently of this moment:

O this Pepi!
Thou hast departed!
Thou art a Glorious One,
Mighty as a god, seated as Osiris!
Thy soul is within thee;
Thy Power ("Control") has thou behind thee;
The *your head, the Misut*-crown is at thy hand ...
Thou ascendest to thy mother, goddess of Heaven
She lays hold of thine arm,
She shows thee the way to the horizon,
To the place where Ra is.
The double doors of heaven are opened for thee,
The double doors of the sky are opened for thee ...
Thou risest, O Pepi ... equipped as a god.

(An illustration in the tomb of Ramses IX suggests that the Double Doors were opened by inclining them away from each other; this was achieved by the manipulation of wheels and pulleys, operated by six gods at each door. Through the funnel-like opening, a giant man-like falcon could then emerge.)

[COMMENT: The parentheses are Sitchin's. According to the Velikovskian Historical Reconstruction, Ramses IX may have been the Pharaoh Inaros, who revolted in 455 BCE against the Persian conquest and occupation of Egypt from 504 BCE, which was the fifth year of Persian King Cambyses II. Inaros/Ramses IX was executed in 449 BCE at the end of the Six-Year Egyptian War. During this half-century there also occurred the epic battles between Persia and Greece at Marathon, Thermopylae, Salamis, Eurymedon and Cyprus. Persia and Greece signed the Peace of Callias in 449, simultaneously with the end of the Egyptian War. This period was 200 years after the "dissolution" and final "departure" of The Cosmic Tree in 687 BCE, so any illustration by Ramses IX would have been based solely upon historic memory, not personal observation. See the chronology in Chapter 18. RS]

With great self-satisfaction at the achievement, the texts announce to the king's subjects: "He flies who flies; this king Pepi flies away from you, ye mortals. He is not of the Earth, he is of the Heaven ... This king Pepi flies as a cloud to the sky, like a masthead bird: this king Pepi kisses the sky like a falcon; he reaches the sky of the Horizon god." The king, the Pyramid Texts continue, is now "on the Sky-Bearer, the upholder of the stars; from within the shadow of the Walls of God, he crosses the skies".

He encompasses the sky like Ra,
He traverses the sky like Thoth ...
He traveleth over the regions of Horus,
He traveleth over the regions of Seth ...
He has completely encircled twice the heavens,
He has revolved about the two lands ...
The king is a falcon surpassing the falcons;
He is a Great Falcon.

(A verse also states that the king "crosses the sky like *Sunt*, which crosses the sky nine times in one night"; but the meaning of *Sunt* and thus the comparison are as yet undeciphered.)

Still sitting between "these two companions who voyage over the sky", the king soars toward the eastern horizon, far far away in the heavens. His destination is the *Aten*, the Winged Disk, which is also called the Imperishable Star. The prayers now focus on getting the king to the *Aten* and his safe arrival upon it. "*Aten*, let him ascend to thee; enfold him in thine embrace," the texts intone in behalf of the king. There is the abode of Ra, and the prayers seek to assure a favorable welcome for the king, by presenting his arrival at the Celestial Abode as the return of a son to his father.

Ra of the Aten,
Thy son has come to thee;
Pepi comes to thee;
Let him ascend to thee;
Enfold him in thine embrace.

Now "there is clamor in Heaven: 'We see a new thing' say the celestial gods; 'a Horus is in the rays of Ra.'" The king — "on his way to Heaven, on the wind" — "advances in Heaven, he cleaves its firmament", expecting a welcome at his destination.

The celestial journey is to last eight days: "When the hour of the morrow comes, the hour of the eighth day, the king will be summoned by Ra"; the gods who guard the entrance to the *Aten* or to Ra's abode there will let him through, for Ra himself shall await the king on the Imperishable Star:

When the hour of the morrow comes ...
When the king shall stand there, on the star
Which is on the underside of the Heaven,
He shall be judged as a god,
Listened to like a prince.
The king shall call out to them;
They shall come to him, those four gods
Who stand on the *Dam*-scepters of Heaven,
That they may speak the king's name to Ra,
Announce his name Horus of the Horizons:
"He has come to thee!
"The king has come to thee!"

Traveling in "the lake that is the heavens", the king nears "the shores of the sky". As he approaches, the gods on the Imperishable Star indeed announce as expected: "The arriver comes ... Ra has given him his arm on the Stairway to Heaven. 'He Who Knows The Place' comes, say the gods." There, at the gates of the Double Palace, Ra is indeed awaiting the king:

Thou findest Ra standing there;
He greets thee, lays hold on thine arm;

He leads thee into the celestial Double Palace;
He places thee upon the throne of Osiris.

[COMMENT: The "shores of the sky" obviously refer to the perimeter boundaries of the Winged Celestial Disk. The "tomcat" had once again climbed up the "Golden Pole" to the "Golden Cage" on top, telling his tales as he travelled, that is to say, generating a lot of low-rumbling noises that were audible to wide-eyed ancient "tourists" on the ground in Alaska, Siberia and Scandinavia. And this reference to "the lake that is in the heavens" reminds one that The Cosmic Tree was also remembered as "Pool of the Sun", "Pool of the Moon", "Pool of Natron" (a word used by Sitchin earlier in this chapter), "Lake of Sa" and "Lake of the Dragons". See Illustrations 6 and 7 of *Slow-Motion Doomsday.* RS]

And the texts announce: "Ra has taken the king to himself, to Heaven, on the eastern side of Heaven ... the king is on that star which radiates in Heaven."

Now there is one more detail left to accomplish. In the company of "Horus of the *Duat*", described as "the great green divine falcon", the king sets out to find the Tree of Life in the midst of the Place of Offering. "This king Pepi goes to the Field of Life, the birthplace of Ra in the heavens. He finds Kebehet approaching him with these four jars with which she refreshes the heart of the Great God on the day when he awakes. She refreshes the heart of this king Pepi therewith to Life."

Mission achieved, the texts announce with glee:

Ho, this Pepi!
All satisfying life is given to thee;
"Eternity is thine," says Ra ...
Thou perishest not, thou passest not away
For ever and ever.

The king has ascended the Stairway to Heaven; he has reached the Imperishable Star; "his lifetime is eternity, its limit everlastingness."

CHAPTER V, THE GODS WHO CAME TO PLANET EARTH
(initial excerpt only)

Nowadays, we take space flight for granted. We can read of plans for permanently orbiting space settlements without blinking an eye; the development of a reusable space shuttle is viewed not with wonderment, but with appreciation of its cost-saving potentialities. All this, of course, because we have seen with our own eyes, in print and on television astronauts fly in space and unmanned craft land on other planets. We accept space travel and interplanetary contacts because we have heard with our own ears a mortal named Neil Armstrong, commander of the Apollo 11 spacecraft, report on his radio — for all the world to hear — the first landing by Man on another celestial body, the Moon:

Houston!
Tranquility Base here.
The *Eagle* has landed!

Eagle was not only the code-name for the Lunar Module, but the epithet by which the Apollo 11 spacecraft was called, and the proud nickname by which the three astronauts identified themselves. The *Falcon* too has soared into space, and landed on the Moon. In the immense National Air and Space Museum of the Smithsonian Institution in Washington, one can see and touch the actual spacecraft that were flown or that were used as backup vehicles in the American space program. In a special section where the Moon landings have been simulated with the aid of the original equipment, the visitor can still hear a recorded message from the surface of the Moon:

O.K., Houston.
The *Falcon* is on the plain at Hadley!

Whereupon the Manned Spacecraft Center at Houston announced to the world: "That was a jubilant Dave Scott reporting Apollo 15 on the plain at Hadley."

Up to a few decades ago, the notion that a common mortal can put on some special clothes, strap himself in the front part of a long object, then zoom off the face of Earth, seemed preposterous or worse. A century or two ago, such a notion would not have even come about, for there was nothing in human experience or knowledge to trigger such fantasies.

Yet, as we have just described, the Egyptians — 5,000 years ago — could readily visualize all this happening to their Pharaoh: he would journey to a launch site east of Egypt; he would enter a subterranean complex of tunnels and chambers; he would safely pass by the installation's atomic plant and radiation chamber. He would don the suit and gear of an astronaut, enter the cabin of an Ascender, and sit strapped between two gods. And then, as the double-doors would open, and the dawn skies would be revealed, the jet engines would ignite and the Ascender would turn into the Celestial Ladder by which the Pharaoh will reach the Abode of the Gods on the "Planet of Millions of Years".

On what TV screens had the Egyptians seen such things happen, that they so firmly believed that all this was really possible?

In the absence of television in their homes, the only alternative would have been to either go to the Spaceport and watch the rocketships come and go, or visit a "Smithsonian" and see the craft on display, accompanied by a knowing guide or viewing flight simulations. The evidence suggests that the ancient Egyptians had indeed done that: they had seen the launch site, and the hardware, and the astronauts with their own eyes. But the astronauts were not Earthlings going elsewhere: they were, rather, astronauts from elsewhere who had come to Planet Earth.

Greatly enamored with art, the ancient Egyptians depicted in their tombs what they had seen and experienced in their lifetimes. The architecturally detailed drawings of the subterranean corridors and chambers of the *Duat* come from the tomb of Seti I. An even more startling depiction has been found in the tomb of Huy, who was viceroy in Nubia and in the Sinai peninsula during the reign of the renowned Pharaoh Tut-Ankh-Amon. Decorated with scenes of people, places and objects from the two domains of which he was viceroy, his tomb preserved to this very day a depiction in vivid colors of a rocketship: its shaft is contained in an underground silo, its upper stage with the command module is above ground. The shaft is subdivided, like a multi-stage rocket. Inside its lower part, two persons attend to hoses and levers; there is a row of circular dials above them. The silo cutaway shows that it is surrounded by tubular cells for heat-exchange or some other energy-related function.

[**COMMENT:** Tutankhamen ruled Egypt during 852-845 BCE. He was the "historical duplication" or "ghost correlation" of King Takelothis. He was a successor of the controversial King Akhnaton and Queen Nefertiti.

These people would have seen The Cosmic Tree with their own eyes. Tutankhamen was the contemporary of King Jehu in Israel and King Joash in Judah. However, Seti I The Great did not rule Egypt until 659-605 BCE. Whether he was born before 687 BCE and actually caught some glimpse of the departing Planet X Nibiru is unknown. It is possible, but not probable, that Seti I ("ghost correlation" of Psammetichus I) saw remnants of The Cosmic Tree with his own eyes. The reader is again referred to Chapter 18. RS]

Above ground, the hemispherical base of the upper stage is clearly depicted in the color painting as scorched, as though from a re-entry into the Earth's atmosphere. The command module — large enough to hold three to four persons — is conical in shape, and there are vertical "peep holes" all around its bottom. The cabin is surrounded by worshippers, in a landscape of date palm trees and giraffes.

The underground chamber is decorated with leopard skins, and this provides a direct link with certain phases in the Pharaoh's Journey to Immortality. The leopard skin was the distinctive garb symbolically worn by the *Shem* priest as he performed the Opening of the Mouth ceremony. It was the distinctive garb symbolically worn by the gods who towed the Pharaoh through "The Secret Path of the Hidden Place" of the *Duat* — a symbolism repeated to stress the affinity between the Pharaoh's journey and the rocketship in the underground silo.

As the Pyramid Texts make clear, the Pharaoh, in his Translation into an eternal Afterlife, embarked on a journey simulating the gods. Ra and Seth, Osiris and Horus and other gods had ascended to the heavens in this manner. But, the Egyptians also believed, it was by the same Celestial Boat that the Great Gods had come down to Earth in the first place. At the city of An (Heliopolis), Egypt's oldest center of worship, the god Ptah built a special structure — a "Smithsonian Institution", if you will — wherein an actual space capsule could be viewed and revered by the people of Egypt!

[COMMENT: Ptah, the Egyptian "god of the handsome face", is the equivalent of the Nibiruan Baron Ninurta, who was active against the usurpers Baron Marduk and Baroness Sarpanit during The Pyramid Wars. He was known in Greco-Roman traditions as Hephaestus, Vulcan or Typhon, in England as Cú Chulainn and in Germany as Niflhel — a word with linguistic similarities to both the Biblical word for giants or "Nefilim" and the name of the German Goddess "Hel", who was actually Nibiruan Duchess Ereshkigal, consort of Duke Nergal in the African Underworld of "partially lighted regions". She was known in Egypt as Serqet and in Greece as Persephone. See Chapters 11 and 12 for more details. RS]

The secret object — the *Ben-Ben* — was enshrined in the *Het Benben*, the "Temple of the Benben". We know from the hieroglyphic depiction of the place's name that the structure looked like a massive launch tower from within which a pointed rocket was poised skyward.

The *Ben-Ben* was, according to the ancient Egyptians, a solid object that had actually come to Earth from the Celestial Disk. It was the "Celestial Chamber" in which the great god Ra himself had landed on Earth; the term *Ben* (literally: "That Which Flowed Out") conveying the combined meaning of "to shine" and "to shoot up in the sky".

An inscription on the stela of the Pharaoh Pi-Ankhi (per Brugsch, *Dictionnaire Géographique de l'Ancienne Egypte*) said thus:

The king Pi-Ankhi mounted the stairs toward the large window, in order to view the god Ra within the *Ben-Ben*. The king personally standing up and being all alone, pushed apart the bolt and opened the two doorleaves. Then he saw his father Ra in the splendid sanctuary of *Het-Benben*. He saw the *Maad*, Ra's Barge; and he saw *Sektet*, the Barge of the *Aten*. ...

End Quoted Sitchin Text

[COMMENT: Mr. Sitchin doesn't specify whether this king was the Pharaoh Piankhi I who ruled Egypt during 776-725 BCE, or his son Piankhi II The Great who ruled during 725-703 BCE. Piankhi I was also known as Sheshonk "III", and it was in the 15th year of Sheshonk III that "The Sun Devoured The Moon" on 15 June 762 BCE at the time of the Great Eclipse and Great Earthquake. This event has been clearly documented and established by traditional historians and astronomers. This year of 762 BCE also fell during the reigns of King Uzziah in Judah and King Asshurdan III in Assyria and was contemporaneous with the Trojan War, which lasted from 770 until 760 BCE. Piankhi II is the equivalent "ghost correlation" of Sheshonk "I" or So. We can only speculate as to what these two Piankhis and their contemporaries actually saw with their own eyes during this crucial 75-year period from 762-687 BCE when the detethering and departure sequence of the "Sampo" visibly played itself out in the overhead sky. See Chapter 18 for full chronological details.

[As for the *Ben-Ben*, it has also been linked to the Egyptian Phoenix Bird which periodically rises from its ashes and is reborn. The following is a short excerpt from a book by D.J. Conway. RS]

http://www.polarissite.net/LegendPhoenix.htm

"The Legend Of The Phoenix"
Magickal, Mystical Creatures By D.J. Conway
Llewellyn Publications, St. Paul, Minnesota, 1996

The Phoenix is known in various forms and by various names throughout the Middle and Far East, the Mediterranean, and Europe, as a symbol of resurrection. The name Phoenix may have come from the Greek *phoinix* and may be related to *phoinos* (blood-red). Although it was an enormous bird, it had certain characteristics of the eagle, pheasant, and the peacock.

[COMMENT: Planet X Nibiru was repeatedly described as "blood-red". RS]

The earliest known Greek reference to the Phoenix was by Hesiod in the eighth century B.C.E. Such Greek and Roman writers as Tacitus, Ovid, Pliny, Herodotus, and Hesiod referred to the Phoenix either as the Arabian Bird or the Egyptian Bird. An extremely gentle creature, it was said to weep tears of incense, while its blood was balsam.

[COMMENT: In the 8th Century BCE, the 700s BCE, The Cosmic Tree, Winged Disk, Thunderbird or Phoenix Bird was still visible at Hyperborea, "beyond the north". Hesiod, as others noted above, surely saw it with his own eyes. RS]

There are two ancient records of first-hand sightings of a Phoenix: one by Pliny, who saw one exhibited in the Roman Forum during the reign of the Emperor Claudius; another by Clemont in the first century C.E.

The Phoenix was a graceful bird, with brilliant plumage and a distinctive tuft of feathers at the back of its head. There are at least three different descriptions of the plumage colors of the Phoenix. One says that the head, breast, and back are scarlet or reddish-gold, and the iridescent wings are many colors. Its feet are a Tyrian purple hue, while its eyes are sea-blue. Another says the body is plum-colored with a scarlet back and wing feathers, a golden head, and a long tail of rose and azure. The third description states that the Phoenix is a royal purple with a golden neck and head. It is possible that these descriptions are of the Phoenix in various stages of its life.

[COMMENT: And also perhaps these various colors refer to the ever-changing appearance of Planet X Nibiru, as it stands stationary atop the North Mountain. RS]

Tradition says that the Phoenix fed only on air, harming no other creature. It lived a solitary life in a far-away land, coming to human-inhabited land only when it was ready to die. The length of a Phoenix's life differs from ancient writer to writer; most believed that it lived for a thousand years.

[COMMENT: And the period of The Cosmic Tree or Thunderbird lasts for 900 years, or a "millennium of the gods", as it were. RS]

When the Phoenix knew its time had come, it flew to Arabia where it gathered myrrh, laudanum, nard, and cassia. Carrying a great load of these fragrances in its wings, the Phoenix flew on to Phoenicia. There, it chose the tallest palm tree and built a nest in it from the essences it had brought. At the next dawn, the great bird faced the rising Sun and sang in a beautiful voice. The heat of the Sun ignited the fragrant spices, and the Phoenix died in its own funeral pyre.

[COMMENT: "Sang himself a boat of copper, with a copper deck provided. ... Still he sang as he was sailing, 'May the time pass quickly o'er us. One day passes, comes another, and again shall I be needed.'" Finnish *Kalevala*. RS]

After nine days, a fledgling Phoenix rose out of the ashes. A few days later, when its wings were strong enough, the young Phoenix gathered the ashes of its parent and flew them to Heliopolis in Egypt. Thousands of ordinary birds accompanied it on its journey. There, the Phoenix put the ashes of its parent on the altar in the Sun temple. Then it flew toward the east and its distant home.

Other writers of the Phoenix story disagree on several points. Some said that instead of flying to Phoenicia with its spices, the Phoenix flew directly to the temple at Heliopolis and built its funeral pyre on the altar there. Others believed that the priest of the Sun temple gathered the spices and prepared the nest for the Phoenix. A few writers recorded that the Phoenix did not rise straight from the ashes, but rather spent three days in a worm-like form before turning into the glorious Phoenix.

The Phoenix never died permanently. Legend says it existed when the universe was created and that it knows secrets of life and reincarnation even the deities do not know.

Humans are fascinated by the sweet song of the Phoenix, and the bird is friendly to humans, although it seldom concerns itself with human affairs.

A similar mythological Egyptian bird was the Bennu, a heron-like bird. The Bennu was born in a spice-lined nest in a sycamore tree. It too made its own funeral pyre in which it died. Its first flight, after being reborn, was accompanied by thousands of ordinary birds. In fact, "Bennu" in Egyptian and "Phoenix" in Greek both mean "date palm". The Bennu was sacred to Osiris and Ra, and a symbol of the Sun and resurrection. It also represented the morning star.

[COMMENT: And so this bird which made its home in a Cosmic Tree was connected to Sirius. The first sightings of Planet X Nibiru will probably be as a new "morning star" rising heliacally just before dawn. RS]

The Egyptian Phoenix was called the "Lord of Jubilees", and was considered to be the *ba* (spirit) of the Sun God Ra. At one point in the Book of the Dead, the deceased says, "I have gone forth as a Phoenix." In Heliopolis, the Bennu was said to live in the benbenstone (obelisk) or in the sacred willow.

[COMMENT: As noted previously with the Sitchin material, the "ben-ben" was a "conveyance" of some sort that "ferried" the Anunnaki and others from the northern Vigrid Plain to "Heaven", *i.e.*, the tethered object "beyond the north". Here the tree is compared to a willow. RS]

Queen Elizabeth I had a Phoenix engraved on her medals; Mary Queen of Scots also used the same emblem. Jane Seymour, who died giving birth to Edward VI, had a Phoenix crest, which her son later used.

In Mesopotamian art, the Phoenix may have been symbolized by the horned and winged solar disk. Ancient bas-reliefs show this winged disk also having tail-feathers, legs, and claws of a bird. Often this winged disk also had horns. The winged disk of Abura Mazdah on a relief at Persepolis distinctly shows this disk with tail-feathers and bird's legs and feet.

[COMMENT: Winged Solar Disk, Phoenix Bird, Thunderbird — and on and on the ancient imagery continues. If we treat all of these olden legends as mere "myths", we do so at our peril. RS]

Alchemists used the Phoenix to symbolize the color red and the successful end of a process, while medieval Hermeticists used the Phoenix as a symbol of alchemical transmutation. The word Phoenix was also used to identify one of the secret alchemical formulae.

The ancient Mysteries used the sign of the Phoenix to symbolize the immortality of the human soul and the great truths of esoteric philosophies revealed only through special initiations. In some ancient Mystery Schools, accepted initiates were referred to as Phoenixes, or those who had been "born again".

[COMMENT: Here this idea of being "born again" has absolutely nothing to do with modern evangelical Christianity. It refers back to the idea that once every *shar* the Planet X Nibiru returns to our view, like a Phoenix Bird, and appears to have been reborn for another millennium. RS]

Chapter 10

The Home Of The Gods

*

The Home Of The Gods By Andrew Tomas
(Berkeley Medallion Books, New York, 1972)

INTRODUCTORY COMMENT
By Rob Solàrion

This book will be excerpted from beginning to end, with my comments interspersed as usual. The chapter is quite long, but I have decided not to abbreviate it. I'll note the page numbers of my paperback edition but shall not include the scores of footnotes. If anyone wishes to know whether certain material is documented by Mr. Tomas, please contact me by email from my website, and I'll look in the book for a footnote. This 1972 edition was the first American printing of this international bestseller. I also have in my library the book *We Are Not The First* by Andrew Tomas, which contains a chapter on Apollonius of Tyana in the context of his visit to the "magical technological palace" of the Indian Sage Iarchas in 26 CE. Following is an "About The Author" from the frontispiece of *The Home Of The Gods*:

"Andrew Tomas, born in St. Petersburgh [presumably Russia, not Florida, RS] in 1913 and now an Australian citizen, has spent most of his life travelling extensively throughout the world. In fact *The Home Of The Gods* was begun in Australia, continued in the Himalayas, Russia and France and completed in London. He is the author of the best-selling book on the riddles of Ancient Science, *We Are Not The First*."

Thus, Andrew Tomas, if still alive, would be 93 this year. His work has had a profound influence on my own. The full title of this book is *The Home Of The Gods: Atlantis, From Legend To Discovery*. Where I disagree with Andrew Tomas is that not *all* of these historical items refer to Atlantis, but to The Cosmic Tree, The World Tree, The Sacred Tree, The Holy Mountain, The Sacred Mountain, Mount Olympus, Mount Zion, Mount Meru, Yggdrasill, Asgard and Hyperborea. Mr. Tomas was using all of his material in an attempt to validate the existence of the Lost Atlantis, with which in and of itself I do not disagree. It has taken quite a number of persistent researchers over the span of more than a century to piece together the "true" solution to all of these legends — and just in time, I might add, for the next Crossover Dreamtime is fast approaching!

*

Page 7

Dedicated to NICHOLAS ROERICH (1874-1947), artist, explorer and philosopher, who wrote this verse about a forgotten race:

We do not know. But they know.
The stones know,
And they remember.
Airships were flying.
Came pouring a liquid fire.
Came flashing
The spark of life and death.
By the might of spirit
Stony masses ascended.
Scriptures guarded wise secrets.
And again all is revealed.

Pages 11-12

PRELUDE

Once a youth stood before the image of Quetzalcoatl in Mexico City, reflecting upon the culture hero's links with legendary Atlantis.

In Los Angeles and Hollywood he frequented public libraries examining chronicles of the conquistadores and legends of the American Indians.

In Japan the man learned of a myth that formerly the earth had been connected with heaven by a bridge, and he thought of prehistoric space travel.

Taoists in China spoke to him about the abode of "immortals of the west" and Shambhala, the city of star-men.

In Australia he found out about Dreamtime, a forgotten epoch in which mankind commuted with the sky-beings.

Recently he beheld the mighty Kanchenjunga on [the] Tibetan border, known as the Five Treasuries of the Great Snow, where secret treasures are believed to be stored from time immemorial.

In India he admired the Himalayas and listened to legends about the underground palaces and treasure-vaults of Nagas, the flying serpents who have brilliant lamps to illuminate their subterranean abodes.

The man lived in the Himalayan village of Manali which, according to Brahmin scriptures, was founded by Manu who had rescued his sages from the Great Flood in a bark.

Then he explored the Khufu Pyramid and asked the Sphinx for the answer to his age-old riddle.

In Australia he tried to solve the mystery of a polished steel cube found in a coal stratum, millions of years old.

At the Bibliothèque de l'Arsenal, near the Place de la Bastille in Paris, he perused eighteenth-century volumes of the 'Astronome du Roi' Jean-Sylvain Bailly. In the Louvre he found the ancient Egyptian Zodiac of Denderah.

Then he studied in the reading room of the British Museum Library in London with its wealth of books and manuscripts of great antiquity.

After that — research work in the Lenin Library of Moscow, meetings with Russian scientists and writers, visits to the Hermitage Museum in Leningrad.

This tale is about the author of this book.

[COMMENT: The Egyptian Zodiac of Denderah is considered to be so mysterious and significant because it depicts the Vernal Equinox in Egypt as being in the Constellation of Leo, thousands of years ago in terms of Precession of the Equinoxes. And in the tomb of Senmut, chief architect of Egyptian Queen Hatshepsut (who ruled 983-971 BCE, during the time of the last Cosmic Tree), in another anomalous ceiling painting the stars and constellations are "upside-down" from what they are today, that is, North and South are reversed. However, this can be quite easily explained if the North Pole were located near the Ruins of Zimbabwe in southern Africa, due "south" of the Pyramids. What is now "south" was at that time "north", and so the heavens were reversed, probably at about the same time that the Denderah Zodiac was fashioned. These "problems" are discussed at length by Dr. Immanuel Velikovsky in *Worlds In Collision*; and the tomb of Senmut is mentioned again below.

[The subterranean abodes of the Indian "Nagas" or "flying serpents" call to mind the *Duat* legends of Ancient Egypt, discussed at length by Zecharia Sitchin in *The Stairway To Heaven.* See Chapter 9. The "bridge" connecting heaven to earth in the Japanese legend simply refers to the legendary Nordic "Rainbow Bridge Bifrost" which connected heavenly "Asgard" with the Earth at our North Pole. It is the Cosmic Treetrunk, the Golden Pole, the Totem Pole. Shambhala was not another name for Atlantis. It was Planet X Nibiru, Winged Disk, Celestial Ship Of The North.

[In the *New Larousse Encyclopedia Of Mythology* (New York 1978), the Japanese bridge to heaven is mentioned briefly on pages 405 and 412: "Formerly earth was linked with heaven by a sort of bridge, *Ama-no-Hashidate*, which allowed the gods to go to and fro. According to the *Tango-fudoki*, one day when the gods were all asleep this bridge or stairway collapsed into the sea. ... Amaterasu decided to send her son Ame-no-Oshido-Mimi down to earth to reign over it as sovereign. But before leaving, the god looked at the earth from the floating bridge of Heaven, saw it was full of disturbances, and refused to go."

[Then on page 415 we find: "As to the stars, M.G. Kato says: 'They never had a prominent place in early Shinto beliefs, although they included the god of evil, Amaatsu-Mikaboshi, "the brilliant-male".' Later on, due to the influence of Chinese and Buddhist beliefs, the Japanese god of stars was identified with the Pole Star, Myo-ken (in Sanskrit, Sudarsana) ... The god Take-Mika-Zuchi, who was sent by the other gods to subjugate Izumo province, is also considered a god of Thunder, who pursued the son of O-Kuni-Nushi to Lake Suwa and conquered him. Aji-Suki-Takahikone, another son of the same god, is also a Thunder god. At his birth he cried and screamed, and they calmed him by carrying him to the top and then to the bottom of a ladder.' In the Japanese mind the ladder is used to get to heaven, so this episode seems to allude to one of the characteristics of the Thunder, which is to come and go between heaven and earth. He was also placed in a boat which sailed between the eighty islands. The boat was the means by which the Thunder god connected heaven and earth." This "ladder" between heaven and earth is yet another reference to the celestial "bridge", and in the Old Testament we have the story of Jacob's Ladder to Heaven, being used by the angels. Such "celestial boats" were also mentioned in Egyptian and other mythologies and, of course, were spaceships. As I observed in the previous chapter, we could just as easily have referred to our modern space vehicles as "boats" rather than "rockets" or "shuttles".

[And on page 402 we read regarding Chinese mythology: "The Chinese Paradise. As we have seen, when the souls of the just are not sent back immediately to a new life by the tenth Yama-King, they go either to the K'un-lun Mountain, the dwelling place of the Immortals, or to the Amitabha Buddha in the Land of Extreme Felicity in the West. The K'un-lun Mountain has a close resemblance to the Olympus of the Greeks, but while the latter situated the dwelling place of their gods in a mountain of their own country, the Chinese place theirs on a fabulous mountain far away from their land and at the earth's centre." By "centre" is meant the Pole Star as the center of both Heaven and Earth. And needless to say, the modern Greek Mount Olympus is only a "symbolic remembrance" of the true celestial Mount Olympus. RS]

Page 26

A 3,000-year-old [Egyptian] Twelfth Dynasty papyrus preserved at the Leningrad Hermitage mentions the "Island of the Serpent" and contains this passage: "After you leave my island, you will not find it again as this place will vanish under the sea waves."

[COMMENT: This passage could have a double meaning. "Island of the Serpent" could also refer to that "island" in the "heavens" above our North Pole — "Lake of the Dragons". And the quote above might refer to the fact that after Planet X Nibiru detethers and departs, it will seem to have vanished beneath the sea. The Twelfth Dynasty was at the beginning of the Egyptian Middle Kingdom, around 1900 BCE when adjusted for the Velikovskian Historical Reconstruction. RS]

Page 27

A Maya codex states that "the sky approached the earth and in one day all perished. Even the mountains disappeared under water." The Dresden Codex of the Mayas shows the destruction of the world in pictorial form. On the chart there is a serpent in the sky with torrents of water pouring out from its mouth. Mayan signs indicate lunar and solar eclipses. The moon goddess, the patroness of death, has a frightening appearance. An inverted bowl, from which gushes a destroying flood, in in her hands.

The *Popul Vuh*, the sacred book of Guatemala Mayas, bears witness to the dire character of the disaster. It says that the roar of fires was heard above. The earth shook and things revolted against man. It rained tar with water. The trees were swinging, houses crumbling, caves collapsing. Then day became black night. The *Chilam Balam* of Yucatán asserts that the motherland of the Mayas was swallowed up by the sea amid earthquakes and fiery eruptions in a very distant epoch. A white Indian tribe, Paria, used to live in Venezuela, in a village with so significant a name as Atlan. They had a tradition of a calamity which had destroyed their country, a large island in the ocean. A perusal of American Indian mythology discloses an interesting fact that over 130 tribes have legends of a world catastrophe.

Can mythology and folklore be used to some extent in filling the numerous gaps in history? Professor I.A. Efremov of the Soviet Union answers in the affirmative. He insists that "historians must pay more respect to ancient traditions and folklore". Efremov accuses scientists in the West of a certain snobbishness when it comes to the tales of the so-called common people.

[COMMENT: To appreciate fully these remarks, the reader has to put this writing into the context of the times in which it was published. Even though Dr. Velikovsky's *Worlds In Collision* was first published in 1950, it wasn't until the early 1970s that it enjoyed international interest and acclaim. I first heard about Dr. Velikovsky from my dear friend Giovanni Yanello in Brooklyn, NYC, in January 1973, when Gianni came to Texas for a visit and we discussed these matters. This was at the very outset of my interest in these topics. Erich von Däniken was all the rage, too, at the time. Later that year Dr. Maxine K. Asher's expedition would "discover Atlantis" off the coast of Cádiz, only to be chased out of Spain as "Cold War spies". Every heartbeat pumped adrenaline. Several of Dr. Velikovsky's major books had not even been published yet. The journal *KRONOS* did not exist. Because of the "atheistic" nature of Communism, a lot of the major opinion in this field was Soviet. Thus, it is not at all surprising to read this quote from Professor Efremov, and I feel sure that Dr. Velikovsky would agree completely. There are "catastrophe legends" all over this world. The ancient people may not have had our modern technology to guide their thoughts, but they certainly had as much "common sense" as we do, if not more. From their human experience they knew the difference between an ordinary hurricane and a "cosmic catastrophe", between just another tornado or earthquake and a "cosmic catastrophe". A "flying serpent" in the sky does not accompany these "predictable" events. And also, this idea that "the sky approached the earth" resembles the theme of the treatise "The Wayward Sun" by Rand & Rose Flam-Ath (*The Velikovskian* Journal, 1997): "But why did the sky fall?" We could speculate about the exact "scientific" meaning of these legends; but when it happens again, and we see it (again?) with our own eyes, I am certain that instinctively from "race-memory" we'll know exactly what was being remembered, because it will have started to happen again. *Call it to mind. "I lived in such a place," etc.* See Chapter 17. RS]

Page 29

Interesting supporting arguments are worth mentioning. In ancient Mexico there was a holiday devoted to a past event in which constellations had assumed a new aspect. It follows that in a bygone age the heavens did not have the same appearance as now.

Martinus Martini, seventeenth-century Jesuit missionary in China, wrote in *History of China* concerning her oldest records. These speak of a time when the sky suddenly began to fall after the earth had been shaken. This is certainly a strong hint at the wobble of the earth for it alone can explain the astronomical phenomena described in Chinese writings.

Two star maps painted on the ceiling of the tomb of Senmouth, Queen Hatshepsut's architect, present a riddle. The cardinal points are correctly placed on one of these astronomical charts but on the other they are reversed, as if the earth had undergone a tilt. In fact, the Harris Papyrus mentions that the earth had turned over in a cosmic cataclysm. The Hermitage Papyrus of Leningrad and the Ipuwer Papyrus also allude to the world having been turned upside down.

[**COMMENT:** The Papyrus Ipuwer is dated from the end of the Middle Kingdom and describes — from the Egyptian perspective! — the Israelite Exodus from Egypt, during which time the Red Sea temporarily "parted" allowing the Israelites to escape but later swallowed up the pursuing Pharaoh's army. Certainly Andrew Tomas was aware of Dr. Velikovsky. RS]

Pages 34-36

CALENDARS FROM ATLANTIS

Across the Atlantic there is another link between ancient Egypt and Peru. Their calendars had eighteen months of twenty days with a five-day holiday at the end of the year. Is this coincidence or tradition from a common source?

The approximate date of the end of Atlantis can be arrived at from the examination of ancient calendars. The first year of Zoroastrian chronology is 9660 B.C. when "time began". This is very close to the date given by Egyptian priests to Solon for the doomsday of Atlantis, or 9560 B.C.

Ancient Egyptians calculated time in 1,460-year solar cycles. The end of their last astronomical epoch came in A.D. 139. Eight solar cycles from this date can be traced back to the year 11,542 B.C. The lunar calendar of Assyrians divided time into periods of 1,802 years. The last one ended in 712 B.C. Six lunar cycles from this date are followed back to 11,542 B.C. The solar calendar of Egypt and the lunar calendar system of Assyria coincided in the same year — 11,542 B.C., when both calendars were presumably created.

[COMMENT: Trying to analyze and compare ancient calendars is a difficult proposition. I have spent countless hours in this endeavor. It's often an exercise in mathematical futility. We need a completely new and scientific world calendar starting 3 or 4 "shars" ago. As for the Egyptian Sothis Period of exactly 1,461 "Orbit-Years", that simply means that there are 1,461 Earth orbits around the Sun during a period of 1,440 calendar years. 1,461 X 360 = 1,440 X 365.25. Therein lies the "secret" of the Egyptian Sothis Period. One *shar* would contain 2.5 Egyptian Sothis Periods, since 3,600 divided by 1,440 equals 2.5. See Chapter 17. RS]

The Brahmins measure time in rounds of 2,850 years from 3102 B.C. Three of these cycles, or 8,550 years, added to 3102 B.C. give us the date of 11,652 B.C.

The Mayan calendar shows that the ancient peoples of Central America had long cycles of 2,760 years. The beginning of one span is traced to the year 3373 B.C. Three periods of 2,760 years, or 8,280 years from 3373 B.C. would take us back to 11,653 B.C. which, within a year is the same date as that of the sages of India.

[COMMENT: These figures look a little suspect to me, but I don't plan to analyze them for accuracy. Suffice it to say that "shars" began in 1588 BCE, 5188 BCE (not a typo, just a coincidence of counting), 8788 BCE and 12,388 BCE. The destruction of Atlantis undoubtedly coincided with one of the earlier dates. The Flood of Noah would have been in 5188 BCE. The Brahman and Mayan counts of 2,850 and 2,760 years, respectively, could refer to those 2,700-year-long periods, like now, when Planet X Nibiru is *not* visible beyond our north (*hyper borea*). RS]

Vatican Codex A-3738 contains a significant chronology of the Aztecs according to which the first cycle continued for 4,008 years ending in a flood. The second of 4,010 years was destroyed by hurricanes. The third era of 4,801 years was closed by fires. In the fourth period which lasted 5,042 years, mankind suffered starvation. The present era is the fifth and it began in 751 B.C. The duration of all the four periods listed in this Codex is 17,861 years; and its beginning is traced to an incredibly distant date of 18,612 B.C.

[COMMENT: The date of 751 B.C. may be slightly incorrect. A lot of new information about early American calendars has been published since 1972. According to John Major Jenkins in his "The Black Road To

The Sacred Tree" available at my website, the current Mayan Calendar "officially" commenced in 679 BCE, eight years after the final departure of Planet X Nibiru in 687 BCE. "Retrograde calculations" such as those noted below are just that: modern projections backwards based upon suppositions. At any rate, both these approximations comfortably coincide with Planet X's last departure sequence dated 762-687 BCE. These references to destructions by flood, hurricane and fire resemble the references in Buddhism to destructions by fire, water and wind. See Chapter 17. RS]

Bishop Diego de Landa wrote in 1566 that in his time the Mayas reckoned their calendar from a date which was about 3113 B.C. in European chronology. They claimed that 5,125 years had passed before this date in former cycles. This would move the origin of the early Mayas to 8238 B.C., or close to the period of [the] Atlantean cataclysm.

[COMMENT: Note the proximity of the date 8238 BCE to the commencement of the *shar* in 8788 BCE. All of these ancient calendars were simply trying to "interpret" the correct length of a *shar*. People everywhere were aware that "after long intervals of time" (as Plato put it in his essay on Alantis) another "cataclysm" of some sort "destroys" the world, and they wanted to be able to warn future generations when it might happen again. And here we are today, pondering these same age-old questions! RS]

Aside from providing a clue as to the dating of Atlantis, a reasonable supposition can be made on the basis of these figures that many thousand years ago mankind possessed a considerable knowledge of astronomy which is usually characteristic of a high civilization.

The longest day in the Mayan calendar contained 13 hours and the shortest 11. In Ancient Egypt the longest day had 12 hours 55 minutes and the shortest 11 hours 5 minutes. These figures are almost identical with the Mayan hours. But what is really puzzling is this — 12 hours 55 minutes is not actually the duration of the longest day in Egypt, but in the Sudan. In an attempt to explain this difference, Dr. L. Zajdler of Warsaw suggests that this time reckoning had come from tropical Atlantis.

Archaeologist Arthur Posnansky of La Paz, Bolivia, speaking of the uncompleted Sun Temple at Tiahuanaco, claims that the construction was suddenly abandoned about 9550 B.C. The date is familiar — the priests of Sais told Solon that Atlantis perished in 9560 B.C.

In the words of E.F. Hagemeister of U.S.S.R. science has this to say about the sinking of Atlantis: "The end of the European Ice Age, appearance of the Gulf Stream and submergence of Atlantis occurred simultaneously about 10,000 B.C."

Not all scientists take a similar stand in the problem of Atlantis. Some discard the theory altogether in spite of all the evidence. Others attempt to put Atlantis in the Mediterranean, Spain or Germany. Needless to say, this is not the Atlantis of Plato and Egyptian scholars which they placed "in front of the Pillars of Hercules in the Atlantic sea".

[COMMENT: That statement is not factually correct. The priests told Solon only that one had to pass through the Pillars of Hercules (Strait of Gibraltar) in order to sail to Atlantis. No distance from there was ever specified. And as I have demonstrated in Chapter 5 of my book *Planet X Nibiru: Slow-Motion Doomsday*, an excellent case can be made that Atlantis is modern Antarctica. It did not sink to the bottom of the sea; it pivoted or "sank" to the bottom of the world (geographical bottom of the seas!) and then froze over. It was "destroyed" in the sense that it was no longer accessible by or hospitable to ancient visitors. It is difficult even for us today to travel to and across Antarctica. It would have been impossible by pre-technological ancient peoples. During the preparation of this book, an email message came to me from a British source, stating that a new theory is under discussion that Ireland is the true Lost Atlantis. Yawn. I have read all of these various "theories" for over 30 years. Atlantis was Antarctica, pure and simple. RS]

In the Egyptian section of the Louvre in Paris I saw a carved design inconspicuously exhibited in a staircase with no tablet. However, I recognized the carving as the famous Denderah Zodiac. Originally this ancient Egyptian relic was part of a ceiling in the portico of the temple of Denderah in Upper Egypt. It was brought to France by Lelorrain in 1823.

For many generations the Calendar of Denderah has remained a baffling riddle to science. The zodiacal signs are arranged in a spiral and the symbols are easily recognizable, but Leo is at the point of vernal equinox. Because of the precession of the equinoxes this would indicate a date between 10,950 to 8800 B.C., or the period during which the Atlantean catastrophe took place.

The Zodiac of Denderah is of Egyptian origin but it may have been engraved to commemorate a distant event — the end of Atlantis and the birth of a new cycle.

[COMMENT: Read that word "cycle" to mean *shar*! And the trouble with trying to determine when a particular Constellation like Leo was aligned with the Vernal Equinox is that the Constellations themselves are not of equal length. The "astrological signs" are all equal to thirty degrees, but not the actual Constellations in the sky. This complex topic will not be considered here. RS]

Pages 44-45

To preserve the products of civilization for an indefinite period against the dangers of devastating wars and geological calamities, nothing would be more effective than underground shelters. This is as true today as it was in the days of Atlantis.

From the story of man's life on the planet many pages have been torn out by the hand of Time. However, legend speaks of a colossal disaster which wiped out an advanced civilization. Most of the survivors became savages. Those who were later rehabilitated by "divine messengers" rose from their primitive state and gave birth to the nations of ancient history from which we ourselves derive our origins. The secret communities of the "Children of the Sun" were small in number but great in knowledge. By means of their high science they excavated a vast network of tunnels, particularly in Asia.

Isolation has been the eternal law of these colonies. Philosophers, scientists, poets, artists, religious devotees, writers and musicians require a peaceful environment in which to pursue their labours. They do not want to hear the tramping of soldiers' boots or the cries of the market place. No one can accuse these philosophers of the vice of selfishness because down through the ages they had shared their wisdom with those who were ready for it. This detachment is of a protective nature. For is not the rule of the fist as strong today as in the times of Caligula? Perhaps, the fist is even more awful in its technological armour.

Lost in the secret valleys between snowy ridges or hidden in mountain catacombs live the Elder Brothers of the human race. Indications of the reality of these colonies came from such widely separated countries as India, America, Tibet, Russia, Mongolia and other parts of the world. Over the expanse of time these reports have appeared in the past five thousand years. Embellished by fancy of the people living in various lands, they contain grains of truth.

Pages 46-51

NORTHERN SHAMBHALA

A Shanghai newspaper in the twenties featured an article by Dr. Lao-Tsin about his journey to a utopia in Central Asia. In a colourful narrative antedating James Hilton's *Lost Horizon*, the surgeon describes his hazardous trek with a Nepalese yogi to the uplands of Tibet. In a desolate mountainous region the two pilgrims found a hidden valley protected from severe northern winds and enjoying a much warmer climate than the surrounding territory. Dr. Lao-Tsin spoke of the 'Tower of Shambhala' and the laboratories which aroused his wonder. The two visitors saw great scientific achievements of the dwellers of the valley. They also watched outstanding feats in telepathy conducted over great distances. The Chinese doctor could have told much more about his stay in the valley if it were not for some promise he had given its inhabitants not to reveal all.

According to the Eastern tradition of Northern Shambhala, where now are found only salt lakes and sands, there was once a huge sea in Central Asia. This sea had an island of which nothing now remains but mountains. In that faraway epoch a great event took place: "Then with the mighty roar of swift descent from incalculable heights, surrounded by blazing masses of fire which filled the sky with shooting tongues of flame, flashed through the aerial spaces the chariot of the Sons of the Fire, the Lords of the Flame from Venus; it halted, hovering over the White Island which lay smiling in the bosom of the Gobi Sea."

On the background of present-day controversy over a cosmic ship crash in Tunguska, Siberia — let us not ridicule this Sanskrit tradition.

In the folklore and songs of Tibet and Mongolia, Shambhala is exalted to a point where it assumes the form of a supreme reality. During an expedition through Central Asia Nicholas Roerich came across a white frontier post regarded as one of the three outposts of Shambhala. To demonstrate how strong the belief in Shambhala was in lamahood, we will quote the words of a Tibetan monk who told Roerich that "the people of Shambhala at times emerge into the world. They meet the earthly co-workers of Shambhala. For the sake of humanity they send out precious gifts, remarkable relics."

After examining the traditions of Buddhism in Tibet, Csoma de Koros (1784-1842) placed the land of Shambhala beyond Syr Daria River between 45 and 50 degrees north latitude. It is a notable fact that a seventeenth-century map (published in Antwerp, Belgium) shows the country of Shambhala.

Early Jesuit travellers in Central Asia, such as Father Stephen Cacella, recorded the existence of an unknown domain called 'Xembala'.

Explorers Colonel N.M. Prjevalsky and Dr. A.H. Franke mention Shambhala in their works. Professor Grünwedel's translation of an ancient Tibetan book *The Path to Shambhala* is an interesting document. However, the geographical pointers seem to be purposely vague. They are of no use to anyone without a thorough familiarity with ancient and modern names of places and monasteries. Geographical indications may be confused for two reasons. Those who actually know of the colonies will never disclose where they are, so as not to disturb the humanitarian work of the Guardians. On the other hand, references to these retreats in oriental literature and folklore may sometimes seem to be conflicting because they allude to communities in diverse localities.

After having studied the subject for many years, I wrote this chapter in the Himalayas, and to me the name 'Shambhala' covers the White Island in the Gobi, hidden valleys and catacombs in Asia and other places, and a great deal more.

[COMMENT: "Geographical indications may be confused" for another, simpler and more obvious reason, not considered by Andrew Tomas: Shambhala no longer exists. Shambhala was just another name for Hyperborea, Yggdrasill, Asgard, Mount Olympus, Mount Zion and Mount Meru — The Cosmic Tree. It is not here anymore. Most writers would never even think to discuss it in this manner, as I am doing. Analyzing this material in terms of The Cosmic Tree is my own unique "contribution" to this whole field of Nibiruology. These secret caverns of "Guardians" may have been related to Shambhala in some way, but they do not represent

Shambhala's true location. Note the reference above that it was thought to be in "northern" latitudes. In Madrid in 2001 Fabrizio Torricelli and Maria Marin published a Spanish-language book on this subject, titled *Shambhala: La Tierra de los Sabios* (Land of the Sages), in which context they discuss Apollonius of Tyana. RS]

Lao Tse (sixth century B.C.), the founder of Taoism, searched for the abode of Hsi Wang Mu, the goddess of the west, and found it. Taoist tradition asserts that the goddess was a mortal thousands of years ago. After having become 'divine' she retreated to the Kun Lun Mountains. Chinese monks insist that there is a valley of great beauty in the range which is inaccessible to travellers without a guide. The valley is the home of Hsi Wang Mu, who presides over an assembly of genii. These may be the world's greatest scientists.

In this connection the sighting of a strange aircraft over the Karakoram (which is an extremity of Kun Lun) by the Roerich Expedition is quite significant. The strange disc may have come from an aerodrome or spacedrome of the 'gods'.

[**COMMENT:** In reading this, it occurred to me for the first time that perhaps the *Duat* described by Zecharia Sitchin in *The Stairway To Heaven* (see Chapter 9), the "hidden place" or "northern port" from which "conveyances" that looked like "tomcats" departed for Yggdrasill, or "climbed up and down the Golden Pole", may still exist near the modern Altai Mountains. After the detethering/departure sequence played itself out and there was no longer a "Cosmic Tree" beyond the north, the *Duat* "conveyance port" continues to operate as an "airport" even until today. Perhaps it is from this secret location of "Shambhala" that some of our UFOs originate. Mongolia is about as far off the beaten path as one can get on this Earth. The trouble here is that this is generally where the *next* North Pole will be located. Was this *Duat* "conveyance port" *deliberately* constructed for logistical reasons near the site of the next North Pole, perhaps even to assist in the electromagnetic beam's tethering process soon to occur again? It is indeed a fascinating idea to contemplate, amongst all the other "mysteries" from eons long forgotten. RS]

From what has been said by now, it is clear how difficult it is to establish a contact with the dwellers of the secret communities. Yet these meetings have taken place more often than reported. The absence of records is explained by the inevitable vow of secrecy which is demanded of the visitors to these ancient colonies for a justifiable reason. The 'Mahatmas' do not wish to be disturbed by curiosity seekers, treasure hunters or sceptics, for they are the custodians of Ancient Science and the guardians of the Treasure of the Ages.

[COMMENT: We can say only "Bravo!" to this. One certainly hopes that "records" are being kept of our planetary history, independently of the periodic "shar-generated" random destructions of various libraries and museums across the rest of the world. Yes, so many mysteries! RS]

It would be appropriate to quote from one of the letters of the Mahatmas themselves inspiringly outlining the scope of their humanitarian activities: "For countless generations hath the adept builded a fane of imperishable rocks, a giant's Tower of Infinite Thought, wherein the Titan dwelt, and will yet, if need be, dwell alone, emerging from it but at the end of every cycle, to invite the elect of mankind to co-operate with him and help in his turn enlighten superstitious man." Thus wrote Mahatma Koot Humi in July 1881.

The origin of these unknown communities is lost in the night of time. It is more than likely that our elders in evolution ordered the exodus from Atlantis of the people of the 'Good Law'.

[COMMENT: To digress a bit into stream of consciousness, people regularly ask me about why we have not actually seen Planet X Nibiru with a telescope if I am so certain that it is coming back. My speculative answer can vary from one conversation to another, but there is a "lingering thought" in the back of my mind, recalled again here by the idea that there is an "emergence" of this "Titan" from time to time. It could be that Planet X Nibiru is cloaked or in another dimension. Thus, we can't see it, even with the Hubble Space Telescope. If it reflects no light, it would not be visible. Perhaps it may suddenly "emerge" or "decloak", or "come crashing through the dimensional fabric of Space", when it is time for Crossover to publicly begin. It could be that this "emergence" from some "other dimension" will cause a booming trumpet-like noise, remembered in our history as Gabriel's Horn or Gjallar's Horn. Personally I truly anticipate that during the commencement of the arrival sequence, there will sound a great "cosmic herald" which will announce Planet X's arrival and which will scare the living daylights out of all of us. Mark my words! Anybody want to bet a little money here, against this possibility? RS]

All the material and spiritual achievements of Atlantis in her splendour may still be preserved in the secret colonies. Though not represented in the United Nations Organization, this tiny republic may be the only permanent state on the planet Earth and custodian of a science that is as old as the rocks. The sceptics would do well to bear in mind that messages from the Mahatmas are still preserved in state archives of certain governments.

In the folklore of Russia there is a myth of the underground city of Kitezh, where justice reigns. The Old Believers, persecuted by the Czarist government, searched for this Promised Land. "Where to find it?" asked the young. "Follow the path of Batu," answered the old, Batu Khan, the Tartar conqueror, [who] had come from Mongolia in his westward drive. The direction meant that the utopia was to be found in Central Asia.

Another rendition of the legend pointed to Lake Sveltloyar in Russia but it has no basis because the lake bottom has been explored and nothing discovered. It seems that the tradition of Kitezh should be placed alongside with that of Northern Shambhala. The same can be said of the myth of Belovodye.

[COMMENT: Obviously, the terms Kitezh and Belovodye are identical with Shambhala, Hyperborea, Yggdrasill, Asgard, *etc.* To my knowledge, the bottom of Lake Baikal, also in this same general area, has *not* been determined. It is the world's only "bottomless" lake. Why this might have seemed "significant" to Andrew Tomas, I do not know; but the next North Pole will certainly be located in the vicinity of Lake Baikal. RS]

In the Journal of the Russian Geographical Society for 1903 there is an article by Korolenko entitled "The Journey of Ural Cossacks into the Belovodye Kingdom". Likewise the West Siberia Geographical Society published in 1916 an account by Beloliudov, "To the History of Belovodye".

Coming from scientific bodies, both of these articles are of extreme interest. They speak of a strange tradition which was circulating among the 'Staroveri' or Old Believers in Russia. An earthly paradise existed somewhere in 'Belovodye' or 'Belogorye' — the land of the White Waters and White Mountains. Let us recall here that Northern Shambhala was founded on the White Island.

The geographical location of this phantom kingdom may be less vague than it appears at first sight. There are many salt lakes in Central Asia, some drying and covered with a white layer. The Chang Tang and Kun Lun are snow-capped.

Nicholas Roerich learned in the Altai Mountains that there was a 'secret valley' beyond the great lakes and high mountains. Many people had tried to reach Belovodye but without success, he was told. However, a few had found it and stayed there for a short time. Two men in the nineteenth century reached the utopia and resided there temporarily. They returned and described wonders about the lost colony but "of still other wonders they were not permitted to speak".

This account has many points of similarity with that of Dr. Lao-Tsin, related earlier.

[COMMENT: This account also has a similarity with the visit of Apollonius of Tyana and Damis of Ninevah to Indian Sage Iarchas, whose "palace" was located in or near the Himalayas, probably in Kashmir or Tibet. There were marvelous "scientific technologies" on display for Apollonius and Damis. Whether we ever get additional and more definitive information about this place, for our own present purposes, depends entirely upon the whims of these "Guardians". RS]

That the people of these secret settlements are science-conscious can be concluded from a story of Roerich about a lama who was returning to his monastery from one of the communities. The monk had met two men carrying a thoroughbred sheep in a narrow subterranean passage. The animal was required for scientific breeding in the hidden valley.

Vatican archives preserve rare reports of missionaries in the nineteenth century which affirm that in times of crises the emperors of China used to send deputations for advice to the 'Genii of the Mountains'. These documents do not show where the Chinese couriers went to, but it could only be to the Chang Tang, Kun Lun or Himalayas.

These records of Catholic missionaries (and a work by Monseigneur Delaplace, *Annales de la Propagation de la Foi*) indicate the belief of the Chinese sages in superhuman beings living in inaccessible parts of China. The chronicles describe the 'Protectors of China' as human-like in appearance but physiologically different from man.

[COMMENT: "Human-like" but "physiologically different" is correct. They were Saurians from Planet X Nibiru. RS]

Pages 51-54

SACRED MOUNTAINS AND LOST CITIES

Many mountains throughout the world are considered to be the abodes of 'gods'. This is especially true of India, where this chapter is written.

Hindus believe in the divine character of Nanda Devi, Kailas, Kanchenjunga and numerous other high peaks. They think the mountains are residences of the gods. What is more, it is not only the peaks that are considered sacred but the bowels of the mountains as well.

[COMMENT: Modern Mount Kailasa is often mistaken for the "mythical" Mount Meru. After Planet X Nibiru departed and Mount Meru was no longer there, the legends became transferred over to Mount Kailasa, a real peak in the Himalayas. So also in Greece. When Mount Olympus disappeared from "beyond the north", the legends and stories got moved to a tall Greek mountain which took the place of the Celestial Mountain and only symbolizes the true, lost Mount Olympus. RS]

Shiva is said to have his seat on Mount Kailas (Kang Rimpoche). He is also known to have descended upon Kanchenjunga, whereas the goddess Lakshmi, on the contrary, is reputed to have ascended to heaven from the peak. In analysing these myths one forms an impression of a two-way air or space traffic that was going on in a distant epoch when gods walked among men.

[COMMENT: Shiva is the Hindu equivalent of Nibiruan Prince Enki (Poseidon/Seth), and Lakshmi is their name for Princess-Royal Inanna (Aphrodite/Hathor). Inanna was Commander of the Baalbek Airport. After her lover Duke Dumuzi (Adonis/Aten-Ra) was appointed as Lord of India, undoubtedly Inanna and Dumuzi flew back and forth to visit one another. See Chapter 13. RS]

Ever since mankind had risen from the state of savagery at the dawn of civilization, there appeared a belief in beneficent, powerful gods. Certain localities on earth and abodes in heaven were allocated to these sky-beings.

In ancient Greece Mounts Parnassus and Olympus were thought to be the thrones of these gods.

According to the *Mahabharata*, Asuras live in the sky while Pauloma and Kalakanjas reside in Hiranyapura, the golden city, floating in space. At the same time the Asuras have subterranean palaces. Nagas and garudas, the flying creatures, likewise have underground abodes. Do these myths allegorically speak of space platforms, cosmic flights and spacedromes on earth?

[COMMENT: Parts of this word "Hiranyapura" are similar to "Hyperborea" — "yapura" is practically identical with "yper", and both words begin with H. This was indeed a "golden city, floating in space" — "beyond the north". And the idea of the "bowels" or "subterranean palaces" of these mountains also being sacred places references the fact that the subterranean parts of the *Duat* or Yggdrasill were also considered "sacred" and "off-limits" to humans who were not "sponsored" to enter these forbidden zones. See Chapter 9 herein, as well as Chapter 7 ("Heaven & Hell") of *Slow-Motion Doomsday.* RS]

The Puranas mention 'Sanakadikas' — the Ancients of Space Dimensions. These beings remain a mystery if the possibility of space travel in remote antiquity is not accepted.

Since interstellar navigation is impossible without astronomy, the statement in the Surya Siddhantas that Maya, a ruler of Atala (Atlan?), received astronomy from the sun-god, seems to indicate a cosmic source of this knowledge.

[COMMENT: The Hindu Goddess Maya is identical with the Greek Goddess Maia. They are the equivalents of Nibiruan Crown-Princess Ninlil, consort of Crown-Prince Enlil. She was also known as Ma'at in Egypt, Majesta in Rome, Myesyats in Russia and Tien-Mu in China. See Chapter 12. RS]

Whether the gods are Grecian, Egyptian or Indian, they invariably pose as man's benefactors, showering upon him useful knowledge and warning him in critical times.

The scriptures of India speak of Mount Meru, the centre of the world. On the one hand, it is identified with Mount Kailas in Tibet; on the other, it is said to rise 84,000 yojanas or 411,600 miles above the earth. Is Mount Kailas a gateway to space which had existed long before the last cataclysm destroyed Atlantis?

[COMMENT: It had been so long since I first read this book that I'd forgotten about this information. It is my hypothesis that Planet X Nibiru tethers itself about 60,000 miles (100,000 kilometers) above our North Pole, about one-fourth the distance to the Moon. This Indian number of 411,600 miles seems quite excessive to me. And purely from the vestigial legends, it would be impossible to know if this number is accurate or merely a human "guesstimation" of the distance to "the golden city". Mount Meru was located at the "centre of the world". Identical ideas can be found in other cultures, and "centre" in this case means Earth's Axis of Rotation "around" Polaris, or Planet X's occultation of Polaris, at the "centre" of the northern sky. RS]

Tales of superior beings residing on certain mountains are scattered far and wide. In American Indian mythology of the Pacific north-west, Mount Shasta in California occupies a prominent place. One legend recounts the story of the Flood. It tells how an ancient hero Coyote ran to the top of Mount Shasta to save himself. The water followed him but did not reach the peak. On the only dry spot, the top of the mountain, Coyote made a fire. When the Flood subsided, Coyote brought fire to the few survivors of the cataclysm and became their culture hero.

[COMMENT: Like Mount Olympus in Greece and Mount Kailasa in India/Tibet, Mount Shasta in California is probably only a "symbolic remembrance" of the vanished Cosmic Mountain. It is curious here that a "coyote" ran up the mountain. Is the American Indian "animal" that climbs up and down the "totem pole" a "coyote"? In Siberia this "animal" was a "tomcat"; in Scandinavia, a "squirrel" named "Ratatosk"; in Central America, a "monkey". For illustrations see my book *Planet X Nibiru: Slow-Motion Doomsday*, as well as Chapter 9 herein. RS]

In these myths we also hear of ancient times when the Chief of the Sky-Spirits descended upon Mount Shasta with his family. Visits of the earth-people to the abodes of the Sky-People are also mentioned.

Mount Shasta myths may refer to actual incidents of the past — the Great Flood, landing of aviators or astronauts, and the establishment of underground shelters inside the mountain. Moreover, this colony may still be alive. There is evidence which supports this supposition.

[COMMENT: Some of this American evidence is provided by William Bramley in his book *The Gods Of Eden.* RS]

After the Gold Rush days in California, in the middle of [the] last century, prospectors reported mysterious flashes over Mount Shasta. These had sometimes taken place in clear weather, showing that they had nothing to do with lightning. Electricity could not account for the flashes because the country was not yet electrified. In more recent times cars on the roads of Mount Shasta have been known to develop ignition trouble without any apparent cause.

When a forest fire swept over Mount Shasta in 1931, a mystery fog appeared which stopped it from advancing. The demarcation line of the fire damage could be seen for many years. It went around the central zone in a perfect curve.

A curious article was featured by the *Los Angeles Times* in 1932. Its author, Edward Lanser, claimed that after interviewing residents in the Mount Shasta area, it emerged that the existence of a strange community on or in the mountain had been known for decades. The inhabitants of the phantom village were white, tall, noble-looking men with close-cropped hair and a band across their foreheads. They were dressed in white robes. Merchants said that the men used to come to their shops on rare occasions. The purchases were always paid for with gold nuggets, well in excess of the value of the goods. When seen in the forest the Shastians tried to avoid contact by escaping or by instantly vanishing from sight. Strange cattle belonging to the Shasta dwellers have appeared on the slopes of the mountain. They were unlike any animals known in America. To add to the enigma, rocket-like airships have been observed over Mount Shasta territory. They were wingless and noiseless, sometimes diving into the Pacific Ocean to continue out on the sea as vessels or submarines.

Is there a shelter of the Sky-People in the heart of the mountain as the old Indian legends say? Did they truly escape from a global deluge in aircraft?

Similar secret communities seem to exist in Mexico. In his book *Mysteries Of Ancient South America* Harold T. Wilkins writes of an unknown people in Mexico which used to barter goods with the Indians. They were supposed to have come from a lost jungle city.

Roerich's report spoke of mysterious men and women from the mountains who bought goods in Sinkiang and paid for them with ancient gold coins. California, Mexico and Turkestan are far apart, yet the tales about the strange people seem to have many points of resemblance.

L. Taylor Hansen in *He Walked The Americas* tells of an American couple who were flying over Yucatán's jungle in their private plane many years ago. Because of fuel shortage they were forced to crash-land. In the jungle they came upon a secret Mayan city camouflaged against survey from the air.

The Mayas live in ancient splendour in complete isolation from the outer world to preserve their hoary culture, which, no doubt, has its origin in Atlantis. The Americans pledged not to reveal the location of their city. After a long stay in Yucatán, the American couple returned to the United States with an extremely high opinion of the moral and intellectual level of the secret inhabitants of Mexico.

In the *Incidents Of Travel In Central America, Chiapas And Yucatán* J.L. Stephens, noted American archaeologist, cites the story of a Spanish padre in 1838-9 who saw from the Cordillera: "A large city spread over a great space, and with turrets white and glittering in the sun. Tradition says that no white man has ever reached this city; that the inhabitants speak the Maya language, know that strangers have conquered their whole land, and murder any white man who attempts to enter their territory. They have no coin, no horses, cattle, mules, or other domestic animals."

The Spanish conquistadores recorded the Aztec tradition of hidden outposts in the jungle with vast stores of treasure and supplies. These reserve bases had become almost forgotten when the invaders landed in Mexico. Verrill writes that "because no one has ever discovered any of these 'lost cities' does not prove that they did not exist or that they may not exist at the present time".

The Quechua Indians of Peru and Bolivia point to an extensive subterranean network in the Andes. Considering the outstanding engineering achievements of pre-Inca master builders, these tales could be true.

Colonel P.H. Fawcett sacrificed his life in search of a lost city which, he thought, could prove the reality of Atlantis. He claimed to have seen the ruins of such a city in South America.

The legends of lost cities, sacred mountains, hidden valleys and catacombs should be examined without any bias as some of these traditions lead to the colonies of Atlantean descendants or even of still earlier races.

Pages 59-60

'AMPHIBIANS' BRING SCIENCE

The spectacular rise of Sumerians from thousands of years of barbarism to a brilliant epoch is puzzling if we discard the myths of some wonderful beings who came as civilizers.

Babylonian tradition speaks of regular visitations of the gods who taught men the arts and crafts. One of these mysterious beings was Oannes, the fish-god.

[COMMENT: The subjects of Oannes, the Sumerians and the Dogons are discussed in great length by Robert K.G. Temple in *The Sirius Mystery* and Murry Hope in *The Sirius Connection.* See Chapters 4 and 5. RS]

Berosus, a Chaldean priest who lived at the time of Alexander the Great, left an excellent record of the activities of Oannes and his comrades. Learned Berosus writes that in ancient Babylon people were like beasts. Then a strange creature emerged out of the Persian Gulf. Its body was similar to that of a fish but under the fish's head there was a human head. The feet were joined, forming something like a fish's tail. Yet this odd creature could speak, although described by ancient Babylonians as "an animal destitute of reason".

Oannes came out daily from the sea to give the primitive inhabitants of Mesopotamia "an insight into letters and sciences and arts of every kind". He instructed the first men of Babylon how "to construct cities, to found temples, to compile laws, and explained to them the principles of geometrical knowledge". The early Babylonians were also taught agriculture and, as Berosus says, — "in short, he instructed them in every thing which could tend to soften manners and humanize their lives."

The chronicle states that since the appearance of Oannes and "other amphibians" — "nothing material has been added by way of improvement of his instructions."

A tale about "amphibians" or "animals destitute of reason" who acted as science teachers, does not make much sense. Oannes was no god because Berosus clearly says that "his voice and language were articulate and human". Where this culture bearer had come from, is a question which can be answered only if we admit the existence of a superior civilization in former times, or else on other planets.

Berosus tells us that Oannes's head was contained within the fish's head. Is this not a good description of a space helmet within which could be seen the face of a man? Feet subjoined to the fish's tail may be a crude definition of the appearance of the lower part of a pressurized suit. How could the primitive people find words to describe these strange visitors except by a comparison with known things?

Whoever the creatures were, the facts speak for themselves as immediately after their visitations, men began to build cities, construct canals and experiment in the realm of abstract thought. It is then that art, music, religion and science were born in Babylon.

[COMMENT: Not considered by Andrew Tomas is the idea that a "Saurian" might be some sort of aquatic reptile, like an alligator, crocodile or sea-snake. In *Flying Serpents & Dragons* R.A. Boulay presents compelling evidence that these Nibiruans are "reptilian" or "saurian"; Sitchin disagrees completely, making them simply a different species of "mammalian humanoids" from another planet. Until we have further proof, this mystery will remain unknowable. But from the standpoint of "common sense", I tend to think that they are "aquatic reptiles" whose behavior resembles what one would associate with an "amphibian". It would be far easier to confuse a "reptile" with an "amphibian" (or vice-versa) than to confuse a "humanoid" with either one of them. This is "basic knowledge", for crying out loud. Ideally, there should be absolutely no doubt in our recorded history about such an important fact as this. It is unfortunate that we have been left in the dark about it. RS]

Dwellers in the Euphrates valley were beast-like before Oannes, but after him they became civilized and reached a high level of intellectual development. By the second millennium before our era, the mathematicians of Babylon were already proficient in algebra and geometry. The astronomers had exact tables in algebra and knew the position of celestial bodies at any time. And it all started with a fish-like 'god' who had come out of the waters of the Persian Gulf.

Pages 61-64

FROM THE LAND OF SUNRISE

Garcilasso de la Vega transmitted to us the history of the Incas. The Sun, the great parent of mankind, in token of compassion sent Manco Copac and Mama Ocllo to teach men the arts of agriculture and women the crafts of weaving and spinning. The people of Peru accepted the Children of the Sun and laid the foundations of the city of Cuzco. Another legend describes bearded white men who came from the east and imparted the blessings of civilization to the natives.

Blood tests on the tissues of five Inca mummies in the British Museum were made in 1952 by B.E. Gilbey and M. Lubran and reported to the Royal Anthropological Institute. Three out of the five mummies possessed traces of Group A, which is utterly foreign to the American Indian. None was rhesus-negative but one had the substances D and c with the absence of C and E. This combination is rare among the Indians. Further, another royal Inca mummy had the substances C, E and c, with the absence of D. This blood sample is very unique, and almost without a parallel on our earth. These overwhelming facts prove that the Inca kings could not have belonged to the original population of South America.

It should also be noted that the Spanish conquistadores heard the Inca courtiers speak a secret language which their subjects could not understand.

A similar tradition exists in Mexico, Guatemala and Yucatán where [respectively] Quetzalcoatl, Kukumatz or Kukulkán is called a god-man. He was a white man with a ruddy complexion and long beard. On his shoulders was a long robe of black linen with short sleeves. Toltecs, skillful craftsmen, builders, sculptors and agriculturalists followed Quetzalcoatl.

The Feathered Serpent, or Quetzalcoatl, arrived from an eastern land and opened in Mexico an era of great prosperity and progress. One version has an interesting detail as to the manner of his arrival. Quetzalcoatl is said to have landed at the spot now known as Vera Cruz in a strange, winged ship. In Codex Vindobonensis he is shown descending to earth from a hole in the heavens.

[**COMMENT:** Certain liberal-minded Christian historians have speculated that after the Hebrew Jesus Christ "ascended into Heaven" from Palestine, he made an appearance in Central America, because this description of a bearded white "god-man" fits the general description of the Christ. It could be suggested just as easily that this "god-man" with the long beard, ruddy complexion and linen garments was Apollonius of Tyana. Apollonius once wrote, "Live unobserved; but if that cannot be, then slip unobserved from life." When Apollonius at the age of 100 decided to "slip away", he sent Damis of Ninevah to Rome with a private letter for their friend Emperor Nerva. After Damis left for Rome, Apollonius entered a Greek cave and was never seen again. One can only "presume" that he "died". In fact, he could have transported himself in some "miraculous" manner to Mexico, to continue his Hermetic work in an entirely different cultural milieu. Sitchin equates the Greek Hermes with the Egyptian Thoth and the Mayan Quetzalcoatl. All three, in turn, are merely equivalents of Nibiruan Prince Nannar, God of Magic. Apollonius was a Master of these secret Hermetic teachings. By contrast, we have absolutely no "official" record that the Hebrew Jesus (Bar Abbas of Nazareth) was familiar with Hermeticism. And Apollonius would have arrived in Mexico from a land to the east. We can never know the truth about this mystery, but it is fascinating to speculate on the possibilities. See my book *Apollonius Of Tyana & The Shroud Of Turin* for full details. RS]

When culture hero's mission was interrupted by enemies, he returned to the coast and set out for the country of Tlapallan on a raft of snakes. Another myth describes how the messenger cast himself onto a funeral pyre. His ashes then flew up and were changed into birds while his heart became the planet Venus. Quetzalcoatl was resurrected and went to heaven as a god. Was his winged ship — a spaceship, and the funeral pyre — its fiery launching?

As civilizer, architect, agriculturist and religious leader Quetzalcoatl left behind him an unforgettable mark on the history of Mexico. He is still greatly venerated in that country.

According to Pedro de Cieza de León, Viracocha of the Incas was a tall white man who came from the land of dawn. He instilled kindness in the hearts of Quechua Indians and revealed to them the secrets of civilization. After completing his mission, he disappeared into the sea. Viracocha's name means 'foam of the sea'. The sentiments of the Indian in regard to the legend of the white demigod have been so strong that even today some Peruvian Indians salute a friendly white stranger with the greeting — "Viracocha".

[COMMENT: Here another intriguing comparison can be made to Apollonius. Flavius Philostratus related in *Life Of Apollonius Of Tyana* that Apollonius' mother's husband, Apollonius Menodotus, was not his true father, who in fact was the Sea-God Proteus, colleague of Poseidon, God of the Sea. And Poseidon, as we know from Sitchin, was the Greek name for Nibiruan Prince Enki. A truly enigmatic Trident of Poseidon (see Chapter 11) is carved onto the cliffs of the Peruvian Bay of Pisco, pointing directly towards the "landing strips" in the Nazca Plain. After visiting the Incas, did Apollonius return to his "heavenly father" in the sea? Is that the reason a mysterious "fish symbol" became associated with early Christianity? RS]

There is considerable similarity between the legends of Quetzalcoatl and Viracocha in America, and the tradition of Oannes, the Fish-Man, in Babylon, geographically so for apart. The mythology of many races abounds in stories about gods who once trod the earth. Fanciful as the myths are, there is no doubt that some of them may be records of actual historical events.

All these apostles of civilization, descending from the sky or emerging out of the sea, plant ready-made culture among primitive tribes. Who were the founders of solar dynasties? They were the last Atlanteans who had been saved from the Great Flood in airships and spaceships as the *Epic of Gilgamesh* suggests.

The British scientist W.J. Perry believed that the Age of the Gods was largely bound up with the Children of the Sun: "The conclusion, therefore, seems forced upon us that the various groups of Children of the Sun throughout the world are derived from one primordial stock." That source may have been the legendary Atlantis.

In the East, India in particular, the guest is a sacred person because gods are believed to have appeared unexpectedly in olden times in the form and attire of man. To retain for themselves the favours of possible celestial visitors, the Hindus treat the guest with veneration and hospitality up to this day, even if he is only human. This tradition goes back for thousands of years to a time when gods trod the earth.

In India I had a few embarrassing moments when standing decorated with tropical garlands, men and women prostrated at my feet to pay homage to a visiting 'god'.

Pages 66-67

Puzzling portrait galleries exist in the caves of the Kimberley Ranges in Western Australia. The aborigines say they were made by another race. The technique of artwork and the employment of a blue pigment not used by the aborigines, attributes the authorship of these drawings to a people of non-Australian origin. The Kimberley rock paintings portray figures with peculiar head-dress or halos but no mouths. In the land of bare-footed natives, the figures are painted with sandals on their feet.

These 'Wandjina' pictures are supposed to represent the first men. It should be noted here that they are depicted with three or seven fingers and toes. The Wandjina are connected with the Rainbow Serpent paintings in the Kimberleys. The Rainbow Serpent is the term for 'Dreamtime' or the prehistoric age.

There is a remarkable affinity between the Tassili rock paintings and those of the Kimberleys. The creatures without mouths may be beings in space helmets. Numerous theories have been advanced to explain the Mouthless Ones, yet none is satisfactory.

[COMMENT: See Chapter 12 wherein I describe the Celtic God Cú Chulainn, who was said to have had seven fingers and toes. So here in these Wandjina paintings. we find something else to bolster that idea. The "Rainbow Serpent" refers of course to the fact that atop the "Rainbow Bridge Bifrost" there are "serpent-gods" living in Asgard, Shambhala, "the golden city floating in space". The peculiar term "Dreamtime" must refer to that period when The Cosmic Tree is tethered to our North Pole. The majestic arrival sequence and what followed must have seemed like a "dream" to surviving humans. In retrospect, yes, it is a "Dreamtime" for us, too. But then again, this idea could refer to something entirely different, for it is impossible to know with certainty. In any case, it is an intriguing word, this "Dreamtime"! RS]

Page 76

In the Andes, south of Lima, Peru, in the Bay of Pisco, conquistadores of the sixteenth century found "the miraculous sign of the Three Crosses", which actually looks more like the trident of Neptune with branches. This engraving in rock is 810 feet high and can be seen from a distance of 12.5 miles.

[COMMENT: See Chapter 11. RS]

The purpose and meaning of this 'Chandelier of the Andes' have remained obscure until Beltran Garcia, the Spanish scientist and direct descendant of Garcilaso de la Vega, offered his theory. He believes that the trident in rock was used by the Incas, or their predecessors, as a gigantic seismograph. In his opinion, it was a pendulum with pulleys and cords to register earthquakes not only in Peru but in the whole world. This explanation may be much nearer the truth than the one brought by the conquistadores. They thought the Sign of the Three Crosses was carved by God to thank the Christians for the conquest of the Americas.

[COMMENT: "Carved by God"? This demonstrates perfectly well how throughout the history of Christianity, there have always existed such arrogant zealots. The true reason for the Pisco Trident is this: Following the Pyramid Wars, the Nibiruans were exhausted from fighting one another and turned their attention to a looming problem of great concern: depletion of the South African gold mines. New sources of gold were discovered in what is now Peru, and Nibiruan Emperor Anu ordered that the gold-mining operations be transferred there. He placed Prince Enki in charge of the new mining operations, to replace the more elderly Duke Nergal. Then the Nazca Plain was constructed as the new Spaceport, since the old Sinai Spaceport had been sabotaged and destroyed by Prince Utu during the Wars. This destruction, incidentally, also took out the Spaceport satellite cities of Sodom and Gomorrah. Enki is the equivalent of Poseidon, Neptune, Satan and Shiva, all of whom are depicted carrying tridents (pitchforks). Undoubtedly, Prince Enki engraved the Pisco Trident in order to "mark his territory", so to speak. However, if the Trident can be seen from 12.5 miles out to sea, then it also could have been designed to "point the way" to the Nazca landing strips, and it does just that! I'm a bit surprised that this detail was overlooked by Andrew Tomas. Again, see Chapter 11. RS]

Page 78

Babylonians knew of the 'horns of Venus'. They wrote of the crescent of the planet. Since Venus is nearer to the sun than the earth, it shows phases like the moon. But the 'horns of Venus' are [not] visible to the naked eye. The burning question is — how could ancient Babylonian priests watch the phases of Venus without a telescope? They were also aware of Jupiter's four large moons — Io, Europa, Ganymede and Callisto. Till the invention of the telescope by Galileo, mankind had known nothing of these satellites. Strictly speaking, Babylonians should have had no knowledge about them, either.

[**COMMENT:** The same rhetoric is advanced also when considering the Dogon knowledge of Sirius B, "Digitaria". It could be that at certain times, from the pristine unpolluted dark skies of the Chaldean deserts, the "horns of Venus" and Jupiter's four largest moons could actually have been seen with the naked eye. (Note: Tomas left the word "not" out of the above sentence. I'm sure that this was a simple typo.) And as is discussed elsewhere in *Osiris, Isis & Planet X*, Sirius B used to be a red-giant star, "redder than Mars". Then it imploded and became invisible. On the other hand, the phrase "horns of Venus" may refer to the "wings" of the Winged Disk, or the "horns" of the "heavenly heifer", thereby having a completely different meaning. This is unknown. RS]

There are only two explanations for these astronomical observations of the phases of Venus and the four major moons of Jupiter conducted in antiquity. The first theory that the priests of Babylon had a telescope sounds somewhat far-fetched, and most scientific opinion does not even entertain it as a probability. However, the British Museum has a remarkable piece of rock crystal, oval in shape and ground to a planoconvex form. It was discovered by Sir A. Henry Layard during the excavations of Sargon's palace at Nineveh. Sir David Brewster suggested that the crystal disc was a lens but most scientists rejected his theory.

The second hypothesis is that in the course of many generations the priests of Chaldea and Sumer had preserved the elements of antediluvian astronomy. It must be borne in mind that the sages of Babylon were not only priests but scientists as well. Their astronomy was closely linked with religion and reserved exclusively for the elect priesthood.

Page 79

The African Dogons, who have a theocratic system and old traditions, know of the dark companion of Sirius situated at a distance of almost nine light-years from the earth, and visible only through a telescope. Likewise, the Mediterranean people possess the knowledge of [the] Pleiades beyond the seventh [star] invisible to the naked eye. Are these folk memories from a vanished science?

In the study of early astronomy the accuracy of the ancients in measuring the parallax of the sun has always been a riddle because this could not be computed with the instruments then in use.

The *Huat Nan Tzu* book (*c*. 120 B.C.) as well a the *Lun Heng* of Wang Chhung [*sic*] (A.D. 82) outline the centripetal cosmogony in which 'whirlpools' solidify worlds out of primary matter. These writings of ancient China give a preview of modern ideas on the formation of galaxies.

Thus we are faced with two alternatives — either to admit the existence of superior astronomical instruments in antiquity, or to assume that the priests of Babylon, Egypt or India were the custodians of a prehistoric science at least ten thousand years old.

[COMMENT: This analogy of the "whirlpool" has also been applied to the "Golden Pole" leading up to the "Golden Cage" on top (the turning millwheel "scattering the dust a hundred versts away") — "the golden city floating in space". RS]

Pages 83-84

The pundits of India also wrote about the infinitely small in such books as the *Surya Siddhanta* or *Brihath Sathaka*. In olden times they divided the day into sixty kala or ghatika, each equal to twenty-four minutes, in turn subdivided into sixty vikala, each equivalent to twenty-four seconds. Then followed a sixty-fold division of vikala into para, tatpara, vitatpara, ima and kashta. After this split-up of time the Brahmins arrived at the smallest unit of kashta, which is equal to approximately 0.00000003 second (a three-hundred millionth of a second). Needless to say, without precision instruments, 'kashta' — as a fraction of the micro second — is absolutely meaningless. We are inclined to conclude that this measurement of time is merely a tradition preserved by the pundits from an advanced technological civilization probably familiar with nuclear physics. In fact, the author made a startling discovery during his stay in India. The 'kashta' (3 X 10[-8 power] second) is surprisingly close to the life-spans of certain mesons and hyperons!

[COMMENT: On the other hand, these may be measurements connected with the *Nibiruan* Day, not the *Earth* Day. One Nibiruan Day = 10 Earth Years, or 120 Earth Months. If their day were then divided into 60 "kala" (or "ghatika"), then one of these "kala" would equal 2 Earth Months. If each of these "kala" were then divided into 60 "vikala", then 1 "vikala" would almost exactly equal 1 Earth Day, which could then be further subdivided into smaller units, as described above. Let's see. A "vikala" was divided into 60 "para". One Earth Day = 1,440 Earth Minutes (24 X 60). Divide that by 60, and you get 24 Earth Minutes for each "para". 24 Earth Minutes = 1,440 Earth Seconds. 1,440 / 60 "tatpara" = 24 Earth Seconds for each "tatpara". And so on. Finally then, such a "kashta" could become the equivalent of 0.00000003 of 1 Earth Day, *not* 1 Earth Second! Even then, it would still be a tiny fraction of a second. The vexing question therefore inevitably arises: why would this matter, for either us or them?! RS]

Page 86

The statue of Memnon in Egypt spoke as soon as the rays of the rising sun fell upon its mouth. The sound issued from the base of the figure. In the words of Juvenal — "Memnon sounds his magic strings." The Incas had a speaking idol in the valley of Rimac. Needless to say, the construction of these statues required a knowledge of physics.

[COMMENT: Apollonius of Tyana and Damis of Ninevah, when travelling up the River Nile on their way from Alexandria to Ethiopia in July 69 CE, heard this statue of Memnon speak at sunrise. See *Life Of Apollonius Of Tyana* by Flavius Philostratus, Book IV, Chapter IV. RS]

Pages 93-95

RELICS CREATE A DILEMMA

Muhiddin Piri Reis, or Admiral Piri Reis (1470-1554), published in Turkey a navigation atlas *Bahriyye* in 1520. His maps with marginal notes, drawn on roe-skin, were discovered at the palace of Topkapi in Istanbul on 9 November 1929 by Halil Edhem, Director of the National Museums.

[COMMENT: In August 1965, I toured the exquisite Topkapi Palace Museum in Istanbul. RS]

In his notes Admiral Piri reveals the story of his maps. In a naval battle with Spain in 1501, a Turkish officer Kemal took prisoner a Spaniard who had been with Columbus on three of his historic voyages. The captive had a set of curious maps.

Christopher Columbus may have known where he was going thanks to these maps. If this supposition is correct, then we can understand the words of his son Ferdinand who wrote in *The Life Of The Admiral Christopher Columbus* that: "He noted down any helpful hints that sailors or other persons might drop. He made such good use of all these things that he grew convinced beyond the shadow of a doubt, that to the west of the Canary and Cape Verde Islands lay many lands which could be reached and discovered."

Among the articles confiscated by the Turks from the Spaniards were the maps drawn by Columbus in 1498, or six years after the discovery of the West Indies. Yet the charts show a complete outline of the continents of North and South America, their rivers, Greenland and Antarctica — all unknown in 1498. The distance between South America and Africa is surprisingly correct.

[COMMENT: There is an excellent book on the subject of the Piri Reis Maps and other ancient maps, if you can find it. It is titled *Maps Of The Ancient Sea Kings* by Charles H. Hapgood (New York, 1966). All of the Piri Reis Maps are pictured in this book. RS]

Professor Dr. Afetinan of Turkey in his book *The Oldest Map Of America* writes: "In the chapter on this 'Western Sea' we read all that is known about the discovery of America at the time. Of this he [Piri Reis, AT] recounts, on hearsay again, how a certain book from the time of Alexander the Great was translated in Europe and after reading it how Christopher Columbus went and discovered the Antilles with the vessels he obtained from the Spanish government. It is quite evident today that Piri Reis came into possession of the map that the great discoverer used."

[COMMENT: That is an interesting little footnote of history which I'd never paid any attention to before. Apollonius of Tyana "inherited" certain documents from Alexander The Great, and many of those were subsequently acquired by Roman Emperor Hadrian in about 120 CE. If Apollonius "magically transported himself" (as he did when he disappeared from Rome and instantly reappeared at a port city south of there, where Damis was waiting for him) from "the east" to Mexico and Peru as "Quetzalcoatl" and "Viracocha", then he may have known in advance from Alexander's private map collection that such "western lands" indeed existed, and perhaps simply decided to tour them by himself, after "slipping away unobserved from life" in Greece. Why not? RS]

Many things are puzzling about the Piri Reis map. Who drew a chart in Columbus's or perhaps even in Alexander the Great's times with the contours of Antarctica free of ice, and how did he do it? After all, it is only in the International Geophysical Year that the continent was sounded through the ice sheet and charted. Greenland is shown as two or three islands. Greenland is buried under 5,000 feet of glaciers, and it is only recently that a French Polar expedition disclosed the fact that Greenland comprises two main islands.

[COMMENT: Yes, indeed! The Piri Reis Map of "Antarctica" (Atlantis!) is most intriguing. Both Antarctica and Greenland pivoted into polar positions only *after* the last Polar Axis Shift in about 1587 BCE. Before that, both islands were unfrozen and mapable. Thus, the only conclusion to be drawn is that these islands were charted *before* they were buried under ice, snow and glaciers. It could be that the maps of Alexander and perhaps even Columbus were *copies* of other original maps from an even greater antiquity, but the original maps *had* to have been drawn *before* 1588 BCE. Simple. RS]

Arlington H. Mallery, an American authority on cartography, asked the U.S. Hydrographic Office to check the enigmatic map. Commander Larsen, on behalf of [the] U.S. Navy, then made a statement: "The Hydrographic Office of the Navy has verified an ancient chart — it's called the Piri Reis map, that goes back more than 5,000 years. It's so accurate, only one thing could explain it — a worldwide survey. The Hydrographic Office couldn't believe it, either, at first. But they not only proved the map genuine, it's been used to correct errors in some present-day maps."

According to Mallery, the archaic chart had a record of every mountain range in Northern Canada and Alaska, including some ranges which the U.S. Army Map Services did not have on their maps. But the U.S. Army has since found them.

The longitude on this map is exact. This is baffling as it is only two hundred years since we learned how to calculate it. Mallery even remarked: "We do not know how they could map so accurately without an aeroplane."

This map demonstrates the existence of science in a faraway epoch which is considered to have possessed none. Did the expedition of Alexander the Great come into possession of papyri from the temple of Sais in Egypt? Its priests definitely knew about America, for, as Plato writes, Solon was told that the Atlantic "is a real sea and the surrounding land may be most truly called a continent".

The arguments set forth in favour of the very ancient origin of the Piri Reis map which Christopher Columbus allegedly possessed, can be substantiated by another startling fact. Space satellites have disclosed that our planet is somewhat pear-shaped. A letter from Columbus is still extant in which he states that the earth is "pear-shaped". Two decades ago we did not know about the strange form of our planet. How did Columbus learn about it?

[COMMENT: This refers to the fact that Earth has an "equatorial bulge" making it somewhat "pear-shaped" instead of being precisely round. RS]

Pages 98-103

SKYSHIPS OF ANTIQUITY

It is quite reasonable to surmise that most of the legends about skyships in olden times are echoes of aviation and astronautics from a former civilization. Despite the strong opposition to the theory of an advanced technology in the dim past, coming from the majority of scientists, there are numerous facts which support this hypothesis.

The *Ramayana* of India contains detailed descriptions of a "vimana" or airship. It was self-propelled by a yellowish-white liquid. The vimana was large — it had two stories, windows, dome with a pinnacle. The airship of antiquity could fly with the "speed of the wind" according to one's skill, and gave forth a "melodious sound". Its control required high intelligence. The craft could travel in the sky or stop and remain motionless in the air.

The vimanas were kept in "vimana griha" or hangars. The ancient records state that the vimana soared above the clouds and from the altitude "the ocean looked like a small pool of water". The aviator could see the "country round about the ocean and the mouths of rivers meeting the ocean".

The archaic planes were employed in warfare by kings and for sport by the "foremost people among pleasure seekers". It is most unlikely that such precise details could be mere fancy.

In China the Emperor Shun, who lived about 4,200 years ago, constructed a flying chariot. Shun is not only the first recorded pilot but also the first parachutist.

Chu Yuan (340-278 B.C.) wrote a description of an air trip in a poem called "Li Sao". As he knelt at the grave of the Emperor Shun, a jade chariot drawn by four dragons appeared. Chu Yuan boarded the craft and flew at a high altitude across China in the direction of the Kun Lun Range. In his aerial journey he observed the earth, unaffected by the winds and dust of the Gobi. Chu Yuan not only successfully landed upon the completion of the flight but also subsequently made another survey of the Kun Lun Mountains from the air.

The founder of the Shang dynasty, the Emperor Cheng Tang (1766 B.C.) ordered Ki Kung Shi to build a flying chariot. The ancient engineer completed the assignment and tested the aircraft in flight, reaching the province of Honan. However, the vessel was destroyed by imperial edict so that the secret of its mechanism would not fall into wrong hands.

The flying machines of ancient China were either a product of scientific experimentation or a memory from a precataclysmic race. As the Chinese had no technology at the time, there is no alternative but to accept the second possibility.

Chu Yuan's flight to the Kun Lun may, perhaps, give us the clue to the source of this technical knowledge in ancient China. The mighty Kun Lun Range is considered by the Chinese people to be the abode of 'gods'.

These skyships were traditionally reserved for the Emperors and Taoist scholars who were supposed to act as intermediaries between the 'genii of the mountains' and mankind.

An indirect proof of our theory that in ancient times man knew of aviation, is found in the presence of the words 'flying chariot' in Chinese vocabulary. When confronted with the appearance of the aeroplane early in this century, the Chinese did not have, like ourselves, to invent a new term — they used the old one — *fei chi* (flying chariot).

In the twelfth year of the Emperor Yao's reign (2346 B.C.) a strange man appeared. His name was Chih Chiang Tzu-yu. He was so skillful an archer that the Emperor named him 'Divine Archer' and appointed him 'Chief Mechanician'.

In the annals of Chinese history he is reported to have ridden a 'celestial bird'. When "carried into the centre of an immense horizon" he noticed that he could no longer observe the rotary movement of the sun. In space, beyond the earth, our astronauts are also unable to see the sun rise or set. Does the ancient record of the 'Chief Mechanican's' flight intimate that man could bridge interplanetary space thousands of years ago?

The great Chinese thinker Chuang Tzu wrote an essay entitled "Travel to the Infinite" in the third century before our era. He relates how he rode on the back of a fabulous bird of enormous size into space to the distance of 32,500 miles from the earth.

According to Taoist beliefs, 'Chen Jen' or perfect men are able to fly through the air on the wings of the wind. They pass on the clouds from one world to another and live in the stars. Teng Mu, a scholar of the Sung dynasty, wrote about "other skies and other earths". Ma Tse Jan, a distinguished physician of old China, after having mastered the philosophy of Taoism, was taken to heaven alive.

In his travels through Tibet and Mongolia Professor Nicholas Roerich saw passages in Buddhist books about "iron serpents which devour space with fire and smoke" and "inhabitants of the distant stars".

Viacheslav Zaitsev in the Soviet magazine *Neman* (No. 12, 1966) writes about the strange stone discs discovered in the district of Baian-Kara-Ula on the border of China and Tibet. They have holes in the centres like gramophone records. A double groove inscribed with hieroglyphics spirals from the centre to the edge of the discs.

Professor Tsum-Um-Nut, with four colleagues, has deciphered the writing in the grooves. However, their discovery was so sensational that at first the Peking [Beijing today, RS] Academy of Prehistory did not allow the Chinese scientists to publish their findings. After the permission had been granted, a book appeared under an intriguing title, *Disc Hieroglyphics Speak Of Spaceships 12,000 Years Ago.*

An analysis of the stone particles from the discs revealed amazing results — they contained a large quantity of cobalt and some other metal. Tested under an oscillograph the discs displayed a peculiar frequency as if they had been electrically charged thousands of years ago.

The carvings of [the] Baian-Kata-Ula discs depict the sun, moon and the stars as well as some strange dots gliding from the sky towards the earth.

[COMMENT: These "strange dots gliding from the sky towards the earth" probably were intended to depict the "rainbow bridge" connecting Earth to Heaven, beyond the North. RS]

Tschi Pen Lao, of the University of Peking, has discovered curious drawings in the mountains of Hunan and on an island in Lake Tungting. Made about 45,000 B.C. these granite carvings portray people with large trunks and cylindrical craft. It is difficult to admit the existence of spacehelmets and spaceships so long ago — yet what other explanation can one offer?

From the study of myths and historical records it follows that men flying heavenward and cosmic guests coming earthward were a reality in a bygone age. Whether these space visitors came from another planet or from a secret Atlantean colony in a remote part of our globe, is largely a matter for speculation. However, there is no contradiction between the two versions if we assume, on the basis of available data, that Atlantis had contact with other planetary civilizations.

[COMMENT: On the other hand, it could be that all of these memories of space travellers reflect periods when Planet X Nibiru is tethered to our North Pole. Certainly Atlantis experienced such a "Golden Age" as this. RS]

In an article, "By the Path of Legends", U. Tkachev, writing in the Soviet magazine *Smena*, stresses the usefulness of imagination in the field of science. Because of the affinity of his ideas with the plot of the present book we will cite it:

"The earth was visited by an expedition of cosmonauts. It is upon the continent of Atlantis that the spaceship landed. Apparently the earth was not their principal base as otherwise their stay would have left more definite traces. Evidently the astronauts possessed such a technology that they could construct satellites with their own peculiar conditions for life. Using these as bases, they reached the earth and other planets in 'planetoplanes'. Presumably they acquainted the Atlanteans with but few branches of culture, none of which could be employed for enslaving the neighbouring peoples because of their immeasurable humaneness. In all likelihood, these were — painting, sculpture, architecture, mathematics and astronomy. Possibly they had visited the earth a number of times and these flights were recorded in folklore as the descents of gods upon the earth. The Atlanteans founded the first state in the history of earth. Their continent sank 11,500 years ago. The principal site of culture perished. Knowledge was gradually lost by mankind. Occasionally ancient science would come to the surface."

Dr. Carl Sagan, an American astrophysicist of the first magnitude, has made interesting conclusions on the basis of mathematical computations. He suggests that if each advanced civilization in our galaxy sends a spaceship once a year by our time reckoning, in the direction of neighbouring stars, the interval between cosmic visits would be equal to about 5,500 years. According to Dr. Sagan's calculations, the explorers from other solar systems are soon to fly over us on their regular inspection tour. Upon landing on earth the cosmonauts would be greatly surprised at the progress mankind has achieved since the first dynasty in ancient Egypt.

[COMMENT: The late Dr. Carl Sagan of Cornell University in Ithaca, New York, was one of the most vociferous critics of Dr. Immanuel Velikovsky. Their "Velikovsky Affair" is legendary amongst certain academicians. Books have been written about it. Dr. Sagan would never have considered the writings of Zecharia Sitchin, Andrew Tomas or me to be anything other than "fanciful mythological imaginings". And Dr. Sagan got his timeframe wrong: the period is 3,600 years, not 5,500 years. RS]

Incidentally, the tradition of Aztecs speaks of a promise of the 'sons of heaven' to return in 6,000 years — that is in our historical epoch.

Dr. C. Sagan believes that "the earth may have been visited many times by various galactic civilizations during geological times and it is not out of the question that artifacts of these visits still exist". The American scientist recommends us not to discard ancient myths which may contain accounts of appearances of space visitors described as 'gods' or 'angels' in scriptures and folklore.

[COMMENT: According to the accompanying footnote, Dr. Sagan expressed this view in an article published by the *Los Angeles Times*. Generally, Dr. Sagan scoffed at any and all reports of visits by extraterrestrials to Earth. RS]

Today the reaction of simple men and women who have never seen an automobile or aeroplane in some isolated part of the world, would be much the same as the one shown by the ancients when confronted with the appearance of a strange apparatus. In the fifties a jeep was dismantled, carried over the 13,400-foot Rohtang Pass in the Himalayas, and reassembled on the Lahoul side. As it descended into a valley, the surprised natives who had never seen a mechanically propelled vehicle, came out to worship this manifestation of supernatural power. When the first plane landed in Ladakh in 1948, the reaction of Tibetans to this flying monster was even more comic — they brought hay to feed it.

[COMMENT: Leonard "Mr. Spock" Nimoy had a television series in the 1970s titled "In Search Of". I watched most of the episodes. In one of them, Mr. Nimoy reported on a tribe of natives in New Guinea who, not understanding the science behind the airplanes flying overhead, built models of these "sky boats" in their villages and worshipped them as "gods". It is the same idea. RS]

K.E. Tsiolkovsky, the Russian pioneer in astronautics, when asked to express his opinion on the probability of interplanetary contacts, said that the visitation of our planet by cosmonauts could have taken place in the past, and would indeed occur in the future.

Faced with the same question in 1930 Professor N.A. Rynin of [the] U.S.S.R. answered that "if we turn to tales and legends of hoary antiquity, we will notice strange coincidences among legends of countries separated by oceans and deserts. These coincidences in legends comprise the visitation of earth by dwellers of other worlds in time immemorial. Why not admit that a grain of truth still lies in the kernels of these legends?"

If beings from other planets paid us visits in a forgotten epoch, then it is clear how fruits and grains unknown to earth were brought by the 'gods' from other 'lokas' or worlds, as Brahmin books claim.

The subject of cosmic contacts in past ages, possibly in the Atlantean era, has been contemplated by men of science. It is assuredly worthy of a serious consideration in this Space Age, when we ourselves are about to explore other planets.

Behind legends can be dimly discerned a distant epoch in which a vanished race might have attained to a high degree of technology.

Pages 108-109

In 1969 the Cosmic-Ray exploration at Giza took an unforeseen turn. It was aptly described by John Tunstall in the *Times* Saturday Review for 26 July 1969:

"Scientists who have been trying to X-ray the Pyramid of Khephren at Giza, near Cairo, are baffled by mysterious influences which are throwing into utter confusion the readings of their Space Age electronic equipment. The equipment was demonstrated with dramatic success at the New York World's Fair and the agreement for the joint U.A.R.-United States pyramid project was signed on 14 June 1966. Between 1966 and early 1967 the cosmic ray measuring equipment was installed in the burial chamber at the base line of Khephren's pyramid."

On his second visit to Giza John Tunstall interviewed Dr. Amr Gohed, the Arab scientist in charge of the installation at the Khephren pyramid. This is his story:

"He showed me the new I.B.M. 1130 machine surrounded by hundreds of tins of recordings from the pyramid stacked in date order. Though hesitant at first, he eventually told me of the complete impasse that had been reached.

"'It defies all the known laws of science and electronics,' he said, picking up a tin of recordings made in October, 1968. He put the tape through the computer, which traced the pattern of cosmic ray particles on paper. He then selected a recording of the next day in October and put it through the computer. But the recorded pattern was completely different. The salient points which should be repeated on each tape were absent.

"'This is scientifically impossible,' he told me. But it is happening before the scientists' eyes.

"After long discussion and many cups of Arabian coffee, I put the ultimate question to Dr. Gohed: 'Has all this scientific know-how been rendered useless by some force beyond man's comprehension?' He hesitated before replying, then said: 'Either the geometry of the pyramid is in substantial error, which would affect our readings, or there is a mystery which is beyond explanation — call it what you will, occultism, the curse of the Pharaohs, sorcery, or magic, there is some force that defies the laws of science at work in the pyramids.'"

This is a tremendously significant development at Giza. Where does this force field come from? What superior science and technology created it? As modern physics cannot neutralize cosmic rays, the prehistoric engineers must have had a superior science.

And so the Treasure of the Sphinx still awaits discovery.

Pages 117-119

SEEK IN THE MOUNTAINS AND THE SEA

Secret storehouses of a prehistoric civilization are located not only in the Mediterranean basin but also in other parts of the world.

For weeks I used to admire Kanchenjunga with its veil of snow, looming high above the horizon. Why is this Himalayan peak called the 'Five Sacred Treasures of the Great Snow'? Are treasures really buried deep within its bosom? Hillmen from Sikkim and Bhutan pay homage to the Five Treasures of Kanchenjunga. Tibetan folklore affirms that the precious things hidden in the innermost recesses of the mountain, have been guarded for centuries.

Nicholas Roerich, the famous painter and explorer, writes in *Himalayas, The Abode Of Light* that the Himalayan foothills contain entrances to subterranean passages leading far below Kanchenjunga. He remarks that a closed stone door has been seen leading to the Five Treasures of the Great Snow but the time is not ripe for its opening.

Also from Nicholas Roerich we learn that there are other secret storehouses in the Himalayas. On the Karakoram pass at an elevation of 19,500 feet, Roerich's groom said that great treasures were buried in the snowy ridge. He remarked that even the lowly ones among the natives knew of vast caverns which contained treasures from the beginning of the world. He wondered if Professor Roerich was aware of the books recording the location and contents of these subterranean vaults. The man of the mountains was asking Roerich why foreigners, who claimed to know so much, could not find the entryway to underground palaces. Then he concluded that the gates to them were guarded by a mighty fire which outsiders could not pass.

These legends of hidden treasures are persistently heard in Asia. The Tibetan epic of Ghessar Khan predicts the opening of the Treasures of the Mountains.

[COMMENT: Indeed, I cannot think of a better location to hide secret treasures than in the forbidding Himalaya Mountains. When I spent a week in Kathmandou, Nepal, in January 1970, the days were warm, but there was lingering snow all over the ground. Kathmandou is the "jumping-off point" for those who seek to climb Mount Everest. One afternoon I took a two-hour tourist flight from Kathmandou out to Mount Everest and the Himalayas. It was some of the most remote but spectacular vistas that I have ever seen in my life. But I had absolutely no desire to be on the ground or climb any of the mountains. I also saw the IMAX film about Mount Everest. You could not pay me any amount of money to climb that mountain, and I have no idea how they actually got that heavy, cumbersome IMAX camera all the way to the top of Everest. I have never been a "mountain person" anyway. I prefer hot, flat desert lands to mountains. I rather enjoy the smaller, more rolling mountain ranges like the Ozarks and the Ouachitas; but as for the Alps, the Rockies, the Himalayas — no, thank you. Somebody else can do that particular investigative research. Incidentally, my Himalaya tourist plane was a French Fokker from the era of World War II. Right after we got back to Kathmandou, the same plane took off on a routine commuter flight to New Delhi. It crashed along the way. Stunned is an understatement when I read about it in the *Paris International Herald-Tribune* the next morning during breakfast at the Royal Hotel. The fourth of my nine Leo cat-lives had just been used up. I'm down to about two cat-lives left by now. RS]

India has a number of secret repositories according to H.P. Blavatsky. She wrote that initiated yogis knew of a vast network of underground galleries which run from cave temples. This engineering feat suggests a high technology in remotest antiquity.

In her travels through Tibet, Blavatsky met Buddhist pilgrims who claimed that in a secluded part of [the] Altyn Tagh Ridge there existed a vast network of galleries and halls housing a collection of several million books. Madame Blavatsky estimated that the British Museum could not have accommodated all the cultural treasures of this underground library.

The locality is described by her as a deep gorge with a small cluster of unimpressive houses to mark the site of what may be the world's greatest library. It is secure from intrusion and nothing will disturb its age-old manuscripts. The entrances are thoroughly concealed and the vaults with books lie deep within the bowels of the earth. It is most unlikely that our world will ever see this fabulous treasure-house of culture. But we can be more optimistic about the treasures of Atlantis buried in Egypt.

[COMMENT: One certainly hopes that this is true. One hopes that in this great world library there is a copy of every single book that has ever been published in the history of this planet. RS]

The sages of the East are in a position to present strange documents which will upset the opinions held by our historians. Blavatsky predicts that some of these manuscripts will soon be released.

Our central topic of secret halls of records left by a former civilization may be hotly debated. Nevertheless, more significant than the writer's theories is the testimony given by Plato, Cicero, Manetho, Josephus, Proclus, Ibn Abd Hokm, Masoudi, and in recent time, by Blavatsky and Roerich.

The stage is now set for a momentous event in world history — the discovery of Atlantean antiquities. The prophetic words of Ignatius Donnelly, the American pioneer of Atlantology, uttered late in the last century, ring now in our ears: "Who shall say that one hundred years from now the great museums of the world may not be adorned with gems, statues, arms, and implements from Atlantis, while the libraries of the world shall contain translations of its inscriptions, throwing new light upon all the past history of the human race, and all the great problems which now perplex the thinkers of our day."

[COMMENT: For additional information on Atlantis, see Chapter 5 of my *Planet X Nibiru: Slow-Motion Doomsday.* RS]

Pages 124-125

IT HAS ALL HAPPENED BEFORE

Civilization is largely the product of human intelligence. From caves to skyscrapers, from boomerangs to space satellites, is a jump achieved by the power of mind. Deprive man of one-half of his present intellect, and the whole social edifice of today will suffer a setback comparable with a planetary disaster. Culture the mind, and you will have a sky-rocketing civilization. Develop the moral nature of humanity, and you will have a utopia to live in.

Intellectual development in human society is not unlike a chain reaction in nuclear physics. Jean Sylvain Bailly, the French astronomer of the eighteenth century, summed up this process like this: "Ideas have been successively gathered together, heaped up; they have mutually engendered each other, the one has led to another. It remains therefore merely to rediscover this succession, to begin with the earliest ideas; the path is traced out; it is a journey that one may make again because it has already been made."

[COMMENT: My own present work reflects these sentiments in a microcosm: Alexander Von Humboldt, Professor William Smith, Sir Paul Brunton, Sir E.A. Wallis Budge, Dr. Whitfield Leggett Russell, Madame Helena P. Blavatsky, U.S. Congressman Ignatius Donnelly, Miss E. Valentia Straiton, Dr. Immanuel Velikovsky, Giorgio de Santillana, Hertha von Dechend, Erich von Däniken, Peter Tompkins, Gerald Hawkins, Andrew Tomas, Robert K.G. Temple, Murry Hope, Zecharia Sitchin, Darío Salas Sommer, William Bramley, R.A. Boulay, Rose and Rand Flam-Ath, prominent amongst others for two centuries, all have led to me; and from me there will inevitably be those who follow on this mesmerizing path to ultimate cosmic truth about the origin and home of our Nibiruan Archons, our "custodial gods" from Planet X. RS]

Behind Copernicus, Galileo and Bruno stood the shades of Pythagoras, Aristarchus, Anaxagoras, Anaximenes and other Greek philosophers. Newton acknowledged his debt to antiquity by saying: "If I have seen further, it is by standing on the shoulders of giants."

But many of these giants of the classic world studied, in their turn, at the feet of Egyptian hierophants. From whom did the wise priests of the Land of the Nile receive their secret philosophic and scientific tradition? From Thoth, who had come from an island in the Western Sea. Thus the fount of learning can be traced to legendary Atlantis.

Much of the New World civilization is an enigma without the Atlantis theory. No race ever built such roads as the Peruvians. They crossed the deepest canyons and pierced the highest mountains by tunnels which are still in use. On ancient asphalt-surfaced roadways modern cars travel today. No people, past or present, ever erected such megalithic structures as the pre-Incan races. No other nation ever has woven, by hand or machine, textiles of the workmanship of ancient Peruvians. No civilization ever had such an accurate astronomical calendar, in which every one of the 18,980 days were individually distinguishable, as the Aztecs and the Mayas.

[COMMENT: One of the reasons for my including the foregoing excerpt was the notation of a "?" that I had placed in the margin of this book years ago next to the mention of "18,980 days". At the time that I first read this book, I was greatly absorbed by the mathematics of ancient calendars, and apparently I'd intended to come back "later" and analyze this figure. Well, here we are. The first thing to do is to divide it by the number of days in one year: 18,980 / 365.25 = 51.964. Thus, this figure refers to the number of individual days in the basic Mayan cycle of "about 52 years" (13 X 4). Despite all that has been written about the Mayan Calendar, I personally have always thought that it is "illogical" because in no way can it be precisely mathematically reconciled with the Precession of the Equinoxes as other ancient calendars can be. The reason for these "peculiarities" in the Mayan Calendar is that they incorporated a Venus cycle along with an Earth cycle. Why Venus? Well, even Dr. Immanuel Velikovsky confused Venus with Planet X Nibiru. Perhaps after Nibiru last departed, the Mayas felt that Venus had somehow "usurped" the orbital position Planet X. But that is only a guess. Surely somewhere there is a "valid" reason as to why the Mayas so venerated Venus, and thus why Dr. Velikovsky was so mistaken. If any reader ever runs across such information, please send me an email. Thanks. See also Chapter 17. RS]

Pages 128-131

Archaeology considers the civilizations of Mohenjo-Daro, Sumer or Egypt as the earliest in history. In fact, science does not recognize any history as such prior to about 5000 B.C. Nevertheless, there may be relics of advanced nations buried in the oceans. If discoveries are ever made to that effect, history will have to make drastic corrections.

The rise of man from simple agricultural states of the valleys of the Nile, Tigris and Euphrates to our present technological era is too short a period for so fantastic an advance, unless man carried with him hereditary traits from another cycle of civilization.

To have progressed from ox-pulled carts to luxurious automobiles and from boomerangs to earth satellites in the short period of six thousand years is truly a miracle. But science has no place for miracles. Unknown factors could account for them. The time is too brief because it represents but a fraction of one percent of the total accepted age of man. The "Caveman to spaceman in twenty-five thousand years" idea may be altogether erroneous.

[COMMENT: Exactly what is meant by "caveman to spaceman in twenty-five thousand years", I do not know. Obviously Mr. Tomas was referring to a phrase in or the title of some book or television program. However, I disagree with this. We have only to look at the swift technological progress of the past century to understand that one invention quickly led to another, and that in turn to another, and so on and so on. Just consider the evolution of Internet technology within the past decade. When I got my first computer in 1994, there was not even a World Wide Web. Email was UNIX-based and cumbersome for someone like I. But today, our computers are not only word-processors and more but also home entertainment centers (based upon old electronic stereophonic and VCR technologies) and clocks (based upon digital watches). We have integrated so many previously unrelated technologies into one machine with our latest, top-of-the-line computers. I can perfectly understand how we could have evolved from a purely non-technological society two centuries ago into our modern world. The discovery of electricity was at the heart of it. The Roman Empire was different from our modern era mainly because they did not discover electricity. Otherwise, their lives were as technologically advanced for *their* time as ours are for *our* time. The Romans had running water, traffic signs, home postal delivery and more. What they didn't have was electricity. How advanced would the world be today if Benjamin Franklin had lived during the reign of Emperor Hadrian, followed by Thomas Edison during the times of Emperor Septimius Severus and Empress Julia Domna Bassianus? Within that span of fewer than one hundred years, would there have been invented (*ipso facto* as in our time) an International Space Station a century later, even before the Council of Nicaea had had time to convene and plunge the Greco-Roman World back into philosophical primitivity?! You decide. RS]

It is anticipated that these views will be condemned by most scientists. Nevertheless, the controversy can be resolved overnight in our favour if one of the Atlantean 'time capsules' becomes accessible to the public. Until then is the author prepared to act as a target for the outbursts of irritated professors.

[COMMENT: As contemporary examples, I cite the unceasingly vitriolic, almost hysterical attacks by ex-Velikovskian C. Leroy Ellenberger on the ideas propounded by Planet X specialist James McCanney, or the denunciations by Professors Carl Sagan and Stephen Jay Gould of the historical achievements of Dr. Velikovsky. (And incidentally, James McCanney's interview is included in the first Planet X Video, as Andy Lloyd and I were presented in the second video. See Appendix A.) Some academicians are simply too slavish and weak-willed to ponder the real meaning of life outside the strict parameters of their entrenched establishment doctrines. Apollonius of Tyana encountered this same sort of "robot behavior" when he engaged the pseudo-philosopher Euphrates in debate at the headquarters of Roman Emperor Vespasian in Alexandria in 69 CE, before Apollonius and Damis departed for Ethiopia and before Jerusalem was destroyed the following spring by General and future Emperor Titus, son of Vespasian and friend of Apollonius. For further details, see my 2005 book *Apollonius Of Tyana & The Shroud Of Turin.* RS]

To summarize: man is more ancient than our academies conjecture. The earth has experienced violent cataclysms, mainly due to axis shifts and falls of huge meteorites. In these geological upheavals great civilizations disappeared without a trace.

Bhagavata Purana, a sacred book of India, speaks of four ages which have passed, each destroyed by the fury of the elements. Our present cycle is the fifth.

Hesiod, an ancient Greek poet who lived in the eighth century before our era, writes about a similar belief in Greece. There were four ages in the past. First, the gods created a golden race. They were mortal yet lived like gods. The second race was silver. They were of inferior intelligence. The next cycle was that of brass. Men were then strong and warlike and they destroyed themselves. The fourth era was the age of heroes whose adventures inspired men ever since. According to ancient Greek lore we are now in the fifth age — the age of iron. They believed we should likewise be destroyed by Zeus as were the other races. Censorinus (b. 238 A.D.) writes that the Greeks thought the world was either inundated or burnt after each epoch.

Ancient Egyptians divided history into three principal eras. First was the kingdom of the gods. In the second period demigods and heroes ruled on earth. Then with their departure, men reigned over Egypt and the world. When classic historians and myths speak of gods and demigods, we do not take them seriously. Yet superior men could have walked the earth in a golden era.

In China the people of Yunnan have preserved the memory of an age when levitation of heavy rocks was a commonplace matter, when all were prosperous and lived long lives. The Pai tribesmen sing of that bygone epoch in these words:

In olden days rocks used to walk,
Is this true or false?
In olden days the rocks could walk,
This is true not false.
At that time the world was all peace,
Do you believe what I say?
At that time the world was all peace,
I believe what you say.
At that time there were no rich or poor,
Do you believe what I say?
At that time there were no rich or poor,
I believe what you say.
At that time people lived hundreds of years,

Do you believe what I say?
At that time people lived hundreds of years,
I believe what you say.

It is easy to call legends mere fantasy and laugh at the traditions of many peoples. It is much more difficult to appraise history in its entirety.

This work has an aim and a moral. Its aim is to call the attention of the public to the startling possibility of a discovery of a hidden depository left by a race now considered mythical. Its moral is contained in the question: "Are we not treading upon the footsteps of Atlantis?"

At the entrance of La Sainte Chapelle in Paris a guide was explaining to us the meaning of various ornaments. Before the panel depicting Noah's Ark and the Flood he delivered an oration on the biblical story of the Flood and concluded: "And so, *mesdames et messieurs*, people and animals began to multiply and, the process is still going on — till the next Flood."

According to Plato, the Atlanteans perished when they were engaged in imperialistic wars. But in a better epoch they loved peace, cherished camaraderie and despised avarice.

[**COMMENT:** Sound familiar, America? RS]

It is only hoped that the modern world may ensure a better destiny for itself than Atlantis. The Catalan poet Jacinto Verdaguer mourned Atlantis in these words:

Atlantis, woe to thy children!
Alas, shall we live to see another dawn?
Our ancestor's words came true one by one:
His Atlanteans, land and gods are no more.

Plutarch in *Isis and Osiris* recorded the opinion and belief of the most ancient sages "that there will become a fated and predestined time when the earth will be completely levelled, united and equal, there will be but one mode of life and but one form of government among mankind who will all speak one language and will live happily".

The discovery of the Treasure of the Ages will completely revolutionize all of our views on ancient history. From the lessons of the past, mankind will learn to avoid the mistakes of the vanished race. Man will then be able to find his proper place on this beautiful planet, his true mission and work towards a glorious future.

Page 136

At the source of all ancient civilizations there always stands a divine bearer of culture. Thoth brought a ready-made culture from a western land. His titles 'Lord of the Overseas' and 'Guardian of the Two Lands' by which he is referred to in the Book of the Dead and certain Pharaonic inscriptions, suggest that he was an Atlantean leader. A significant myth about the god Thoth says that it was on wings that he transported the gods to the east — to "the other side of the lake Kha". Airlift of a cultural *élite* from Atlantis to Egypt?

[COMMENT: This "Lake of Kha" is just another name for the "Lake of Sa" or "Pool of Natron" or "Pillar of Thoth". See Illustration 6 of *Planet X Nibiru: Slow-Motion Doomsday.* The "two lands" were Planet Earth Tiamat and Planet X Nibiru. Thoth was/is Hermes ("Salute, Hermes!"), Prince Nannar of Planet X, member of The Council Of Twelve. Even Darío Salas Sommer (pseudonym, John Baines) writes about Thoth's ancient influence on Egypt in his book *The Stellar Man.* One wonders, in passing, if Señor Sommer is keeping up with the latest in this field of research. Or is he simply laughing all the way to the bank from his retirement villa in Santiago de Chile? I hope that I don't seem too cynical here, but I have good reason to feel this way. Enough said. RS]

Pages 138-142

MYTHS PROVEN TRUE

The Mansi tribe in the Arctic tundra of Siberia has a legend. Long, long ago a firebird lived with the ancestors and it was so warm that giant trees grew and queer animals grazed. But someone stole the firebird and severe cold and winds set in. The strange trees and animals perished.

This is not a myth but a scientific fact as fossils of prehistoric trees and animals are found in [the] Siberian tundra. Verbal accounts, handed down from generation to generation, can often preserve an amazing degree of accuracy.

[**COMMENT:** Giant trees and queer animals? Did these animals look like tomcats, squirrels, monkeys, bulls or coyotes? All of the above? None of the above? "The strange trees and animals perished" when The Cosmic Tree last detethered and departed. In 1588-87 BCE when Planet X Nibiru arrived last time and the North Pole shifted from the North Atlantic Ocean to the Arctic Ocean, the change was so swift and the Siberian cold so sudden that whole animals froze to death instantaneously. Even the food in their stomachs was undigested because of the abrupt polar cold. Some of these animals were so perfectly preserved that when discovered centuries later and defrosted, their meat was still fresh enough to be eaten by people in the former USSR. See also Chapter 2 of *Slow-Motion Doomsday.* RS]

We have talked a great deal about myths in this book. The average person thinks they are a by-product of fancy. However, this is not always the case. Folklore, as the collective memory of the human race, contains many records of past events, often coloured by ancient story-tellers and unavoidably distorted because of the passing of legends from one generation to another. But not infrequently traditions are actual fossils of history. It is utterly unscientific to discard mythology as a collection of fables. A reality of yesterday is a myth today. The world we live in will be hardly more than a myth itself ten thousand years from now. In that distant future, sages will engage in polemics about the mythical character of legends connected with our vanished civilization.

Until about two hundred and fifty years ago the cities of Herculaneum and Pompeii were nothing but a myth. After their excavation and discovery the two cities became history. When I saw Pompeii, the city seemed to be only asleep.

Among the more fabulous stories of Herodotus is the tale of a distant country where griffins guard a golden treasure. Soviet archaeologists have discovered that country. It is Altai, or Kin Shan in Chinese, meaning the golden mountain. Gold mines have been located there since ancient times. In the valley of Pazyrka scientists have found remains of a high culture. Rich decorations prominently display the griffin. Thus a vague myth about griffins guarding the gold has ceased to be a mere legend.

[**COMMENT:** There is more to this than Mr. Tomas realized. The "griffins guarding the gold" were the Saurian guards at the various "checkpoints" in and around the *Duat*, or "Hidden Place" where the "celestial chariots" landed and departed, climbing up and down "the golden pole" to "the golden city floating in space": Hyperborea, situated beyond the Rhipaean Mountains, which were in turn accessible from the Altais. See Chapter 9 herein and also Chapter 8 of *Slow-Motion Doomsday.* RS]

Although the mountain stronghold of Petra, lost in the desert south of the Dead Sea, was described by Eratosthenes, Pliny, Eusebius and others, in time it became a legendary city. It was only at the dawn of the nineteenth century that Burckhardt gained entrance into the gorge and beheld an edifice carved out of solid rock, an amphitheatre and numerous caves. Once again a fable was turned into fact.

When Heinrich Schliemann started excavations on the mounds of Hissarlik in Asia Minor in 1870 in search of the legendary city of Troy, professors thought he was mad. Yet the *Iliad* of Homer was right — Troy was no myth. Schliemann found ruins of a city which was even more ancient than Troy itself. Eventually Troy's remains were identified to the triumph of Heinrich Schliemann.

The story of Diego de Landa, written in 1566, about the Sacred Well of Sacrifice in Yucatán into which human victims and jewellery were thrown, has been regarded by historians as a mere tale. Then in the nineteenth century an American diplomat and archaeologist E.H. Thompson discovered the well of Cichén-Itzá and validated the old Indian legends.

Six hundred years ago a Chinese ambassador, Chow-Ta-Kwan, submitted to his emperor a report of a fantastic walled city, the hub of a thriving kingdom, south of China, completely lost in the jungle. When the document was published in 1858, Western scholars dismissed it as fiction. Before long a French naturalist — A.H. Mouhot — stumbled upon the remains of Angkor Thom in Indo-China. The description of the legendary jungle city by the mandarin surprisingly corresponded to the actual panorama of Angkor Thom.

[COMMENT: During October 1969, I borrowed a Citroën Deux Chevaux from an Embassy colleague in Saigon. I drove through the Tay Ninh semi-warzone in western central South Vietnam to the Cambodian border and crossed as a tourist. Prince Sihanouk was still in power. The Khmer Rouge were only a minor nuisance, and the Americans had not yet invaded. Crossing that border was like stepping back in time, going from darkness into light. Cambodia was a veritable paradise on Earth back then. I spent over a week driving around Cambodia, including Phnom Penh, Sihanoukville, Angkor Thom and Angkor Wat. I stayed at Phnom Penh's elegant old colonial French Hotel Metropole and gambled at the posh, Monte Carlo style riverboat casino moored in front of Prince Sihanouk's palace. The Phnom Penh International Film Festival was in full swing. At that time remote Angkor was not a popular tourist destination. I had the whole Angkor complex practically all to myself for two full days, driving through the jungles, not seeing another vehicle, smoking "Cambodian Red Bhanga" (legally available in all the farmers' markets), taking photographs, listening only to the animal sounds emanating from the immense, old trees. You wouldn't believe some of the photographs that I still have of this magnificent temple and palace complex! By now the place is probably swarming with tourists. And I brought a large grocery sack full of Cambodian Red back to Vietnam. It lasted me well into the next year after I returned stateside in January 1970. It was still an easy matter to smuggle drugs across borders during those innocent, halcyon days, especially when travelling on an official or diplomatic passport. And then, of course, there still remains all that old "gossip" of the drug smuggling that transpired on the diplomatic "Air America" flights that U.S. Saigon Ambassador Ellsworth Bunker made between there and Kathmandou, Nepal, to visit his wife Clare Booth Luce, who was stationed in Nepal as the U.S. ambassadrix! Aie! Old secrets best left forgotten! Let me add also as an aside that I was one of only three American men in the history of the Vietnam War who risked making that drive from Saigon to Phnom Penh. We three "daredevil explorers" (including Charles Benoit, whose car I used) became Saigon legends in that respect, at least amongst the Americans. The French businessmen and rubber plantation scientists regularly made this road trip, however, and they had assured me that it was not nearly so dangerous as the Americans exaggerated it to be. But skittish Americans *always* exaggerate danger, even until our present day with its vague "terrorist" threats. RS]

When Marco Polo returned to Europe with tales of black stones found in China which burned and heated daily baths, the Venetians of his day only laughed. First of all, no stones could burn, secondly how could anyone in the world afford such a luxury as a hot bath daily? The reader has, no doubt, recognized the reference to coal in Polo's story. His accounts of black oil of the Caspian available in large quantities from the bowels of the earth, were also ridiculed. What were amusing tales to the citizens of Venice, are now scientific facts familiar even to children.

At times it is difficult to ascertain where myth ceases and history commences, or where history ends and myth begins. There is a tendency in scientific circles nowadays to regard mythology and folklore as sources of history. Dr. Carl Sagan, a prominent United States astrophysicist, has successfully proved this point by referring to the voyage of La Perouse to the north-west of America in 1786. The legends of Indians who saw the ships of the navigator, contain amazingly accurate details as to the actual appearance of the French fleet which had visited the lands of these tribes. This shows how an actual event can be preserved in the memory of the masses by verbal transmission from one generation to another.

[COMMENT: Dr. Immanuel Velikovsky, not Carl Sagan, was the pioneer in interpreting myth in terms of history and science. In fact, today's *The Velikovskian* Journal styles itself as "A Journal of Myth, History and Science". RS]

Guatemalan Indians have interesting legends which date to the sixteenth century. When miraculous appearances of gods and their way of life were closely examined by the University of Oklahoma, it became apparent that the mythological beings were none other than the Spanish invaders.

Allowance must be made for inaccuracies, distortions and exaggerations which creep into a tale transmitted through the centuries. Nonetheless, it may contain a kernel of truth and a chronicle of life in bygone epochs. In this light, we should not cast aside legends speaking of a highly advanced civilization of the past which perished in a planetary catastrophe.

Present-day science is gradually reiterating the wisdom of the ages. Have we not proven the correctness of the formula of Democritus — "in reality there is nothing but atoms and space"?

The children of ancient Greece were taught that the earth was a globe floating in infinite space. Their teachers knew about the relative sizes of the sun and the moon, and their approximate distances from the earth. Philosophers delivered lectures in forums about the Milky Way as a conglomeration of stars, each a sun in itself. In colonnaded temples learned men in tunics and togas spoke of life on other planets.

Almost two thousand years later the schoolchildren of Europe were taught that the earth was flat, the centre of creation, and that the stars were holes in the firmament. What right have we then to look down upon the sages of the classic world who possessed more wisdom than the theologians of the Dark Ages?

The tradition of the ancients in regard to a treasure hidden thousands of years ago, is not a myth. If we but take it as a working hypothesis, a great discovery could be made in this century.

Its impact on our life may be stronger than imagined. Evidence of a sudden geological cataclysm that had destroyed Atlantis, will necessitate adjustment in a science which admits no abrupt catastrophes on a planetary scale. History, with so many missing chapters, will gain an undistorted picture of the story of mankind. Our sociology will find out what social and economic systems had existed in the pre-cataclysmic world, and how they had developed — a fact of utmost importance in the modern conflict of ideologies. Archaic instruments or machinery constructed on principles unknown to us, might set our science on a new track. The acquaintance with the beliefs of the vanished race will show the growth of human consciousness. The discovery of a new world in time can be equated with the discovery of an inhabited world in space — both can set mankind in upheaval.

Great revelations have been made by questioning the accepted opinions of the times. Roger Bacon has well diagnosed the causes of human error in his *Opus Majus*: "For every person, in whatever walk of life, both an application to study and in all forms of occupation, arrives at the same conclusion by the three worst arguments, namely, this is a pattern set by our elders, this is the custom, this is the popular belief: therefore it should be held."

Like our predecessors we still live in a mentally conditioned society in which every departure from the recognized mode of thinking is regarded as a revolt against the idols of the time. But thousands upon thousands of people nowadays are beginning to think for themselves on all subjects. To them this book will be more than fiction.

[COMMENT: When this book was published in the 1970s, the best decade of my life in many ways, indeed more and more people were beginning to think for themselves. That changed in the 1980s, and today we are worse off intellectually than at any time that I can remember. But at least now, we have the Internet, making available every idea under the Sun to anyone who seeks to know about it. The contemporary, pathetic, mainstream American "intellect" (if one can call it such) hit rock bottom years ago and now is hopelessly mired in vacuous mindlessness. RS]

Part II

Chasing The Centuries

Chapter 11

Brief History Of Planet X Nibiru

*

This essay was first written in 1996 and subsequently revised in 2003. This latest version has been modified again for inclusion here. Several graphics accompany this chapter. For these and additional color graphics, you can consult this page at my website.

http://www.slowmotiondoomsday.com/nibiru.html

By way of preface, let me state that that this is my own personal theory, albeit it based upon the pioneering research of both Zecharia Sitchin and R.A. Boulay. Mr. Sitchin and Mr. Boulay disagree on the nature of these "gods" (whether they are "humanoid" or "saurian"), and in that debate I fully support the contentions of Mr. Boulay. A list of Mr. Sitchin's books and others is appended below, and a copy of Mr. Boulay's first-edition of *Flying Serpents & Dragons*, with my commentary, can be read at my website in both English and French.

http://www.slowmotiondoomsday.com/flyingserpents.html

Also, the use of royal titles for the hierarchy of these "gods" (Emperor, Prince, Duchess and so forth) is purely of my own invention and is intended to convey to the reader the sort of royal relationships that exist amongst these Nibiruan ruling entities.

After the discovery of the Planet Pluto in 1930 CE, astronomers soon noted that earlier theories regarding the hypothetical influences of such an unknown planet on the orbits of Planets Uranus and Neptune were not validated by the existence of only Pluto. So eventually in the 1970s after computers were becoming commonplace, a computer-generated model of this "Planet X", as it was called, was created. It was determined that Planet X would have to be at least five times bigger than the Planet Earth in order to cause these "perturbations" at Uranus and Neptune. They also calculated the length and shape of its orbit around the Sun as well as the number of years necessary to complete such an orbit.

In 1983 with NASA's cooperation a group of astronomers began a comprehensive survey of the sky with the Infrared Astronomical Satellite (IRAS). In the fall of that year the IRAS discovered several moving objects in the vicinity of this solar system, including 5 previously unknown comets, a few "lost" comets, 4 new asteroids and "an enigmatic comet-like object". Headlines read "Giant Object Mystifies Astronomers" and "Mystery Body Found in Space". In *The Washington Post* were the following headline and story: "At Solar System's Edge Giant Object is a Mystery — A heavenly body possibly as large as the giant planet Jupiter and possibly so close to Earth that it would be part of this solar system has been found in the direction of the Constellation Orion by an orbiting telescope called the IRAS. So mysterious is the object that astronomers do not know if it is a planet, a giant comet, a 'protostar' that never got hot enough to become a star, a distant galaxy so young that it is still in the process of forming its first stars, or a galaxy so shrouded in dust that none of the light cast by its stars ever gets through. 'All I can tell you is that we don't know what it is,' said Gerry Neugebauer, chief IRAS scientist."

The United States Government squashed the story immediately. For some arcane top-secret reason apparently, the government doesn't want to alarm or panic the general public by any "premature" disclosure of this discovery. Why? Because a race of reptilian super-beings inhabits that planet, and common knowledge of this fact could have people screaming in the streets. Pay attention to this history.

Long, long ago, approximately 500,000 years back into the past, perhaps even longer than that, our Planet Earth/Tiamat was a quite different place than it is today. "Tiamat" was this planet's original name. The term "Earth", from the Greek "Gaia", is only a more recent designation. Herein, this planet will be referred to as Planet Tiamat.

A half a million years ago, Tiamat was not located in Space where it is today. It orbited farther out from the Sun, in between the paths of Mars and Jupiter. Mars was orbiting at a distance much closer to the Sun than now and was quite habitable, with a temperate climate and liquid water. This fact has been verified numerous times by NASA and other astronomers. What those scientists do not acknowledge is the fact that Mars had a different orbital pattern than today. Establishment schools do not embrace a cataclysmic evolution of this solar system, at least not publicly.

Then, too, Tiamat's system was closer to the Star Sirius (or Sothis, as the ancient Egyptians called it). This solar system and the Sirius planetary system are part of a unit, or so I have read somewhere before. The two neighboring systems, at only 8.3 light-years distance, are gravitationally connected to one another, a new fact that is now beginning to gain widespread consensus from the scientific community. Probably our "Sirius-Sol Regional Sector" revolves around the Central Sun Alcyone in the Pleiades Cluster, as part of what might be termed the "Pleiades Quadrant", since ancient legends in the Americas and Pacific Islands mention "gods" who arrived here from the Pleiades. This greater quadrant revolves around the Galactic Center, in the direction of the Stars of Sagittarius, once every 200,000,000 years or so. What is significant about our present-day epoch is that certain great cycles relating to orbital alignments within the Pleiades Quadrant and between this Quadrant and the Galactic Center are starting to repeat themselves, and there's nothing that we can do about it. It's happening, folks! Prepare yourselves for the unexpected, for what I have termed "Slow-Motion Doomsday"!

But returning to the history of Nibiru and Tiamat, our planet had a much colder environment back then than it does today. The humanoid population, whom our paleontologists and the like call "Neanderthals", were hardier and hairier than we are. They lived in caves to take advantage of the natural internal warmth of Tiamat. These early Tiamatians may have been directly descended from or seeded by hoary ancestors in the Pleiades, but this "Pleiadian Connection", if any, will not be explored in this book.

This solar system was internally stable 500,000 years ago. But an unanticipated, strange event occurred. After red-giant Osiris (Sirius B) imploded, one of the larger planets from the Sirius System was ejected from that system into our direction. This Planet Nibiru ("Planet of the Crossing") was unwittingly captured by our Sun and thrown into an extremely elongated comet-like orbit, lasting 3,600 of our current years (one Nibiruan *shar*), with its aphelion at the Oort Cloud, the very boundary of the Sun's gravitational field, or perhaps even farther away than that. Perhaps it even transits back and forth between our system and the Sirius System. This we do not know. It is approximately the size of our Planet Neptune. It is populated with a reptilian or "saurian" super-race governed by an élite aristocracy called "Nefilim". The general population is known as "Anunnaki" (Sumerian) or "Anakim" (Old Testament). As a species, they also refer to themselves collectively as "Igigi". At the time of this ejection-capture event, the Planet Nibiru was being ruled by Emperor Alalu and Empress Lilitu.

During its ejection and subsequent capture by our Sun, the Planet Nibiru slowly began to suffer atmospheric deterioration. The governing "Council of Twelve" headed by Emperor Alalu met in emergency session and concluded that in order for their planet to survive, a heat-shield of gold dust would have to be constructed to protect their atmosphere and prevent a cooling off of the planet, which would have been disastrous for a reptilian-based species constantly dependent upon external heat sources for bodily warmth. They began an immediate exploration of their new solar system. A fleet of spacecraft was dispatched to Tiamat and other planets in an attempt to search for gold.

On the other hand, it is entirely possible that this heat-shield had been constructed in anticipation of the negative consequences of the implosion of red-giant Osiris. Certainly these Igigi had eons to ponder this ineluctable event. Following the implosion and their ejection from that system, this heat-shield could have suffered damage, requiring these repairs. The scant remaining Sumerian records of this period do not discuss these heat-shield details.

Commander and Crown-Prince Anu (the Sirian known in old Sumerian legends as "Oannes"), along with his two sons Enki and Enlil and daughter Ninkhursag, landed in what is now the Persian Gulf and first went ashore in what is modern-day Kuwait. Eventually they established a spaceport at Sippar in Mesopotamia, as well as a nearby headquarters complex known in our mythology and religion as E-din or Eden. Guided by Neanderthal Tiamatians, they found abundant sources of gold on Tiamat and successfully created or repaired their Nibiruan heat-shield. They soon realized, however, that this heat-shield required periodic maintenance, in turn forcing them to always maintain a contingent of Anunnaki gold-miners here on Tiamat, constantly shipping more gold home to Planet Nibiru.

After the passage of several Nibiruan orbits around their new Sun, perhaps about 26,000 Tiamatian Years, as their planet's orbit matured, Nibiru passed perilously close to Tiamat and its primeval Moon Kingu. One of Nibiru's host of moons and moonlets crashed into Tiamat in what is now the Pacific Ocean, blowing the lost continent of Lemuria to smithereens, leaving its remnants floating about in Space in what we now so casually refer to as the "Asteroid Belt", and knocking Tiamat and Kingu out of their old orbits and into newer ones closer to the Sun. Tiamat and Kingu finally settled into their current positions, and in all of this chaos the Planet Mars got displaced to a new orbit farther away from the Sun, eventually causing that planet's surface to freeze and die. On the up-side, however, was the fact that this cosmic catastrophe made Nibiru's orbit finally stabilize, so that no more events of this awful magnitude have subsequently occurred.

During this upheaval, Crown-Prince Anu/Oannes and his consort Crown-Princess Antu/Apas became enraged by some of Emperor Alalu's actions and decisions, so the next time that Emperor Alalu and Empress Lilitu visited Tiamat to check for gold-mine damage that might have resulted from the collision, Anu and Antu staged a *coup d'état* and proclaimed themselves Emperor and Empress. They banished deposed Emperor Alalu and Empress Lilitu to Tiamat and forbade them from ever returning to Nibiru. Alalu and Lilitu, believe it or not, may still be alive and living in a palatial underground spaceport in the Grand Teton Mountains near Yellowstone National Park, surrounded by massive hordes of gold and precious gems, and operating an enormous transmitter device (the source of the Taos Hum?) that allows them to stay in communication with Nibiru and its space patrols. It should be emphasized as well that the Nibiruans are not the typical bug-eyed "Greys" that we hear about so often in the media.

CURRENT MEMBERS OF
THE RULING COUNCIL OF TWELVE

Emperor Anu & Empress Antu
Crown-Prince Enlil & Crown-Princess Ninlil
Prince Enki & Princess Ninki
Prince Nannar & Princess Ningal
Prince Utu & Princess-Royal Inanna
Prince Ishkur & Queen Ninkhursag

To digress for a moment at this point, the saurian inhabitants of the Planet Nibiru are anywhere from 10-20 feet (3-6 meters) tall. They have elaborate cranial hair, often multicolored, but very little, if any, body hair, although some of the males do have mustaches and beards. Many of the males have goat-like horns on their heads, and many of the females are winged. They do not sweat and have no body odors, which was one reason they didn't let the Tiamatians do their gold-mining for them. They thought that the noisy, hairy Tiamatian Neanderthals stank, and they didn't like to be around them. They kept to themselves. They can have up to seven pupils in each eye, seven fingers on each hand, and seven toes on each foot. They regularly wear clothing made of pure goldleaf, and their diet mainly consists of elixirs which provide them with their necessary longevity nutrients without their having to consume many solid foods. They seeded Tiamat with many of our current fruits and vegetables, however, and they deserve credit for that. See my companion essay "Nibiruan Physiology" available at my website.

http://www.slowmotiondoomsday.com/physiology.html

Time passed. And the Anunnaki gold-miners got restless, despising their drudgery in the mines, which were primarily located in what is now South Africa, under the command of Nefilim Duke Nergal and Duchess Ereshkigal. The mysterious spiral Ruins of Zimbabwe were possibly a part of this gold-mining consortium. Eventually the Anunnaki miners revolted and refused to do anymore mining. Alarmed, Emperor Anu summoned Queen Ninkhursag to the throne room. Queen Ninkhursag is Nibiru's Chief Medical Officer and Geneticist. Emperor Anu beseeched Queen Ninkhursag to come up with a cloned hybrid Nibiruan-Tiamatian to act as a slave gold-miner. She accepted the challenge and eventually created a crossbred male "savage" from the egg of a Tiamatian female and the sperm of her brother Prince Enki. She referred to this hybrid as "Adamu". At first, only males were produced, cloned one just like the other. A group of Igigi females was ordered to serve as incubation vessels for these "mass-produced" Adamu clones, and these female Anunnaki became known as "Birth Goddesses".

Life went merrily along. Nibiru had its gold; the Anunnaki males were freed from the mines; the Adamu clones went into full-scale production and became excellent gold-mining slaves. But as always happens both here and on Nibiru, a "hitch" developed. It usually does, you know? The Birth Goddesses got fed up with sitting around pregnant all the time with little

Adamu clones, so they revolted and refused to continue with the incubation process. Emperor Anu immediately had Queen Ninkhursag brought back to the throne room and this time ordered her to create a "female Adamu" that could mate with the males and produce their own clones. By now, for Queen Ninkhursag, this was a piece of cake. She created the female "Eva" prototype along with an absolutely perfect Adamu clone. These two were then allowed to romp around naked like pets in the botanical garden of the Eden Palace, in hopes that they would produce a child, a little "bundle of joy" slave gold-miner, or PERFECT ROBOT!

But alas — they were infertile. Just as two mules cannot produce another mule, requiring instead that a horse mate with a donkey, these two hybrid clones could not reproduce. At this point Emperor Anu was beside himself. He had to have gold! So he ordered Queen Ninkhursag to undo the hybridization. She asked her brother Prince Enki to travel to the Eden Palace to do the job for her. Prince Enki had the hybrids Adamu and Eva ingest a chemical substance of some sort that caused them to revert back to more of a Tiamatian than a Nibiruan animal, about two-thirds Tiamatian and one-third Nibiruan. They immediately shed their reptilian outer-skins and began to mate. Realizing that he'd been dealt a major blow, in that as a result of the partial dehybridization he'd lost some of his power over these new creatures, Emperor Anu banished the Adamu and Eva from forever re-entering the Eden Palace grounds.

Cro-Magnon Human was born. It was approximately 20,000 years ago. Neanderthal Man was slowly but inexorably going extinct as a result of Tiamat's closer proximity to the Sun and its warmer climate. By 10,000 BCE, Neanderthals no longer existed, leaving Cro-Magnon to dominate Tiamat.

In his book *The Rainbow Conspiracy*, Brad Steiger writes in connection with the Philadelphia Experiment that U.S. President Franklin Delano Roosevelt met with some "nearly human aliens" in the 1930s. These aliens had a greenish tint to their skin. In order to mingle unnoticed here on Tiamat, they use a bleaching solution to lighten their skin-tone. And then there are all those old drawings of "gods" from India, nearly human-looking "gods" who have a bluish skin-tone. Did you ever look closely at the skin color and quality of anole lizards (faux chameleons)? Their skin is oh-so-smooth and silky! And green! According to a report on TV in August 1996, medical researchers, looking into better ways to develop medicine-delivery skin-patches for sick people, have discovered that snakeskin is extremely similar to human skin. They are developing new medicine-delivery skin-patches from plain old snakeskin. Why is this important? Well, R.A. Boulay in his book *Flying Serpents & Dragons* writes in connection with our Nibiruan ancestors that the snake was a correlative creation that came into existence during the dehybridization process of the Adamu and Eva. What unnecessary or useless reptilian characteristics that had to be "shed" during dehybridization rematerialized as a "serpent". Thus, one might contemplate the following: if an anole or a chameleon can change colors, can a Nibiruan Igigi do the same?

What happened next is our recorded history. Nibiru once again came too close to Tiamat, unleashing a worldwide catastrophe of floods and earthquakes, recounted in The Bible as the Flood of Noah. This time the Eden Palace and Sippar Spaceport themselves were inundated and destroyed. Nefilim Prince Utu was ordered to rebuild the Spaceport in what is now the Sinai Peninsula. Life on Tiamat eventually got back to normal, but then "The Pyramid Wars" erupted.

One of Emperor Anu's favorites was the Princess-Royal Inanna, whom he appointed as ruler of what are now known as India, Pakistan and Nepal. Her Hindi/Sanskrit title is Lakshmi. She is still worshipped there today. Her lover and consort was the most handsome man on Nibiru, the Duke Dumuzi. Duke Dumuzi became involved in a quarrel with Baron Marduk, resulting in the outbreak of the Pyramid Wars. Princess-Royal Inanna and Duke Dumuzi began a protracted power struggle with Baron Marduk and his consort, the Baroness Sarpanit. In the process Duke Dumuzi was murdered by Baron Marduk; and Prince Utu in cahoots with Princess-Royal Inanna blew up the Sinai Spaceport along with the satellite R&R cities of Sodom and Gomorrah. The South African goldmines fell into disarray when Duke Nergal and Duchess Ereshkigal allied themselves with Baron Marduk and Baroness Sarpanit. Once again there was chaos amongst the Nefilim Ruling Council.

See Chapter 13, as well as Zecharia Sitchin's *The Wars Of Gods And Men.*

Emperor Anu was forced to rebuild the Nibiruan Spaceport, this time putting it under the command of his son Prince Enki. Prince Enki and his consort Princess Ninki were sent to what is now the area around Lake Titicaca, Peru, where they rebuilt the complex on the Nazca Plain. Massive new sources of gold and tin (for making bronze) were found in the surrounding Andes Mountains, so the South African mining operation was moved to Lake Titicaca.

And that brings us up to their previous arrival about 3,600 years ago, or one *shar* ago. The last time that their planet approached Tiamat was in 687 BCE, almost 2,700 years ago, when they once again headed out for their long winter hibernation at the Oort Cloud or beyond. But they have continued to maintain a secret presence here.

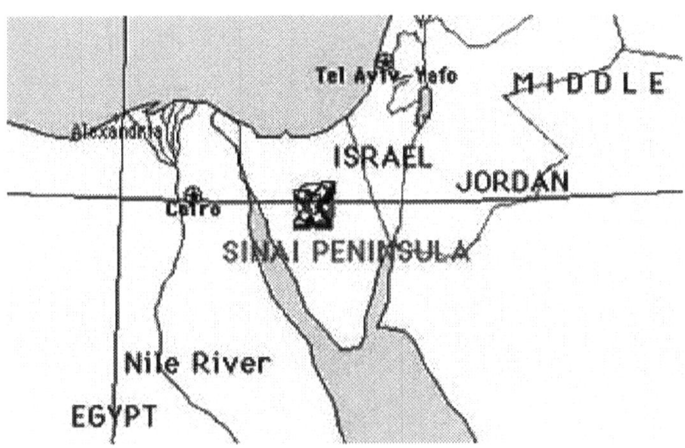

The Nibiruan Sinai Spaceport was located due-east of
The Great Pyramid of Egypt.
The Giza Pyramids pointed the way to the Sinai landing strips.

The blast that destroyed the Spaceport was felt over a large area,
even extending into southern Israel.

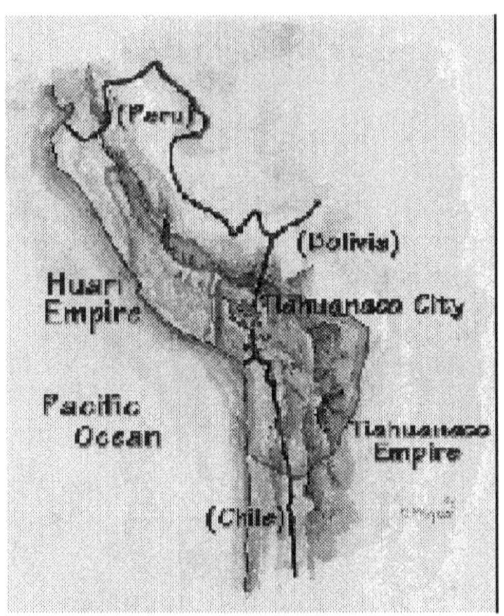

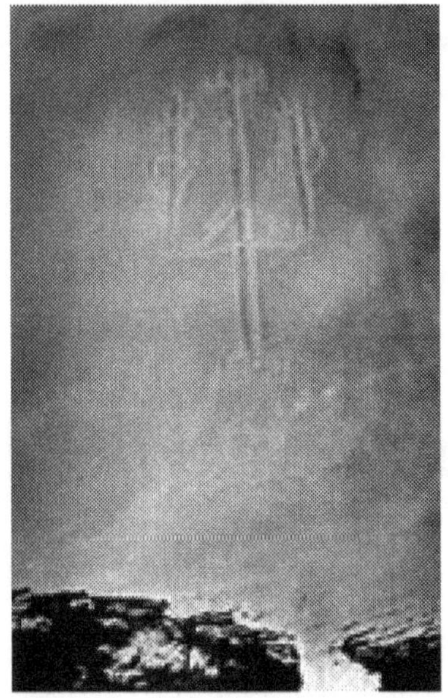

This Trident of Poseidon, which is carved into a cliff at the Bay of Pisco, Peru, points directly towards the Nazca Spaceport.

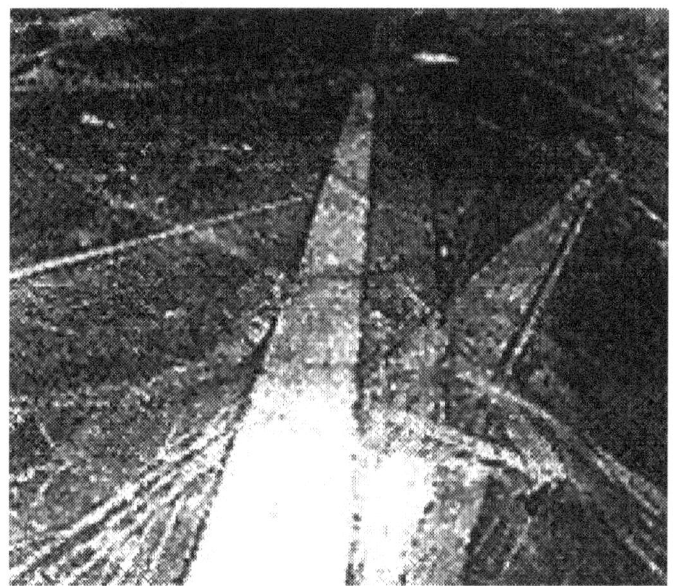

The Nazca Plain, Peru

Much has been written about these "runways" at Nazca.
Clearly they were not "engineered" by primitive Incas.
Those "experts" who tell us this must take us for fools.
These runways and other designs are visible only from the air.

According to William Bramley in *The Gods Of Eden*, in addition to their underground installation beneath the Grand Teton Mountains, these Nibiruan "custodial gods" have other underground facilities in South America, the Saudi Arabian Empty Quarter and the Himalaya Mountains, to name three of them. And we are told by some "occultist" writers that there's another underground chamber just southwest of the Great Pyramid of Egypt, that the Nefilim built in connection with the Sinai Spaceport; but there is a debate of sorts about which extraterrestrial race actually controls access to that most sacred, ancient chamber, which some have speculated contains the hidden records of the Lost Atlantis. The answer to that question, as well as many others, will undoubtedly be forthcoming in the future.

Where is Planet X Nibiru today? Well, it's still out there. Occasionally some astronomer will spot it and refer to it as an enigmatic cosmic object, like an "Object Kowal" or a "mini-galaxy". Our governments definitely suspect that it's there but may be keeping this knowledge officially under wraps, away from the general public to avoid causing any unnecessary early panic over its inevitable return.

One of the things that none of us knows about Planet X is the exact position of its aphelion. Its periodic perihelion is its Earth-synchronized orbit over our North Pole for one-fourth of its total transit, 900 years, one of its "seasons", as it were. But where does it go after that for the remaining 2,700 years of its *shar*? It could be as close as the Planet Jupiter or as far away as the Sirius System.

In *Enuma Elish* we are informed that the "Planet of the Crossing" first becomes visible in the "station of Jupiter". In *Worlds In Collision* Dr. Velikovsky described how a "comet-planet" arrived here from Jupiter at the time of the Israelite Exodus from Egypt and the Santorini Cataclysm, but he concluded that this event represented "The Birth Of Venus" from an explosion at Jupiter, which ejected this new cometary planet into the Solar System. Is Planet X thus "cloaked" right now underneath the vast clouds of Jupiter? We have only speculations about what is beneath Jupiter's clouds. Our singular attempt to send a satellite to penetrate those clouds ended in failure, as the satellite burned up in the Jovian upper-cloud atmosphere. We shall never know what is underneath those clouds until we can actually see it with our own eyes. Jupiter is certainly a big enough place to "warehouse" Planet X Nibiru during its "off-season" from Earth. Does it just suddenly burst forth from Jupiter every 3,600 years and take us by surprise? Is

Jupiter's mysterious Giant Red Spot a "portal" of entry and exit for Planet X Nibiru's Jovian "season of rest"?

But its aphelion could be located near the Oort Cloud at the outer boundary of our Solar System, which would result in an extremely elongated oval-like orbit, as Sitchin has postulated. In this case it would spend 2,700 years transiting from "Hyperborea" to the Oort Cloud and back. Then again, perhaps its aphelion is in the Sirius System, and it "crosses over" from one system to the other twice in every *shar.* It would remain here for 900 years, then transit back to Sirius over a second period of 900 years, where it would remain for 900 years, before taking another 900 years to return to Earth's "Hyperborea", which would be its perihelion from our perspective but its aphelion from the perspective of the Sirius System. We do not have answers to these questions, but for additional information the reader is referred to Chapter 12 ("The Day Of The Lord") of *Planet X Nibiru: Slow-Motion Doomsday.*

In his 2005 book *Dark Star, The Planet X Evidence* (see Appendix A), Andy Lloyd theorizes that Planet X is in orbit around a brown-dwarf star that is a binary companion to our Sun. This "Dark Star" has an orbit completely outside the known planetary Solar System, but on its approach towards Pluto or even Neptune, its satellite Planet X Nibiru periodically becomes visible to observers on Earth. The word "nibiru" can mean "ferry" or "ferryman" as Andy has pointed out on pages 184-185, citing a reference in Appendix 39 of *Hamlet's Mill* by Giorgio de Santillana and Hertha von Dechend (Boston, 1969), quoted here from pages 431-436: "The plain meaning of *nibiru* is 'ferry, ferryman, ford' — *mikis nibiri* is the toll one has to pay for crossing the river — from *eberu*, 'to cross'. ... There have been other identifications (including even a comet!) — the summer solstice, or the celestial North Pole. ... This is all very nice so far, and certainly not without highest interest, but do we know meanwhile what Nibiru, 'ferry, ferryman, ford', was supposed to be? We know it not. ... Nibiru is only one case among many, but it is a rather significant model case for proving that no concrete problem is going to be solved as long as the experts of astronomy are too supercilious to touch 'mythical' ideas — which are firmly believed to be plain nonsense, of course — as long as historians of religion swear to it that stars and planets were smuggled into originally 'healthy' fertility cults and naïve fairy tales only 'very late' — whence these unhealthy subjects should be neglected by principle — and as long as the philologists imagine that familiarity with grammar replaces that scientific knowledge which they lack, and dislike."

Andy's personal interpretation of "ferry" and "ferryman" is included in Appendix A; but to carry his idea of a "Dark Star" one step further, perhaps there is indeed a "Dark Star" located between here and Sirius, and this Dark Star is what "ferries" Planet X Nibiru from one system to the other, giving it "slingshot momentum" to traverse this distance at a high rate of speed. If Planet X Nibiru is the "Planet of the Crossing", then this Dark Star would be the "Ferryman of the Crossing". Again, we are simply confronting a scientific impasse here, but for the record our Sun is the third largest star within a distance of 10 light-years, exceeded in size only by Sirius A and Alpha Centauri A. Thus, our Sun "Sol" would have the third-strongest gravitational effect in this immediate region of the Galaxy and could exert a tremendous influence upon a local interstellar "rogue planet" like Nibiru.

http://www.solstation.com/stars/sol.htm

As noted, Planet X Nibiru arrives from the station of Jupiter. Also, the Mayas recorded that it arrives from the "dark rift" of the Milky Way in the Constellation of Sagittarius. Combining both ideas, we could presume that it arrives from the station of Jupiter when the station of Jupiter is aligned with the Galactic Center at the "dark rift" in Sagittarius. Jupiter's next transit of Sagittarius lasts from 2 December 2007 through 4 January 2009, and it will be directly aligned with the Galactic Center on 22 December 2007. This date conveniently coincides with the December Solstice and the "High Noon" midpoint of this current Nibiruan Day, which began on 21 December 2002 and ends on 21 December 2012. See also Chapter 17. And I want to express my appreciation to my friend Mark McHugh for his assistance with this astronomical data, which accurately updates the more generalized information in *Slow-Motion Doomsday.*

On 29 June 2001 NASA announced the discovery of mysterious "microlensing rogue planets" between Earth and globular cluster M22 which is seen from here in the Constellation of Sagittarius.

"Wandering Mystery Planets" (NASA Science News)
http://science.nasa.gov/headlines/y2001/ast29jun_1.htm?list553051

Are these "wandering mystery planets" Planet X Nibiru and its accompanying "host" of planetoids and moons? As Jupiter transits Sagittarius, it will go into alignment with M22 around the middle of 2008, about the time of the June Solstice. Will that be the point at which Nibiru rapidly becomes visible in the station of Jupiter during Jupiter's transit of Sagittarius? It seems from history that suddenly it's just up there, like a golden miniature second sun with a lashing, fearsome cometary tail trailing along behind it and its host of satellites, before hovering, suspended like a jewel over the Tiamatian North Pole for a "millennium of the gods". The Cosmic Tree. The World Tree. The Sacred Tree. The World Mountain. Mount Olympus. Mount Meru. Mount Zion. Hyperborea, the North Country, "land beyond the mountains where the North Wind rises"!

"My cloak I shed in the light of my reflection. My grand entrance I make after the serpent appears. Then they will know me." Old Greek Riddle from the Isle of Samos, Birthplace of Pythagoras.

The material presented here in only the tip of the iceberg of our available knowledge of Planet X Nibiru and its stormy interaction with Planet Earth-Tiamat. If you would wish to obtain additional information, you are referred to the following books, including all of the Sitchin material in ten volumes consisting of more than 3,500 pages total.

Genesis Revisited by Zecharia Sitchin
Divine Encounters by Zecharia Sitchin
Of Heaven & Earth by Zecharia Sitchin
The Earth Chronicles by Zecharia Sitchin
1 — *The Twelfth Planet*
2 — *The Stairway To Heaven*
3 — *The Wars Of Gods And Men*
4 — *The Lost Realms*
5 — *When Time Began*
6 — *The Cosmic Code*
7 — *The Lost Book Of Enki*

Flying Serpents & Dragons by René A. Boulay
The Gods Of Eden by William Bramley
Dark Star, The Planet X Evidence by Andy Lloyd
The Sirius Mystery by Robert K.G. Temple
The Sirius Connection by Murry Hope
Chariots Of The Gods by Erich Von Däniken
Gods From Outer Space by Erich Von Däniken
Gold Of The Gods by Erich Von Däniken
The Home Of The Gods by Andrew Tomas
We Are Not The First by Andrew Tomas

Not Of This World by Peter Kolosimo
In Search Of Ancient Mysteries by Alan & Sally Landsburg
The Cosmic Conspiracy by Stan Deyo
Fingerprints Of The Gods by Graham Hancock
Gods Of The New Millennium by Alan F. Alford
The Stellar Man by John Baines (Darío Salas Sommer)
The Rainbow Conspiracy by Brad Steiger
The Hidden History Of The Human Race by Michael A. Cremo & Richard L. Thompson

Many other books with similar themes are also available.

Chapter 12

Names Of The Gods

*

In my opinion, one of the unique and most fascinating ideas of Zecharia Sitchin in *The Earth Chronicles* was his comparison of the Nibiruan "custodial gods" (as William Bramley refers to them) with personages in our various mythological traditions. In his *Flying Serpents & Dragons* R.A. Boulay pursued this idea further. Sitchin equated Nibiruan Crown-Prince Enlil with the Egyptian Osiris, the Greek Zeus, the Teutonic Thor and the Hindi Indra. He equated Prince Enki with the Egyptian Seth and the Greek Poseidon, and Prince Nannar with the Egyptian Thoth, the Greek Hermes and the Mayan Quetzalcoatl. Reading such comparisons (and totally agreeing with them, I might add), I was immediately struck with the logical implication that if *some* of these "custodial gods" equaled personages in various mythologies, then *all* of them would have comparable equivalents in these same and other mythologies.

Thus, in 1994-95, after having finished *The Earth Chronicles*, I began an investigation into several mythological traditions to determine the names of all the gods and goddesses in relationship to the ruling élite of Planet X Nibiru. Primarily I used the following reference sources: *A Dictionary Of Greek And Roman Biography And Mythology* by Professor William Smith (3 Volumes, London, 1890); *New Larousse Encyclopedia Of Mythology* (London, New York, 1978); *The Gods Of The Egyptians* by Sir E.A. Wallis Budge (2 Volumes, New York, 1969); *Ye Gods!* by Anne S. Baumgartner

(Secaucus, New Jersey, 1984); and *Mythology* by Edith Hamilton (New York, 1961).

An abbreviated version of this information has been available at my websites ever since, but only recently did I decide to revise and expand it and include it here, as a more permanent record than a mere file on the Internet. I had fully intended to go back to all the original references and try to double-check some of the vaguer of these identifications; but once I attempted to embark upon this task, I found myself mired in confusion and abandoned the effort. Therefore, the comparisons which follow are mostly identical with the original comparisons that I made a decade ago. Looking over all of them again in preparation for this book, I concluded in hindsight that probably about 75% of the comparisons are valid. Nobody is an expert on *all* the world's mythological traditions, and to my knowledge no one has ever even attempted such a feat as this before. I consider my work here to be "pioneering" in that regard, and it would be easy for interested authorities in these various mythological traditions to refine my data. And I would most welcome any improvement and commentary on it!

When this information was first published by Barbara Marciniak in her *Pleiadian Times* in 1995, she received letters from people who were critical of some of my comparisons. She received a long letter from a woman in Mexico who bitterly complained about my Mayan comparisons, saying that I had confused Toltec names with later Aztec names, putting them into the same list, whereas they should have been placed into separate lists. In the Indian mythological traditions there are also clearly two sets of names of the gods. Here I have attempted to correlate them into an older group (Sanskrit) and a later group (Hindi), but again I may have inadvertently confused the two groups. And some of the gods in the Hindi category, such as Mitras and Ahriman, also show up in ancient Persian mythology. As for the Chinese and Japanese names, since there are numerous gods and goddesses in both traditions, neither of which is a field of expertise of mine, I make only a few suggestions, leaving more accurate comparisons to authorities in those schools.

As I researched one mythology after another, certain, as it were, "plot-line elements" began to emerge from culture to culture. There are always several "olden gods", one of whom fathers a family of quarrelsome offspring. Two brothers fight for control of Earth. One brother is usually a "god of thunder and lightning", whilst the other is a "god of the sea". There are always a kind and loving "great mother goddess" who is sometimes credited with the creation of humanity, a mighty "god of war", and a gorgeous "goddess of sex, love and beauty". Fearsome "gods of the underworld" permeate each mythology, and generally there's at least one set of twins, a male and a female. The further this research progressed, the easier it became to match "characters" across the various traditions.

Also, as a trained linguist, I immediately looked for any linguistic similarities amongst the mythologies. Just as today, for instance, we have the female names Mary, Marie, Maria, Marya, Mariam, and the male derivative Mario; or the male names Paul, Paulo, Paolo, Pablo, Paulus, Paulick, Pavlov, Pavel, and the female derivatives Paula and Pauline in our modern languages — so also do we find similarities in the international names of the custodial gods and goddesses. Take Empress Antu as an example. She was also known in Mesopotamia as Apsu, in the Levant as Apas, in Rome as Ops, in India as Apo, in Mayan as Akna, and in Celtic as Danu or Anu.

Three other significant linguistic comparisons should be noted here. In the Teutonic sections of the charts, Hel is the name of Duchess Ereshkigal, the consort of Duke Nergal who is known in Teutonic as Midgard or Loki. They were rulers of the "underworld" or "partially lighted regions" of "Hell", in southern Africa opposite the northern Night Sun. (See Chapters 6-7 of *Slow-Motion Doomsday*.) There is also the Teutonic name Niflhel, which I have equated with Baron Ninurta; and one of the "mythical" regions of the Teutonic world was called Niflheim. These words Niflhel and Niflheim greatly resemble the Biblical term Nefilim for those "giants" who came to Earth from the heavens, *i.e.*, the inhabitants of Planet X Nibiru. See Genesis 6:1-5.

N F H L
N F H M
N F L M

In 2003 a strange "star map" was discovered in Germany, and this discovery was featured in the January 2004 issue of *National Geographic* Magazine, page 76: "Bronze Age Map Of The Stars". This article begins as follows: "Buried on the Mittelberg hill near the town of Nebra in 1600 B.C., the sky disk shines at sunset in central Germany. The disk, which tracks the sun's movement along the horizon, contains the oldest known depiction of the night sky and may have served as an agricultural and spiritual calendar." Noteworthy is the fact that this "Nebra Sky-Disk" is dated at about 1600 BCE, or exactly one *shar* ago, and the words Nebra and Nibiru are linguistically identical.

N B R
N B R

Finally, in the Finnish creation epic *Kalevala*, there is a "Maiden of Northland" named Annikki. She is the sister of the mythical hero Ilmarinen. This "Northland" was certainly "Hyperborea", yet another reference to the "North Mountain" or "North Country". But even more important is the linguistic similarity of the words Annikki and Anunnaki. Clearly the maiden's name was derived from the general term Anunnaki for the inhabitants of Planet X, whose station is "beyond" the North Pole.

A N N N K
A N N K K

This linguistic radical also resembles the Egyptian word Ankh, a symbol that predated the Christian cross and which is an "amulet" of sorts depicting The Cosmic Tree. The word Anunnaki is also identical with the Old Testament "tribe" of Anakim, who are mentioned several times during the period of the last Cosmic Tree. See, for example, Numbers 13:22-33, Deuteronomy 1:28 and Joshua 15:13-14.

The physical characteristics and abilities attributed to these gods and goddesses in the various mythological traditions often read like fantasy, and one wonders how much of it is actually true and how much is fictitious legend. Provided here are two examples which I found particularly intriguing, those of the Celtic Cú Chulainn and the Mayan Tezcatlipoca.

In the *New Larousse Encyclopedia of Mythology* on page 233, we find the following description of a "god" named Cú Chulainn from the Celtic mythological tradition, the equivalent in my opinion of the Nibiruan Baron Ninurta: "In his normal state Cú Chulainn is pictured as a young man with well-defined physical attributes. He had seven pupils in each eye, seven fingers on each hand and seven toes on each foot. His cheeks were multi-coloured, yellow, blue, green and red. His long dark hair was of three tints, dark close to the roots, red in the middle and lighter in colour towards the tips, suggestive of the practice of the Celts of smearing their hair with a thick wash of lime. Rich and gorgeous jewellery adorned him, a hundred strings of pearls on his head, a hundred golden breast ornaments. Far different was Cú Chulainn in his battle-frenzy when his body was seized by contortions. He turned round in his skin so that his feet and knees were to the rear and calves and buttocks to the front. His long hair stood on end and on the tip of each hair was a spot of blood or a spark of fire. From his open mouth spurted fire and from the top of his head a jet of black blood rose mast high. One eye receded far back into his skull while the other protruded onto his cheek. Finally, on his forehead appeared the 'hero's moon', a strange inexplicable sign. When in this state Cú Chulainn's fury was uncontrollable, and he needed to be plunged into three vats of cold water before he could be pacified."

When a theory is inherently correct, then all of the loose evidence will easily fall into place like clockwork. Nowhere is this more apparent than in the writings of Dr. Immanuel Velikovsky. In the "Sky Watch" daily column from the University of Texas McDonald Observatory, published by *The Dallas Morning News*, the following item appeared on 28 July 2004.

http://www.as.utexas.edu/mcdonald/mcdonald.html

"The Big Dipper is visible every night of the year, circling the North Star, Polaris. To the Aztecs, the Big Dipper personified the god Tezcatlipoca, HE WHO CAN ENTER ALL PLACES. He reigned over the cardinal directions as well as the night."

In the *Kalevala* there are numerous references to the Great Bear (Big Dipper) that rests in the branches of The Cosmic Tree, or Sampo. This Constellation is associated with Planet X Nibiru. The phrase "he who can enter all places" mystifies me, but perhaps it refers to the superiority of these "gods" over mere humans, or perhaps it refers to the "other place" occupied by Planet X when it is not tethered here. Googling, I found:

http://www.pantheon.org/articles/t/tezcatlipoca.html

Tezcatlipoca
By Micha F. Lindemans

"Tezcatlipoca was the Aztec god of night and all material things. He carried a magic mirror that gave off smoke and killed enemies, and so he was called 'god of smoking mirror'. He was god of the north. As lord of the world and the natural forces, he was the opponent of the spiritual Quetzalcoatl, and sometimes appeared as a tempter, urging men to evil. Punishing evil and rewarding goodness, he tested men's minds with temptations, rather than trying to lead them into wickedness. He was also god of beauty and war, the lord of heroes and lovely girls. He once seduced the goddess of flowers, Xochiquetzal, wife of the god Xochipilli, because such a lovely goddess was a good match for him, being a handsome war-like god. Yet he appeared most frequently as a magician, a shape-shifter and a god of mysterious powers."

In my accompanying charts, I have equated Tezcatlipoca with Nibiruan Baron Marduk who attempted to stage a *coup d'état* against Emperor Anu and Empress Antu; but other personality traits of this god bring call to mind aspects of Apollo and Hermes. The term "god of the smoking mirror" is quite intriguing. Is that analogous to, or from where we obtained, our English expression "smoke and mirrors"? What sort of "smoking mirror" could this be? Could it refer to the powerful MEs that Baron Marduk and his wicked co-conspiratorial consort Baroness Sarpanit stole from Anu and Antu?

The reader can Google for much more information about all the names found here in my charts, where only the 24 most significant of the custodial gods and goddesses are examined; and the question inevitably arises as to exactly why there are so many different names for them in all the mythological traditions, some names of which are completely dissimilar from culture to culture. The answer can perhaps be found in Sitchin's work. After Queen Ninkhursag successfully created the Cro-Magnon prototypes, the Adamu and the Eva, and determined that they were fertile and able to procreate the planet with generations of offspring, undoubtedly all of these "slaves" spoke the same language, probably that of the Nibiruan Overlords themselves. But then these Adamus and Evas began to progress technologically apart from their creators. They began to envy the scientific superiority of those who could create a Cosmic Tree "beyond the north" and began to construct their own "stairway to heaven" at Babel, Mesopotamia. Possibly fearing future competitors, these Nibiruan Overlords moved at once to destroy humanity's futile achievement, scatter people to the far ends of the Earth, and teach them new languages to prevent their further communicating with one another. This "legend" of the Tower of Babel is preserved in our Old Testament.

But today, in great part as a result of our Internet, millions of people communicate daily primarily in the English language. We are easily able once again to share ideas on a worldwide basis. Thus, following the forthcoming Crossover and new Cosmic Tree, we should be alert to their negative intentions and not fall prey to the illusions that they'll certainly attempt once again to foist upon us in order to perpetuate their now very tenuous control of our existence. They need us for whatever reason they may still have, and by now we *know* that they need us. We should never let them take advantage of us again.

Of note for the record, when I first completed my chart in 1995, I had matched the Hebrew/Levantine God Yahweh with Nibiruan Crown-Prince Enlil, as there seemed to be no other logical pairing. Subsequently in reading R.A. Boulay's *Flying Serpents & Dragons*, I found that Boulay had also equated Enlil with Yahweh. Then in 1996 Sitchin published *Divine Encounters*, to which he appended an Endpaper titled "God, the Extraterrestrial". The first sentence reads: "So, who was Yahweh?" He then proceeded to analyze several of the more important Nibiruan Overlords (Enlil, Enki, Ishkur, Ninurta, Nannar, Marduk) to determine which one might be Yahweh. He could make no valid comparisons and ultimately decided that Yahweh is not amongst the Nibiruan pantheon but exists on a higher plane than either we or they. Personally I thought that this was a royal "cop-out" on Sitchin's part, and I suspect that he had ulterior motives for coming to this conclusion.

The charts on the following four pages are image files, which were the most convenient and efficient way for me to present them in this book. They include 12 gods and 12 goddesses denoted by (m) for male and (f) for female. Current members of the Council of Twelve are indicated by an asterisk (*).

NIBIRUAN	Alalu (m)	Lilitu (f)	Anu (m)*	Antu (f)*	Enlil (m)*	Ninlil (f)*
SUMERIAN	Teshub	Kali	Anshar	Ninmah/Apsu	Asshur	Ninlil?
EGYPTIAN	Shu	Tefnut	Geb/Seb	Nut/Neith	Osiris	Ma'at
GREEK	Uranus	Gaea	Kronos	Rhea	Zeus	Maia
ROMAN	Caelus	Terra	Saturn	Ops	Jupiter	Majesta
LEVANTINE	Elohim?	Lilith	Leviathan	Apas	Yahweh	Mariam
SANSKRIT	Tvashtri	Prithivi?	Kasyapa	Aditi	Indra	Indrani
HINDI	Mazda	Sarasvati	Brahma	Apo	Vishnu	Maya
CELTIC	Partolón	Tailtiu	Eochaird	Danu/Anu	Dagda	Gwydian
TEUTONIC	Ymir	Audumla	Buri/Mannus	Tuisto	Thor/Donar	Jørd
SLAVONIC	Byelun	?	Svarog	Mati-Zemlya	Dazhbog	Myesyats
FINNO-UGRIC	Jumala	Louhi	Ukko	Ahka/Rauni	Ilma	Ilmatar
MAYA	Tonatiuh	Tzinteotl	Akanchob	Akna	Tlaloc	Chalchiuhtlicue
INCA	Monan	?	Inti/Punchau	Quilla	Catequil/Pillan	Cuycha
CHINESE	Pao-Yu	Ching-Te	Yu-Huang	?	Lei-Kung	Tien-Mu
JAPANESE	Izanagi	Izanami	Susanoo	Amaterasu	Ninigi	Kono-Hana-Sakuya

NIBIRUAN	Enki (m)*	Ninki (f)*	Nannar (m)*	Ningal (f)*	Ishkur (m)*	Ninkhursag (f)*
SUMERIAN	Anshar	Nintud/Apsu	Sin	Ninsha	Adad	Ninmah
EGYPTIAN	Seth	Nephthys	Thoth	Sesheta	Horus	Isis
GREEK	Poseidon	Libya	Hermes	Leto	Ares	Hera
ROMAN	Neptune	Africa	Mercury	Latona	Mars	Juno
LEVANTINE	Satan/Lucifer	Demeter?	Kristos?	?	Aria	Fauna?
SANSKRIT	Agni	Svaha	Sudhanvan	Saranya	Mitra	Surya?
HINDI	Ahriman/Shiva	Ganga	Buddha	Yasodhara	Mitras	Hariti
CELTIC	Manannán	?	Lug/Llew	?	Angus/Og	?
TEUTONIC	Hoenir	?	Oden/Wodin	Sif	Tiw/Tyr/Ve	Edun/Bestla
SLAVONIC	Ogon	Marzanna	Volga	?	Perun	Dennitsa
FINNO-UGRIC	Ahti	Vellamo	Vogul	Vizanya	Arianrod	Mader-Akka
MAYA	Itzamna	?	Quetzalcoatl	Xilonen	Huitzilopochtli	Coatlicue
INCA	Viracocha	Cocha	Pihuechenyi	?	Pachacamac	Pachamama
CHINESE	Lung-Wang	T'ien-Hou	Han-Hsiang-Tzu	?	Kuan-Ti	Nu-Kua
JAPANESE	O-Wata-Tzu-Mi	Haigi-No-Kami	Shinatsu-Hiko	?	Bishamonten	Hariti-Kismojin

NIBIRUAN	Utu (m)*	Inanna (f)*	Dumuzi (m)	Shala (f)	Nergal (m)	Ereshkigal (f)
SUMERIAN	Shamash	Ishtar	Eshmun	Ninsha	Irrigal/Gaga	Ereshkigal?
EGYPTIAN	Harpocrates?	Hathor	Aten-Ra	Uatchet	Anubis	Serqet
GREEK	Apollo	Aphrodite	Adonis	Hecate	Hades	Persephone
ROMAN	Helios	Venus	Priapus	Luna	Pluto	Prosepina
LEVANTINE	?	Astarte	Tammuz	Hestia/Vesta?	Beelzebub	Pyrtania
SANSKRIT	Rudra	Ushas	?	?	Varuna	?
HINDI	Rama	Lakshmi	Sri Krishna	Sivaha	Yama/Siva?	Parvati
CELTIC	Nuada/Lud	Brigit	Finn	?	Culann	?
TEUTONIC	Ull/Magni	Friga/Frigg	Modi	?	Midgard/Loki	Hel
SLAVONIC	Varpulis	Kupala	Yarilo	?	Erisvorsh	?
FINNO-UGRIC	?	?	?	?	Tuoni	Tuonetar
MAYA	Xochipili	Tlazolteotl	Bachlum-Cham	Chikomecoatl	Mictlantecuhtli	Mictanchihuatl
INCA	Tamendonare	Chasca	Huachlepenyi	?	Supai/Annwn	Nina
CHINESE	I (letter i)	Chih-Nii	?	?	Yen-Wang-Yeh	?
JAPANESE	?	Benzaiten	?	?	Emma-Hoo	?

NIBIRUAN	Marduk (m)	Sarpanit (f)	Ninurta (m)	Bau (f)	Nabu (m)	Aya (f)
SUMERIAN	Merodach	Ninharrak	Ningirsu	Gula	Nabo	Kingu
EGYPTIAN	Amon-Ra	Nut-Bast	Ptah	Sekhet	Bakha	Mehurit
GREEK	Belus	Anchinoe	Hephaestus	Athena	Dionysus	Artemis
ROMAN	Bel	Achiroe	Vulcan/Typhon	Minerva	Bacchus	Diana
LEVANTINE	Ba'al	Belit	Mikael?	?	Bassareus	?
SANSKRIT	Vritra	?	Yavishtha	?	Soma	Prisni
HINDI	Karttikeya	Tara	Ananta	Manesa	Haoma	Sita
CELTIC	Belenus/Fal	?	Cú Chulainn	?	Ossian	?
TEUTONIC	Balder/Frey	Gerda	Niflhel	Nerthus	Thyrm	Freyja/Njørd?
SLAVONIC	Volos/Vlas	Rusalka	Krukis	Zorya	Sadko	Devana
FINNO-UGRIC	?	?	Sar-Akha/Rot	?	?	?
MAYA	Tezcatlipoca	Meztli	Chin/Xipetotec	?	?	Xochiquetzal
INCA	Ngurvilu	Nandecy	Urcaguary	Cuna	?	?
CHINESE	?	?	?	?	Fu-Hsing	Ch'ang-O
JAPANESE	?	?	Shina-Tsu-Hiko	Shina-To-Be	?	?

Chapter 13

Vengeful Birth Of Lord Hellespontiacus

*

This short piece was first written for Halloween 1996. I mixed facts from mythological sources with Zecharia Sitchin's identifications of the "gods", and then I wrote it like a short story. Please indulge me here.

After Emperor Anu and Princess Inanna ended their love affair, she cruised around the Sirius Sector in her personal Vimana spacecraft. Eventually she rekindled her never-ending passion for ultra-handsome, ultra-risqué Nibiruan Adonis Duke Dumuzi — she "the Miss Universe", he "the sexiest man alive". Their torrid interstellar romance was "the talk of the Sector". This was at a period in time prior to Inanna's elevation in rank to Princess-Royal and the Council, preceding the opening salvos of the Pyramid Wars.

Duke Dumuzi was the son a Nibiruan couple named Phoenix and Alphiboea. After he was born, Inanna noticed him and was captivated by the sheer beauty of this infant. So, she kidnapped him, hid him in a wooden chest and gave the chest to Queen (now Duchess) Ereshkigal for safekeeping. But when Ereshkigal realized what a treasure that she had in her possession, she refused to give Dumuzi back to Inanna.

Inevitably, the dispute was brought before Emperor Anu. After hearing all the arguments, Anu decided that for four months of every year Dumuzi could be by himself, for four months with Queen Ereshkigal, and for four months with Princess Inanna. Of course, Duke Dumuzi chose to spend the four months that he had for himself with voluptuous Inanna. Their ensuing love affair lasted until his murder by Baron Marduk, setting off the first of the three Pyramid Wars.

Duke Dumuzi, also known as Tammuz, was a "lunar god", as Inanna was a "lunar goddess". He was known as "the shepherd of the flocks of the stars". According to some, this idea is closely bound up with the passion of Tammuz (as Adonis) for Ishtar (as Aphrodite), because of the relationship between them and the Phases of the Moon. Dumuzi was also the "Ferryman of Souls to the Land of the Dead" — the psychopomp and a symbol of supreme power, since flocks are representative of cosmic forces. This would indicate that Duke Dumuzi was under the immediate command of King (now Duke) Nergal, who was also known as Hades/Pluto, the one in charge of the "lakes of fire" at the South African gold smelters; Dumuzi was probably the man responsible for rounding up a "flock" of Adamus, herding them onto "celestial boats", and "ferrying" them down to the "Underworld". And this could also explain why soon after Dumuzi's murder, Inanna hightailed it to South Africa to enlist the support of Nergal and Ereshkigal in her war with Baron Marduk.

Before proceeding further, here is some genealogical information about these personages. Nergal is the eldest brother of Inanna; Utu, her elder brother. Their father and mother are Prince Nannar and Princess Ningal. Prince Nannar, in turn, is the son of Crown-Prince Enlil and Crown-Princess Ninlil. Enlil is the son of Anu and Antu. So Inanna is the great-granddaughter of Anu and the sister-in-law of Ereshkigal, the close confidante of the Empress.

Princess Inanna was Nibiruan Airfleet Commander at the Baalbek Airport. She was in charge of all air traffic on Tiamat. Her brother Prince Utu was Nibiruan Space Commander at the Sinai Spaceport. If you think of these individuals in terms of astronaut rankings, Anu is Admiral and Antu is Vice-Admiral. All the others would be subordinate to them.

Because Inanna's work as Airfleet Commander kept her stationed primarily at Baalbek, Duke Dumuzi was left in charge of her Shangri-La Palace of Lakshmi in the Land of Indra. He flew back and forth a lot on business. During his absences, Inanna began a clandestine love affair with her brother Prince Utu. As Air and Space Commanders, they were together a lot at work. This secret *liaison dangereux* went on for many years. One can just imagine them as a couple in disguise, going out on the town in the Spaceport Satellite R&R Cities of Sodom and Gomorrah, to mingle with the natives, to stir things up a little bit and give everybody some juicy gossip. The jaded sybarites of Sodom thought it was "fun" to "mate with a son of God" or "know an angel". There were lots of unwanted hybrids in the Negev Desert. The Philistine "giant" who fought King David was undoubtedly one of these crossbreeds. Don't forget also that during the Pyramid Wars, Space Commander Utu was the chief ally of Airfleet Commander Inanna, even blowing up the Sinai Spaceport and Satellite Cities to assist her cause against the villainous Baron Marduk and his co-conspiratorial consort, the wicked Baroness Sarpanit.

Now here's a little background information on Prince Utu. As we know from Zecharia Sitchin's books, Utu is the equivalent of the Greek God Apollo; thus, his parents in Nibiruan history would be Nannar/Hermes and Ningal/Leto. However, Homer reported that Apollo was the son of Zeus/Enlil and Leto/Ningal and was the twin brother of Artemis/Aya. But Herodotus wrote that the Egyptians believed that Apollo was the son of Queen Ninkhursag/Isis and Baron Nabu/Dionysus, an odd combination to say the least. In all of the Greek, Roman and Levantine mythological traditions, Inanna / Ishtar / Aphrodite / Venus was the lover of both Utu / Apollo / Helios and Dumuzi / Tammuz / Adonis / Priapus.

Prince Utu resided in "Hyperborea", which was also known as "The North Country" and "the land beyond the mountains where the North Wind rises". Every once in a while this hidden Land of Hyperborea will surface again in some moldy reference volume from the back of the library shelves. Where was Hyperborea? It certainly wasn't the North Polar Zone, or even Britain as some have suggested, so it must have been another ethnic name for the Planet X Nibiru, stationed like a gigantic "Heaven" above our North Pole, visible to all but those in the "Underworld" (below the Equator). Its planetoids and moonlets revolve around it, painting patterns of pastel colors in the firmament above us. It must be a spectacularly magnificent sight to see, especially during an intense Aurora Borealis display.

Prince Utu's official emblem was that of a god holding a bow and arrow, the supreme "averter of evil". Away from his post at the Spaceport, however, Prince Utu was a most cultured man in other ways. He was known as "the god of music and song", and it is said that he invented the flute and the lyre. Of all the Gods in the Greek Pantheon, the Greeks ranked Prince

Utu/Apollo the highest and gave him the status of a "sun god". Some historians even feel that Prince Utu represented in himself the best of the mind of the Greeks and that their civilization could not have risen to the heights of culture that it did without the assistance of Prince Utu.

The most beautiful of the extant representations of Prince Utu are the Apollo of Belvedere at Rome, which was discovered in 1503 CE at Rettuno, and the Apollino at Florence. In the Apollo of Belvedere, the god is represented with commanding but serene majesty; sublime intellect and physical beauty that are combined in a most wonderful manner. The forehead is higher than in other ancient figures, and on it there is a pair of locks, whilst the rest of his hair flows freely down his neck. The limbs are well-proportioned and harmonious, the muscles are not worked out too strongly, and at the hips the figure is rather thin in proportion to the chest.

Once more, Duke Dumuzi had to fly to Indra on routine business, leaving Inanna and Utu alone again. By the time that Dumuzi returned, Inanna found herself unexpectedly pregnant. It is unclear who the father was, probably either Dumuzi or Utu; maybe Inanna herself didn't know for sure. But this time their cover was blown, and gossip about the illicitly incestuous Inanna-Utu romance spread quickly amongst the ruling Council of Twelve and the Nibiruan aristocratic élite.

Empress/Vice-Admiral Antu was one of the first to be informed about it. She was furious! Not only were there already too many good-looking royals running around all over the place, but Inanna's condition would put her out of commission for a time, away from her responsibility as Airfleet Commander.

Undoubtedly, the Empress gossiped about this turn of events with her friend Queen Ereshkigal. Ereshkigal and Inanna were rivals from several perspectives, another of which, it might be added, was that whereas Inanna was the ultimate in feminine celestial beauty, elderly Ereshkigal was a typical "old crone" or "witch". It is unclear from the historical records whether this ugliness resulted from biological design by her progenitors or was purely an accident of Nibiruan genetics. But she was jealous of Inanna and would jump at any chance to "get back at her". You can be certain that old Queen Ereshkigal put in her "two cents' worth". She certainly goaded Empress Antu on, into what quickly became the Empress' secret plot of vengeance against promiscuous, haughty Inanna, her consort Anu's former lover.

The highest-ranking male and female on the Council of Twelve and on all of Planet X Nibiru possess a certain magical ability that allows them to "wield a sword" over everybody else, keeping all the others in line with a threat of bizarre chastisement for any "wrongdoing". The Emperor and Empress, using their magical MEs, can physically touch another Nibiruan, on the arm, for example, and "will" that something should happen to that individual, for good or ill.

Empress Antu summoned Princess Inanna to the throne room, ostensibly to ooh-and-aah over the impending arrival of a beautiful new royal. During this meeting, Antu touched Inanna and willed her wish: that Inanna's newborn bastard would be the ugliest creature this side of the Pleiades. Lovely Inanna, of course, was totally unsuspecting.

When it came time for reptilian Nibiruan Princess Inanna to lay her egg, she flew to what is now the Dardanelles Strait of Türkiye, formerly called the Hellespont, that narrow oceanic divider of Europe and Asia, about a day's drive southwest of modern Istanbul. That part of Türkiye was then at the southernmost stretches of the Land of the Amazons and Lesbos, a land which snaked around the northern Turkish seacoasts from the Sea of Marmara and the Black Sea to the Caucasus Mountains of present Armenia. These Amazons and Lesbos were probably the legendary "Birth Goddesses". They most likely supervised certain large, safe hatcheries for soon-to-break-forth little Nibiruan royals and Adamus.

Young Lord Hellespontiacus finally hatched out of his shell. He was a child of extreme ugliness with unusually large genitals. His gorgeous mother Inanna was horrified. But his name was given to the Hellespont; and he eventually became known as "the promoter of fertility" both of the vegetation and of all animals connected with an agricultural life; and in this capacity he was worshipped as the protector of flocks of sheep and goats, of bees, the vine, all garden produce, and even of fishing. Like other divinities presiding over agricultural pursuits, he was believed to be possessed of prophetic powers and is sometimes mentioned in the plural.

Lord Hellespontiacus was represented in carved images, mostly in the form of *hermae*, with very large genitals, carrying fruit in his garment and either a sickle or cornucopia in his hand. His ultimate fate is unknown, buried safely away from our modern view, another sordid, scandalous skeleton in the long-forgotten dusty closets of the inscrutable Planet X Nibiru.

Chapter 14

Saturn & The Flood

*

"On Saturn And The Flood"
By Dr. Immanuel Velikovsky
KRONOS Journal (Volume V, Number 1), Fall 1979
Glassboro, New Jersey

Editor's Note: This essay is based upon a lecture given by Immanuel Velikovsky at the University of the New World, Valais, Switzerland, in 1971. It was first transcribed by his secretary, Mr. Jan Sammer, who supplied both additional textual material (offset by brackets) as well as certain relevant referential data. Dr. Velikovsky's lecture was derived from his own unpublished manuscript — *Saturn And The Flood* — which was first conceived and written in the early 1940s. The present article is thus a condensation of a much larger work currently being prepared for publication. — Lewis M. Greenberg

[COMMENT: The Velikovskian presumption has always been that these "cosmic cataclysms" were isolated, one-time events, not cyclical in nature, despite all of the overwhelming evidence from around the world that cycles of destruction periodically occur. [JS] indicates an insertion by Jan

Sammer. Footnotes are indicated by {braces} and are placed at the end of the article. RS]

*

Worlds In Collision comprises only the last two acts of a cosmic drama — one that occurred in the middle of the second millennium before the present era; the other during the eighth and early part of the seventh century before the present era. Prior to the events described in *Worlds In Collision*, Venus — following its expulsion from Jupiter — was on a highly eccentric orbit for a period of time measured certainly by centuries, perhaps millennia, before its near-encounters with the Earth.

[COMMENT: With all due respect to Dr. Velikovsky, I do not accept his idea about "The Birth Of Venus" from the Planet Jupiter, leading to the worldwide catastrophes that occurred at the time of the Israelite Exodus from Egypt and the Santorini Cataclysm. The event that precipitated this global disaster was the previous arrival sequence of Planet X Nibiru, starting on 21 December 1588 BCE. RS]

While the actual beginning of the drama is shrouded in the mist of grey antiquity and difficult to pinpoint with exactitude, there is a point at which a clearer picture emerges. This is the time when the two giant planets — Saturn and Jupiter — approached each other closely. Possibly they were close for a long period of time, passing near one another as they traveled along orbital paths quite dissimilar to those of today.

Saturn and Jupiter are so often associated in cosmological history that sometimes I even considered the possibility that they may have constituted a double star system, of which there are many in the universe. I said that Saturn and Jupiter were stars, though today we know them as planets. Actually, in *Worlds In Collision*, in the last chapter, I also used the word "star" in referring to the two giant planets. There I wrote, with respect to the future, that "some dark star, like Jupiter or Saturn, may be in the path of the sun, and may be attracted to the system and cause havoc in it".{1} At the time it was said that they were planets, not stars, while today it is known that Jupiter and Saturn, too, are star-like, producing several times the amount of heat they receive from the Sun.{2}

Today Jupiter moves on an orbit of twelve terrestrial years and is about half a billion miles away from the Sun, whereas we are some ninety-three million miles distant. Saturn is much farther: it is the next planet beyond Jupiter, approximately another half billion miles outside Jupiter's orbit. They are presently not of the same size or volume. Jupiter is more than three hundred times more massive than the Earth, but Saturn only ninety-five times. In volume Jupiter is about thirteen hundred times that of the Earth, whereas Saturn is only about eight hundred times that of the Earth. Today Jupiter is actually more massive than all the other planets, Saturn and the rest, put together.

The cosmological thought of ancient peoples conceived of the history of the Earth as divided into periods of time, each ruled by a different planet. Of these the epoch of Saturn, or Kronos, was remembered as a time of bliss, and it was made to precede the period during which Jupiter was the dominant deity. Insofar as I could understand the physical events that affected the globe in times preceding the Middle Kingdom in Egypt, I was able to explain them as the results of a disturbance in which both Jupiter and Saturn participated. Various peoples witnessed the events and described them, as a celestial-human drama in different forms: the Greeks, for example, had Jupiter-Zeus, the son of Saturn-Kronos, dethrone his father and banish him, and take his place to become the supreme deity. In Egyptian folklore or religion the participants in the drama are said to be Osiris-Saturn, brother and husband of Isis-Jupiter. And it is not that the wife dethrones the husband, nothing of the kind — there is, instead, a fight going on in the sky in which some body, described as Seth, attacks Osiris and kills, actually dismembers him; and after this Isis travels in search of the dismembered parts of Osiris. You see how the two dramas are hardly at all alike. I believe that my long experience in interpreting dreams and associations of my fellow men probably was of help to me to see similarities where the similarities were not easily seen.

An Egyptologist, one of thc most prominent Egyptologists of the last forty years (he died several years ago), Sir Alan Gardiner, wrote — and I read it twice in his writings{3} — that he could not understand who Osiris was. Osiris occupied an extremely important role in the religion, folklore, and rites of Egypt. But who was he? Was he a king who had been killed? — Gardiner could not figure it out. He did not understand that Osiris represented a planet, Saturn, as did Tammuz in Babylon. Sir James Frazer, author of *The Golden Bough*, describes in the volume *Adonis, Osiris, Attis* the great lamentations and crying for the fate of Tammuz. Similar rites were observed in Egypt for Osiris; and it should be understood that these lamentations were actually for Saturn, because the time of Saturn — the

Golden Age of Saturn, or Kronos — came to it end when the supreme god of that period, the planet Saturn, was broken up.

[COMMENT: Dr. Velikovsky did not believe in "ancient astronauts", particularly any from Planet X Nibiru. He interpreted *all* of these legends purely in cosmic terms, whereas some of them indeed recall real events involving the Nibiruan hierarchy. Tammuz and Adonis were other references to Nibiruan Duke Dumuzi, who was murdered by Baron Marduk during The Pyramid Wars, as described by Zecharia Sitchin in *The Wars Of Gods And Men*, Volume 3 of *The Earth Chronicles*; see also Chapter 11 herein. And needless to say, the last Golden Age of "Saturn" was that time during 1587-687 BCE when Planet X Nibiru was tethered to our North Pole. Thus, whilst reading this article, one must think in terms of Planet X Nibiru when the Planet Saturn is mentioned. RS]

I have already discussed the statement, contained in the Tractate Brakhot of the Babylonian Talmud, which points to the celestial body Khima as the source of the Deluge; and I have shown why Khima is to be identified with Saturn.{4}

[Hindu sources also provide information which links the planet Saturn with the Deluge. This catastrophe is said to have taken place during the Satya yuga, in the reign of Satyavrata, who is usually identified as Saturn. Actually, it becomes apparent that the whole epoch named Satya yuga was the Age of Saturn as well as of the Deluge. Sir William Jones, who occupied himself mainly with comparative linguistics and with Hindu lore, expressed this very thought. He wrote that the Satya yuga meant the Saturnian Age, and that this was the Age of the Flood.{5} JS]

Also in the Mexican codices it is said that the first world age, at the end of which the Earth was destroyed by a universal deluge, and which was therefore called "the sun of water" or Atonatiuh, was presided over by Ce-actl, or Saturn.{6}

The ancient sources all point to Saturn; but how did Saturn cause the Deluge? What did really happen?

Suppose that two bodies, such as Jupiter and Saturn, were to approach one another rather closely, so as to cause violent perturbations and huge tidal effects in each other's atmospheres. As a double star, or binary, they might interact to the extent that, under certain conditions, the interaction of the members of such a pair will lead to a stellar explosion, or nova. A nova is thought to result from an instability in a star, generated by a sudden influx of matter, usually derived from its companion in a binary system. If what today we call Jupiter and Saturn are the products of such a sequence of events, their appearance and respective masses must formerly have been quite different.

Such a scenario would explain the prominence of Saturn prior to its cataclysmic disruption and dismemberment — it must have exceeded Jupiter in size. At some point, during a close approach to Jupiter, Saturn became unstable; and, as a result of the influx of extraneous material, it exploded, flaring as a nova which, after subsiding, left a remnant that the ancients still recognized as Saturn, even though it was but a fraction of the size of the celestial body of earlier days.

[COMMENT: Compare these ideas with those of Andy Lloyd in his *Dark Star, The Planet X Evidence* regarding a local binary system. See Appendix A below. This idea that Saturn became a "nova", only to explode and leave but a "fraction" of its former self could be transposed onto the Sirius System and refer to the implosion of red-giant Osiris (Sirius B) into its "remnant" neutron-dwarf. There is absolutely no evidence to support the supposition that Saturn and Jupiter used to be a "binary subsystem" within this greater Solar System. Furthermore, if indeed there exists some "confusion" of Planet X Nibiru with Saturn, since Saturn is the farthest planet that is visible to the naked eye, when Planet X Nibiru last departed from the Inner Solar System, it may have travelled in the direction of Saturn, at which point it disappeared from view, leaving the two confused by later generations. RS]

[Jupiter, stripped of its outer envelope, was seen as a new creation and received a new name — Horus in Egyptian texts, Zeus in Greek mythology. Thus did Isis-Jupiter give birth to Horus, who was also Jupiter; and similarly Rhea, Kronos' companion, produced Zeus. JS]

In Saturn's explosion much of the matter absorbed earlier was thrown off into space. Saturn was greatly reduced in size and removed to a distant orbit — the binary system was broken up and Jupiter took over the dominant position in the sky. The ancient Greeks saw this as Zeus, victorious over his father, forcing him to release the children he earlier had swallowed, and banishing him to the outer reaches of the sky. In Egyptian eyes it was Horus-Jupiter assuming royal power, leaving Osiris to reign over the kingdom of the dead.

[COMMENT: Dr. Velikovsky and Jan Sammer cannot logically presume, as they seem to have done, that Jupiter was identified with *both* Isis and Horus. Isis was the mother of Horus. This clearly demonstrates the perplexity that results when Planet X Nibiru is not considered as part of the historical scenario. RS]

My conclusion that, as a result of its interplay with Jupiter, Saturn became a nova,{7} I found confirmed in many ancient sources, in which Saturn is regularly associated with brilliant light; but I was led to this idea first of all by a certain clue contained in the Biblical account of the Deluge. The story as found in the book of Genesis starts with these words: "And it came to pass after seven days, that the waters of the Flood were upon the earth."{8} It is not explained, after seven days of what? Some words seem to be missing here from our text of the Old Testament. It is clear, however, that Isaiah refers to the same seven days in his description of the messianic age to come, when "the light of the moon shall be as the light of the sun, and the light of the sun shall be sevenfold, as the light of the seven days. ... "{9} This memory of the seven days of light preceding the Deluge{10} is a most important indication of the physical cause of the catastrophe. The intense light, filling the sky, points to a source in a nova within the solar system.

[COMMENT: Not necessarily. RS]

If, as all evidence indicates, the nova was in fact Saturn, we may obtain an estimate of the Earth's distance from the source of the illumination in the following way: The light from Saturn's explosion probably reached the Earth in a matter of minutes, practically simultaneously with the beginning of its nova phase; but the waters followed seven days later. Though ejected in the same catastrophic disruption, the Saturnian gases or filaments took a week to reach the Earth. If we can estimate the initial ejection speed of this material{11} and fix with some approximation the length of the day at that time, it may be possible to get an idea of how far removed the Earth was from the focus of the cataclysm. It is conceivable that the Earth was, at that time, a satellite of Saturn, afterwards possibly becoming a satellite of Jupiter.

[COMMENT: There is absolutely no evidence that Earth was ever a satellite of Saturn or Jupiter. It periodically becomes a "satellite" of Planet X Nibiru when it is tethered to Nibiru's South Pole, but the two revolve in unison around the Sun, probably at a precise rate of 360 days per Earth Year. RS]

With the end of the seven days of light the Earth became enveloped in waters of cosmic origin, whether coming directly from Saturn — and Saturn is known to contain water{12} — or formed from clouds of hydrogen gas ejected by the nova, which combined, by means of powerful electrical discharges, with the Earth's own free oxygen. There are definite indications of a drastic drop in the atmospheric oxygen at the time of the Deluge — for instance, the survivors of the catastrophe are said in many sources to have been unable to light fires. The Midrashim and other ancient sources describe the waters of the Flood as being warm;{13} in addition the waters may have been rich in chlorine, an element which in combination with sodium forms common salt. Marine geologists are unable to trace the origin of the huge amounts of chlorine locked in the salt of the earth's oceans, the Earth's own rocks being rather poor in this element and incapable of supplying it in the needed quantities. Chlorine may thus be of extraneous origin; being a very active element, it could possibly be present in some different combination on Saturn.

[COMMENT: From the first time that I read this in 1979, long before my "awareness" of Planet X Nibiru, I had been skeptical about the Saturnian origin of these floodwaters. If indeed this chlorine came from an external source at the time of a dissolution of The Cosmic Tree, then it came from Planet X Nibiru, not Saturn. RS]

The effects of nearby supernovae on the biosphere have been the object of intensive study by geologists in recent years,{14} in an attempt to account for abrupt changes in the history of life on this planet. Sudden extinctions were followed by the appearance of new species, quite different from those preceding them in the stratigraphic record. In a relatively brief interval whole genera were annihilated, giving way to new creatures of radically different aspect, having little in common with the earlier forms they replaced.{15} Thus, over the past two or three decades, many geologists and paleontologists have found themselves increasingly drawn to the view that the observed abrupt changes in the biosphere, such as that which marked the end of the Mesozoic and is thought to have brought with it the extinction of the dinosaurs,{16} among other animal groups, could best be explained by the exposure of the then living organisms to massive doses of radiation coming from a nearby supernova. The radiation would annihilate many species, especially those whose representatives, whether because of their large size or for other reasons, were unable to shield themselves from the powerful rays; at the same time, new organisms would be created through mutations, or macro-evolution".{17} Animals would suffer much more severely than plants — on plants the principal effect would be mutagenic.{18}

[**COMMENT:** And could not a nearby nova, such as that of Sirius B, have been the cause of some of these extinctions and mutations of species? This mention of dinosaurs also reflects the fact that Dr. Velikovsky believed that some of the dinosaurs survived the cataclysms of earlier epochs and lived on into historical times. These leftover dinosaurs then became the "dragons" of legends, such as the dragons seen by Apollonius and Damis in their travels to India. And what is the so-called "Loch Ness Monster" if not a leftover dinosaur? This proposal was one of Dr. Velikovsky's most controversial *vis-à-vis* "establishment" science, and paleontologist Stephen Jay Gould of Harvard University was one of Dr. Velikovsky's harshest critics; but I personally have no problem with it. RS]

After the Deluge many new forms of life came into being, especially plant life. Thus it happened that Saturn was later called a god of vegetation. Frazer in his *Golden Bough* considered Osiris and Tammuz to be nothing more than vegetation gods — so strong was Saturn's connection with the new forms in the plant kingdom that appeared following the Deluge.

[**COMMENT:** But, Osiris is also connected to Sirius B and thus in turn to Planet X Nibiru. And in "real life", Crown-Prince Enlil/Osiris and Duke Dumuzi/Tammuz/Adonis were completely different individuals. RS]

There is one important phenomenon which the supernova theory does not explain, however, namely the geological upheavals that accompanied the great extinctions. The Midrashic sources relate that, during the Deluge, all volcanoes erupted;{19} and other ancient accounts assert the same. Changes took place in the lithosphere as well as in the biosphere. Most pronounced, however, were the changes in the hydrosphere — the volume of water on the Earth was called by the ancients "the sea of Kronos"{20} — indicating that it came to be only after the Deluge.

The memory of these stupendous events survived for millennia and vestiges of the cult of Saturn persist even till today. One of these memorials is the feast of light, celebrated in mid-winter: Hanukkah or Christmas, both stemming from the Roman Saturnalia. These are all festivals of light, of seven days' duration, and they commemorate the dazzling light in which the world was bathed for the seven days preceding the Deluge; in their original form these festivals were a remembrance and a symbolic re-enactment of the Age of Saturn. It was said that in that age there had been no distinction between masters and servants — thus in Rome, for the duration of the Saturnalia festival, the household slaves were freed, and were actually waited on by their masters. Also the statue of Saturn which used to stand in the Roman Forum was for a time released from its bonds. This statue, which had bands around its feet, represented the planet Saturn with its rings — it was understood that it was Jupiter that had bound Saturn with these bonds after he had overthrown Saturn. Astronomers are unable to explain their origin, but they must have formed in that event in which Jupiter disrupted Saturn.

[COMMENT: *Something* certainly "disrupted" Planet X Nibiru from its stationary position as The Cosmic Tree, but this "something" was probably not the Planet Jupiter. What it truly was is anyone's guess right now. As for the "festival of light", since Winter Solstice is automatically the darkest time of year in the Northern Hemisphere, this festival could have been designed with the simple, mundane idea of illuminating the environment at its time of greatest nighttime darkness. And, as I have proposed in *Slow-Motion Doomsday*, Planet X Nibiru was a "Night Sun" in the Northern Hemisphere, so during its presence there was never any *total darkness* in the Northern Hemisphere; and there is no reason at all why it can't also have "rings" similar to those of Saturn. Since its arrival seems to coincide with our Winter Solstice, is it any wonder that our Christmas trees are "miniature models" of The Cosmic Tree, "the golden pole with the golden cage on top", or "star" atop the tree? See:
http://www.slowmotiondoomsday.com/christmastrees.html
Christmas trees are clearly symbolic representations of The Cosmic Tree. The Catholic Church condemned Christmas trees as "pagan" icons until the 1600s, when it finally relented and officially allowed this ancient tradition to persist, no longer considering it to be "sinful". RS]

There is evidence that the ancient Maoris of New Zealand were also aware of the rings around Saturn. They called the planet Parearau, which means "her band quite surrounds her".{22}

Saturn was the chief deity of, among other peoples, the Phoenicians and the Scythians — in cuneiform sources the Scythians are called Umman-Manda, or "the people of Saturn". The Phoenicians used to bring human sacrifices to the planet, calling it Moloch, or "king". Usually children were the victims, consumed by Moloch, as Saturn had devoured his own children. Porphyry records the persistence in some cities of the Greek world of human sacrifices to Saturn well into Roman times.{23]

[The worship of Saturn was also reflected and perpetuated in political institutions in many parts of the world. In ancient Egypt the ruling king was identified with Horus, or Jupiter, as the earthly analogue of the reigning heavenly power. Upon his death he came to be regarded as Osiris, or Saturn, a departed but still venerated heavenly body.

[In the Chinese concept of kingship, which persisted till the early years of this century, the Emperor was the earthly representative of the ruling planet Saturn. Ssu-ma Ch'ien, the great Chinese historian of the second century before the present era, in his treatise on *The Rulers of the Heavens* wrote that Saturn is the planet of the Sovereign, or the Emperor. As Saturn occupied the central place in the sky, even so the Emperor was at the center of his realm on Earth. Thus Saturn came to be connected in Chinese thought with the pole star, because the pole star marks the "center of the sky around which the entire heavenly vault revolves — it was considered the most favored place.{24} Saturn was regarded as the most important celestial object, corresponding to the Emperor, and thus placed in the center of the sky. So we see that the Chinese monarch was not a "sun-Emperor" — he was a Saturn Emperor.{25} What could have moved the Chinese to put such stress on the importance of Saturn if this planet was always only a faint, sluggishly moving point of light in the starry ski? Saturn's role in the Chinese idea of government, preserved in its political institutions almost till our own days, lets us get a glimpse of the planet's importance in the past as a dominant celestial body. JS]

[COMMENT: Dr. Velikovsky, here again via Jan Sammer, is simply mistaken. The Chinese Emperor was *not* a "Saturn Emperor". He was a "Planet X Nibiru Emperor"! RS]

*

FOOTNOTES

1. *Worlds In Collision*, Chapter 9, Section "The End".
2. [D. McNally, "Are the Jovian Planets 'Failed' Stars?", *Nature* 244 (August, 1973), pp. 424-426; R.F. Loewenstein, *et al.*, "Far Infrared and Submillimeter Observations of the Planets", *Icarus* 31 (1977), p. 315. Cf. *Astrophysical Journal* 157, pp. 169ff. — Jan Sammer. Also see *Science News*, Vol. 109 (Jan. 17, 1976). pp. 42-43; *American Scientist*, Vol. 63 (Nov.-Dec. 1975), p. 638; *Science News*, Vol. 116 (Sept. 15, 1979), p. 181; *Pensée IVR* I (May, 1972), p. 12 under the entry JUPITER. — Lewis M. Greenberg]
3. *Journal of Egyptian Archaeology* 46 (1960), p. 104: *Egypt of the Pharaohs* (Oxford Univ. Press, 1961), p. 424.
4. "Khima and Kesil" in *KRONOS* III:4 (1978), pp. 19-23.
5. ["On the Gods of Greece, Italy and India" in *Asiatick Researches* I (1799), p. 234. Cf. E. Moor, *The Hindu Pantheon* (1864), p. 108. — Jan Sammer. Also see *Larousse World Mythology*, ed. by Pierre Grimal (N.Y., 1965), p. 244. — Lewis M. Greenberg]
6. E. Seler, *Gesammelte Abhandlungen*, Vol. II, p. 798. [Also see *Mythology of the Americas* (N.Y., 1968), pp. 180-181. Lewis M. Greenberg]
7. [Cf. the remarks by William Mullen in *Pensée IVR* III (Winter, 1973), p. 14 — "Velikovsky has suggested that as a result of disruption Saturn went through a short nova-like phase in which its light would have obscured everything else visible from earth; the deluge followed shortly thereafter." — Lewis M. Greenberg]
8. *Genesis* 7:10.
9. *Isaiah* 30:36.
10. Similar memories are to be found in Babylonian and Hindu sources; an intense light flooded the Earth just prior to the Deluge.
11. [The usual range of the velocities is between 1,300 and 2,500 km/sec. See *Science News*, Vol. 110 (October 16, 1976), p. 251. — Jan Sammer]
12. [See T. Ferré, "A Record of Success" in *Pensée IVR* I (May, 1972), p. 23 under the entry SATURN. Velikovsky correctly claimed that "Saturn contains (or consists of) water. ... The Saturnian rings consist of ice." Pioneer 11 indicated that Saturn's core is "wrapped in a compressed blanket of such materials as water, methane and ammonia extending to about 0.23 percent of Saturn's radius (0.23 Rs)". Furthermore, "many researchers have assumed that the ring particles are composed largely of water ice, and the new data seem supportive" (*Science News*, 9/15/79, p. 181). — Lewis M. Greenberg]

13. See sources in L. Ginzberg, *The Legends of the Jews* (Philadelphia, 1925), Vol. V, p. 178.
14. [The first proponent of the supernova hypothesis was O.H. Schindewolf in his *Der Zeitfaktor in Geologie und Palaeontologie* (Stuttgart, 1950); see also idem, "Ueber die möglichen Ursachen der grossen erdgeschichtlichen Faunenschnitte" in *Neues Jahrbericht der Geologie und Palaeontologie*, Abh. 10, pp. 457-465; V.I. Krasovsky and I.S. Shklovsky, "Supernova explosions and their possible effect on the evolution of life on the Earth" in *Dokl. Ac. Sci. USSR* 116 (2): pp. 197-199; L.J. Salop, "Glaciations, Biologic Crises and Supernovae", *Catastrophist Geology*, Vol. 2, no. 2 (1977), pp. 22-41. Jan Sammer]
15. See N.D. Newell, "Revolutions in the History of Life," *Geological Society of America Special Papers* 89, pp. 68-91.
16. But see my article in *KRONOS* II:2, "Were All Dinosaurs Reptiles?" (Nov. 1976), pp. 91-100. [Cf. Adrian J. Desmond, *The Red-Blooded Dinosaurs* (N.Y., 1976), especially pp. 184-196, 224-225; also see "Cosmic Radiation Blast Linked to End of Dinosaurs" in *New York Times*, 5/30/79, p. A20. — Lewis M. Greenberg]
17. See my comments in *Pensée IVR* IV, "The Pitfalls of Radiocarbon Dating" (Spring-Summer, 1973), p. 13 — " ... in the catastrophe of the Deluge, which I ascribe to Saturn exploding as a nova, the cosmic rays must have been very abundant to cause massive mutations among all species of life ... " In 1963, in a letter to H.H. Hess, I suggested that "tests should be devised for detection of low energy cosmic rays emanating from Saturn, especially during the weeks before and after a conjunction of Earth-Jupiter-Saturn" (see *Pensée IVR* II (Fall, 1972), p. 28; *Velikovsky Reconsidered*, "H.H. Hess and My Memoranda" (N.Y., 1976), p. 49). Besides cosmic rays, I have suggested that Saturn emits X-rays (see *Pensée IVR* I (May, 1972), p. 23). [Also see "Long-period X-ray transients" in *Science News*, Vol. 110, Oct. 16, 1976, p. 250. — Lewis M. Greenberg]
18. K.D. Terry and W.H. Tucker, "Biologic Effects of Supernovae" in *Science* 159 (1968), pp. 421-423.
19. *Sefer Hajashar.*
20. See for example Plutarch, *Isis and Osiris*, Chap. 32; Clement of Alexandria, *Stromata*, Vol. 8, p. 360; Aristotle, fragment no. 196.
21. For a possible explanation of the mechanics of the formation of the Saturnian rings, cf. H. Friedman, "Cosmic X-ray sources: A progress report" in *Science* 181 (3 August 1973), p. 396.
22. E. Best, *The Astronomical Knowledge of the Maori, Genuine and Empirical*, New Zealand Dominium Museum Monograph (Wellington, 1922), p. 35.

23. Porphyry, *On the Abstinence from Animal Food*, transl. by Th. Taylor (Centaur Press, U.K., 1965), p. 81 (II.27), p. 102 (II.54).
24. [Cf. similar assertions in Proclus, *In Platonis Rem Publicam* II.213.4f.; Eusebius, *Praeparatio Evangelica* IV. 1.4; Nonnos, *Dionysiaca* 41.350. — Jan Sammer]
25. [Cf. Giorgio de Santillana and Hertha von Dechend, *Hamlet's Mill* (Boston, 1969), p. 129. — Lewis M. Greenberg]

Chapter 15

Gates Of Hades

*

This essay was first written in March 2001, and the original is still available at both my websites, with 28 mostly color graphics, some of which I have included here in grayscale. Although revisions of this essay have been made for purposes of this book, I have not modified the original website version. The reader is referred to the following URL.

http://www.slowmotiondoomsday.com/gates.html

Enroute to Cappadocia in January 1998, I stopped in Greece for a week to visit a man whom I'd met on the Internet. I'll refer to him as Yannis. He lives in Athens, but his large family owns an historic, 200-year-old "castle" in Mina, a small village south of Sparta and Areopolis in the Peloponnesus region. At an Internet mailing list to which Yannis and I had been subscribed, he had been informing me and the rest of the group about strange phenomena which were occurring in the southern Peloponnesus, including the notorious "Gates of Hades", the primary motivation for my visit. At my website are two maps showing the region which I am discussing. These small maps will not be reproduced here, but you will notice on any map of Greece that three finger-like peninsulas extend southwards from the bottom of the Peloponnesus. The Gates of Hades are (is) situated at the extreme southern end of the middle peninsula, the southernmost tip of the continent of Europe.

Yannis at a motorcycle competition near Sparta

Yannis' family castle at Mina, north of the Gates of Hades

There is a lighthouse at the southern tip of this middle peninsula; and when one looks due-south from this lighthouse, one is looking in the direction of western Ancient Egypt, or eastern modern Libya. This area is a part of the Mediterranean Sea where several ocean currents converge, resulting in numerous shipwrecks against this rocky coastline over the centuries. Thus, in fairly recent times, the Greek Government built this lighthouse to warn ships not to approach too closely to these potentially dangerous convergent currents. Since there is no roadway from Tainaro, the nearest village, to the location of the lighthouse, nor any feasible spot at which to dock a cargo ship below the lighthouse, undoubtedly all of the materials used to construct this lighthouse had to have been brought manually overland from Tainaro. Additional information about this will follow later in the narrative, but I could point out here that this convergence of treacherous ocean currents is similar to the convergence of two primary Pacific Ocean currents at Carmel, California.

Southeast of Sparta is the ancient Spartan port city of Gytheion, from where one can drive southwesterly to Areopolis. Secondary roads lead farther south to Tainaro and the lighthouse. Just south of Areopolis is the "Dirou Grotte" or Diros Cave, also discussed below. Yannis' family castle is located a bit north of Gerolimen. The distance from Areopolis to Gerolimen is 26 kilometers, or about 16 miles.

There are only a half-dozen windows in Yannis' Mina Castle, which is a four-story structure. On the ground-floor level, there's an old olive press that was pulled by donkeys, along with a storeroom for tools. I could tell by looking at this old olive press, from all the accumulated dust, that it had not been used in years. The second floor contains the main living area, with additional sleeping quarters on the third level. The uppermost "battlement" is a single large, lookout tower room with windows facing in all directions, from which one could fire on any approaching "enemies". The walls of this castle are about 2-feet thick, made of solid rock. As I visited there in January, it was extremely cold in this castle during the nighttime. In fact, I have never been so cold whilst sleeping in my whole life. I had two small electric space-heaters next to my bed, and I slept in my clothes underneath three blankets — and I was still cold! Yannis and I talked about this, of course, and he said that usually the family does not visit the castle in wintertime, but he added that in the heat of the summer, the castle feels "naturally air-conditioned". At night, when I was sleeping there, I was thinking about prisoners in ancient dungeons in castles like this, prisoners who had no electric space-heaters or blankets; and it literally made me terrified of such a prospect.

After a couple of brief stopovers in Madrid and Venice, I arrived in Athens on Sunday afternoon, 11 January 1998. Yannis met me at the airport, and we drove to his three-story home in the north Athens Petroypolis District. We had a marvelous dinner with his large family and then partied until the wee-hours at a posh north Athens nightclub, the Club Apollos, a large split-level, California-style club and restaurant with patios on a mountaintop overlooking the City of Athens — very Los Angeles in every respect. Monday afternoon we departed for the Peloponnesus and arrived in Sparta at sunset. Sparta is a lovely and bustling little city. Its main boulevard through the middle of downtown is lined with citrus trees. We parked near a newsstand so that Yannis could make an inquiry about something. Even in cold January there were oranges hanging on these boulevard citrus trees; and when Yannis was gone, I amused myself by picking a couple of oranges and eating them.

From Sparta we drove to nearby Gytheion, the ancient port city for the Spartans. Gytheion has a completely enclosed harbor that could be easily patrolled by even an ancient navy. It was the main "safe-haven" port for the ancient Spartans. Anthony and Cleopatra once sailed into Gytheion for a clandestine romantic rendezvous. And why not? Gytheion is one of the most beautiful small cities that I have ever seen in my travels. Unfortunately, Yannis and I arrived after dark that Monday (and on the return trip, at twilight), so I have no photographs of this breathtaking little seaport. There is a "corniche" that runs alongside the harbor with sidewalk cafes. On the other side of this corniche avenue are numerous hotels, restaurants and classy shops. We ate Greek seafood dishes on both occasions — *molto delizioso!*

We arrived at the castle around 10 PM. After unloading the car and getting settled in, I used Yannis' laptop computer to send some messages to our Internet mailing list about my travel activities on my way to Tyana, Cappadocia. Whilst I was working at the computer, Yannis was outside on the castle terrazzo, looking for UFOs over nearby Devil's Mountain.

Adjacent to this castle is some of Yannis' family property, which includes a large olive grove with about 2,000 trees. This part of Greece is "Kalamata Olive Country", where all of the world's genuine Kalamata Olives are grown. There were literally millions of Kalamata olive trees around here.

They are fairly short trees, and the olives resemble cherries hanging from a cherry tree. Yannis' late uncle (his mother's older brother) used to live at the castle and depended upon these olives as his source of income. Since his death, the Athens family now rents the olive grove to someone else who pays them a commission based upon the amount of olive oil that he produces each year. I picked a few of these olives right off the tree, and they were very bitter to taste. I found out later that the reason that olives are cured in brine is to eliminate this bitter taste from the commercial product. I was most impressed by this experience and subsequent tidbit of information, as olives and olive oil are two of my very favorite food products.

The local animals deserve some mention, also. The "Dog-Star" Sirius was the guardian star of Ancient Sparta. Today there are dogs running loose everywhere, and they are some of the most exotic and splendid dogs that I have ever seen. Once when Yannis and I stopped for gasoline, we noticed a couple of large dogs standing up on their hind legs, embracing each other with their forelegs and kissing each other, almost like humans! It was a sight that neither one of us had ever seen before. Around Yannis' castle there were numerous wild cats that climbed around on the castle ledges and sunned themselves on rooftops. It is the custom in Yannis' family that when someone visits the castle, they place little pieces of bread out for these cats to eat. I thought that this was an unusual practice, since I had never heard that cats would eat bread; but these cats ate it up. Also, one night when we were arriving back at the castle along a nearby country lane, we confronted a herd of young donkeys that were out on the prowl together like pack of errant schoolboys. Suffice it to say that if you ever visit Greece, you should make it a point to include a few days in the Peloponnesus, if possible at the Tsitsiris Castle Hotel. This is some of the most spectacular scenery on the planet!

In the previous photograph of Yannis' castle, the view is looking north along the adjacent country lane. If one were to walk northwards past the castle and on about the length of a football field, one comes to a path that leads into a neighbor's olive grove. A couple of men were picking olives in the distance. Yannis led me down this pathway for perhaps a couple of kilometers, where we arrived at a peculiar sort of rock pile next to the path. We climbed over these rocks and hiked across a rustic area until we reached a gigantic hole that precipitously sloped down into the ground. It is a "bottomless pit", approximately 20 feet (about 7 meters) across, that apparently reaches down into a watery cave. People have thrown dead animals into that pit, for instance, only to find them washed up on the nearby seacoast. We dared not approach closer than about a meter to this pitch-dark hole. I hurled a large rock into the hole, but it never hit bottom. We looked around for something other than a rock to throw in, and we found an old tin can near the rock pile. So I threw that into the hole. We could hear it clanking around inside as it bounced off the walls of the pit, but we did not hear it hit the bottom. Yes, very spooky. I was a bit surprised that this pit wasn't fenced off. If that were in America, it would be surrounded by barbed wire and warning signs to keep people from accidentally falling in.

Around the cove from Yannis' castle was Devil's Mountain, a flat promontory jutting out from the coast. A local legend has it that a "Demon" lives inside the mountain and periodically causes some trouble for people. The Tsitsiris Castle is located about halfway between Yannis' castle and Devil's Mountain. According to Yannis, numerous UFOs have been sighted flying over Devil's Mountain, hence the name. In fact, Yannis' grandmother, an elderly woman whom I met in Athens at the family dinner party, said that she and two of her sons (uncles of Yannis) had seen a UFO over Devil's Mountain one time when they were driving along a local country road. Yannis' grandmother is about as "normal" an old lady as there is, and certainly she was not lying to me, although she seemed a bit embarrassed or shy when I started questioning her intently about this "encounter". But whilst Yannis and I were at the castle, we spotted no UFOs, unfortunately.

During this time, Yannis and I drove to visit the Diros Cave between his castle and Areopolis (where we went to eat our meals at a home-style local restaurant). Going north, when you turn left off the main country road from the castle to Areopolis to visit the Diros Cave, you'll notice little "dark spots" along the coastline. These are entrances to other "underground waterway passages" or "water caves" in this area. Diros is the only one of these local "caves" that is open for tourist purposes. A German explorer, we were told, has been attempting to map all of these 7,500 "karstic" (seashore) caves and link them together. The problem with such a project is that only the Diros Cave has electric lighting. When one attempts to enter these other caves, they are pitch-black inside, so nobody has ever been able to map the entire network of caves, which Yannis told me are probably ultimately linked underground to the main Gates of Hades cave entrance, just south of Tainaro, west of the lighthouse.

Devil's Mountain in the background, as seen from the Mina Castle

General panoramic view from behind the Tsitsiris Castle

Typical coastal landscape between Mina and the Gates of Hades

Several thousand of these caves have already been explored and mapped, and many of them are open to visitors. But of all these, only about 28, including the Diros Cave south of Areopolis, are very well-known to the public. The Gates of Hades is not listed amongst them.

In ancient times the area around the Gates of Hades was thought to be one of the entrances to the Underworld, where Herakles descended in quest of the dog Kerberos. Its general maritime location was the scene of naval action in the Second World War. In this area of Greece there are numerous military communications facilities, but the rocky terrain of the land has made foreign invasion and occupation practically impossible.

As for the Diros Cave, one pays an admission fee and boards a small rowboat. Besides Yannis and me, there were two other tourists in our rowboat. The boatman used a pole to propel the boat along the cave waterway, in much the same manner that Italian gondoliers negotiate the canals of Venice. At the family dinner party on Sunday, both Yannis' grandmother and mother said that they had once toured the Diros Cave out of curiosity and that it had scared them to death, so to speak. Yannis' mother said that she never wanted to make such a claustrophobic trip again. The tour lasted about half an hour. Most fascinating! When we exited the cave, there was an artificial "dock" of sorts, from which we had to climb up a stairway to reach ground-level again. Even the Diros Cave has not been fully explored and lighted yet. Thus, it is not surprising that so very little is known about the entire extent of all these underground cave waterways that permeate the Peloponnesus.

It was so deathly quiet inside that Diros Cave. As the boatman took us along the cave's waterway, I was thinking to myself how that back in ancient times, if one had wished to explore such a frightful water-cave, one would have had to use a primitive torch to light the way. The quietness of the tour, the gentle splashing sound of the water against the sides of the rowboat, created an eerie feeling, as if we were being steered by Styx into the ancient "Underworld" of Hades.

The following is quoted from a website, the source of which I no longer have.

"The name Hades can lead to some confusion, for it was used by the ancient Greeks both for the god who ruled the Underworld and for the kingdom of the Underworld itself. Though it was the world of the dead, the Greek

Hades was not like the later idea of hell, a place where the damned go to suffer eternal torment. It was a place where all the dead — good or bad — journeyed, guided there by the messenger god Hermes. Only when they arrived was their fate decided. Some, particularly those who had offended the gods, did suffer, but those who had been wise and kind, and those who had achieved brave deeds, could lead an afterlife of great happiness. Over all such matters ruled the god Hades, a stern but at the same time always just king.

"Hades himself does not feature strongly in Greek legends because once he had been established as Lord of the Underworld, he seldom left it. Once or twice, when a nymph caught his fancy, he ventured out in his chariot with its sinister-looking black horses, and on one brief visit to earth he seized Demeter's daughter Persephone. But generally, Hades stayed out of sight in his own kingdom.

"The kingdom of Hades, however, plays a most important part in Greek legends. Many of the Greek heroes, together with other gods, visited it for one reason or another while they were still living. A favourite task for a god to set a mortal was to go to Hades and bring back some object or token of the visit. It needed great ingenuity (or the help of a magic spell) for a mortal to get both into Hades and out again.

"In the very early days, it was believed that Hades lay far to the west, beyond the horizon where the river Oceanus, which encircled the earth, began. Later, some stories contained descriptions of dark caverns and long, gloomy passages which led down to the Underworld from districts on the mainland of Greece such as Thesprotia in the west or from across the Aegean Sea in Asia Minor. But wherever they entered, the dead could always rely on Hermes to show them the way."

Searching again for this URL was futile, although similar information is available at many other websites. Thus, I have no doubt about its authenticity, and so I have included it. The above information is an excellent summary of the Greek Hades. And the mention of ancient caverns and gloomy passages into the Underworld is quite significant, I think.

As an aside, let me also add here that Apollonius of Tyana roamed around this same area of Greece during his lifetime. He was granted permission to enter a cave of Hermes called the Cave of Trophonius. It is said that Apollonius retrieved the Emerald Tablet of Thoth/Hermes from this cave and that this valuable text, inscribed on green stone, found its way to the personal imperial library of Emperor Hadrian at Rome around the year 110 CE. Did Apollonius visit the Diros Cave? Did Apollonius enter the Gates of Hades? Today, of course, nobody can know.

On Wednesday afternoon Yannis and I drove south to the Gates of Hades. About halfway between his castle and Gerolimen, one can turn right, or west, onto a narrow unpaved country road that leads to the Tsitsiris Castle Hotel, which was closed for the winter off-season. It is built in the same style as the castle pictured above, but it is about twice that size, quite sprawling and difficult to photograph.

South of the town of Vathia, one reaches Mianes. There is a road sign just south of Mianes where the highway forks and leads a couple of more miles down to its dead-end at the tiny village of Tainaro, the last line of the sign in the accompanying illustration. Just above Tainaro there are the words *Pule tou Ade* which is Greek for "Gates of Hades". This is only a minor tourist attraction now since it is way off the beaten path. When one reaches the village of Tainaro, one has to leave one's vehicle in a large public parking lot near the Psychomantium of Poseidon. From this parking lot, there is a trail that leads down a cliff to the seashore below, perhaps 100 feet (about 30 meters) down, where there is a public boat-dock and beach. We visited briefly with a German woman who was camping out on the beach with her family.

The Peloponnesus is a popular destination for German tourists. Around the time of World War II, German Führer Adolf Hitler heard that these Greek cave passages were linked with other underground caveways that stretched all the way to Berlin. He dispatched secret agents to Greece to explore this possibility, but nothing ever came of the project, other than that the Peloponnesus itself became a popular holiday and retirement destination for Germans. Just down the country lane from Yannis' castle, a new house was under construction by a German family.

The Gates of Hades also did not go unnoticed by the famous late French undersea explorer Jacques Cousteau. Cousteau knew all about this legend and requested permission from the Greek Government to conduct oceanic research around the lighthouse area. The permission was initially granted; but a few days before the French diving was to get underway, the Greek Government reneged on the deal and denied the request, the reason being that this was a "sensitive military area" and off-limits to such exploration.

One can only speculate about the true motives behind this government action.

Road Sign to Tainaro and the Gates of Hades

Psychomantium of Poseidon at Tainaro

This is an ancient floor mosaic in the ruins of Poseidon's Palace.

Poseidon's Ancient Launch Pad?

According to legend, the Greek God Poseidon had a palace complex here. This round stone "Psychomantium" temple marks the spot at which there used to be a passageway from the surface down into "Hades" itself. This "hole", as it were, has now been covered by concrete slabs, so there is no modern danger of anyone's falling into a "bottomless pit" here, to disappear forever, as would happen to one who fell into the pit near Yannis' castle.

There is a most enigmatic spot, just next to Poseidon's Palace. Note the bare round area amidst all the literally millions of rocks of all sizes from pebbles to beachballs. Apparently this is a totally "natural" phenomenon; nobody actually "faked" this effect, supposedly. Yannis speculated that it indicates an ancient rocket launch pad, Poseidon's private backyard spaceport, if you will, with the downward thrusting of the rocket engines scattering these stones into a bare round circle. You be the judge for yourself.

Curving up from the lower left of the "launch pad" you can see a faint pathway through the stones. This pathway takes one over the hill in the distance, from where one hikes on down to the lighthouse and Gates of Hades.

If one were to explore the Gates of Hades itself, one would have to obtain a boat at Tainaro and travel around the lighthouse promontory to a point opposite Tainaro in order to enter the cave. Several adventurers have attempted to explore the interior of the Gates of Hades, and they were never seen again. It would almost require a Jacques Cousteau-style expedition to undertake such a perilous exploration.

To photograph the Gates of Hades, I was standing on the west side of the lighthouse, looking north. On the far horizon you can see the Devil's Mountain from the opposite side, the southern side. The smaller headland that juts out from the shore in the middle of the picture is the back side of the Gates of Hades. In other words, if one were to take a boat from here to the Gates of Hades, one would have to travel around that headland and enter the cave on the other side, approximately below the point of the hill. To reach the Gates of Hades by foot from here would be practically impossible because of all the large boulders that litter the landscape. This area of Greece is definitely the rockiest place that I've ever seen on Earth. One could theoretically reach it by foot, I suppose, but it would be an arduous undertaking, in that one would have to backpack all one's water and other essentials. I certainly wouldn't want to try it myself.

This is the modern lighthouse at the ancient "end of the world". The view is looking southeastwardly in the direction of Egypt. The shoreline entrance to the Gates of Hades would be directly behind my back, as I took this lighthouse photograph.

Gates of Hades

The actual Gates of Hades cannot be seen, or photographed, from the lighthouse or any other geographic location visible from the country roads. However, I have seen one of Yannis' personal photographs of the Gates that was taken from a small boat near the entrance to the cave. It looked much the same as the triangular shoreline "karstic" cave that I photographed near Mianes, shown here, lower-left center.

Back in Athens on Friday, Yannis and I spent the day driving around, seeing some sights and his favorite places. I suggested that we visit what is known as "Plato's Olive Tree", which was said to be the oldest living tree in Greece. Plato had sat in meditation underneath this olive tree 2,400 years ago. Since Yannis had never seen it himself, he readily agreed. We made some inquiries and discovered that the tree was located in the botanical garden of a city museum. This museum was conveniently situated on a main thoroughfare, and we had no trouble locating it and parking the car on a side-street.

We walked in through the museum gate; and since I was wearing my cowboy hat and carrying a camera, I looked like a typical foreign tourist, so two of the security guards immediately walked over to ask if they could assist me with something. Yannis told them that we wished to see Plato's Olive Tree. One of the guards smiled and beckoned us to follow him. We walked into the botanical garden at a corner of the museum complex. The guard pointed to a small building in the center of the garden and told us that Plato's Olive Tree had died a couple of years ago, to their embarrassment, and they'd built that little stucco building around it to hide it from the public.

Yannis and I looked at each other in amazement! Say what?! Without missing a beat, I walked over to the gated entrance to the little building and quickly took a couple of photos of the dead tree inside. The guard shouted at me and said something loudly to Yannis in Greek. Yannis turned to me and said, “It is forbidden to take any photographs, the guard told me.” I played the dumb American tourist, smiled and nodded at the guard. But I already had my photographs, which turned out to be not so good because I’d neglected to use a flash.

At any rate, Yannis and I left the museum, rather in a state of bewilderment. He had heard absolutely nothing in the news about the death of Plato’s Olive Tree. That evening at dinner with his family again, we related what had happened. His family members were just as surprised as we were about it. None of them had heard a single word. Then Yannis’ father stated solemnly, “They say that when Plato’s Olive Tree is dead, it will signal the end of the world.”

Chapter 16

Dulce Report

*

During October of 2005, I had the fortunate opportunity to visit and tour parts of New Mexico, accompanied by a woman whom I'd met years earlier in the Peace Corps. I visited Roswell (where I checked out the UFO Museum), Ruidoso, Albuquerque, Santa Fe, Taos (where I heard the notorious "Taos Hum" at the Rio Grande Gorge Bridge, shown in the back-cover photograph), Dulce, Echo Canyon Amphitheater and Española. A few photos are included here, but for a full report of this trip with over 80 archived color photographs, the reader is advised to go to my website.

http://www.slowmotiondoomsday.com/enchantment.html

For many the name "Dulce" is instantly recognizable. If you google for < dulce+ufo >, you'll bring up 247,000 websites. Pictured on the next page, to the right of the light poles, is the nefarious Archuleta Mesa, where the reptilian aliens from Outer Space supposedly have an underground base of at least seven levels, the lowermost of which extend out under the sleepy, little Apache Indian village of Dulce itself. Horrific experiments are reportedly being conducted, especially at levels six and seven. For example, they are said to be trying to determine how to extract the "soul" and "spirit" from a human being and "transplant" them into some of their reptilian volunteers. Perhaps this is part of their "training process" for reptilian spies who infiltrate human society, by masquerading as humans.

Archuleta Mesa (looming up to the right)

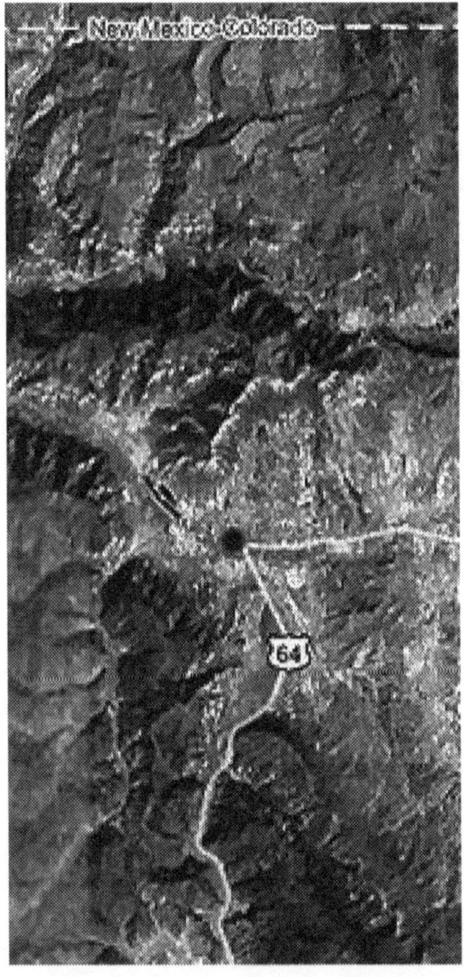

Dulce Map (Dulce is the black dot in the center.)

UFO Museum, Roswell, New Mexico

Music Store across the street from the UFO Museum

Downtown Roswell Nightclub

Rob Solàrion & Roswell Street Lamp

Perhaps the reader will recall from earlier in this book how Brad Steiger related that these reptilian aliens attempt to "disguise" their greenish skin by bleaching it. These antagonistic aliens do *not* seem to be our friends.

The word "dulce" is Spanish for primarily the English word "sweet", but it can also mean "fresh" or "pleasant" or "agreeable". Probably the Spanish *conquistadores* thought that Dulce was a "pleasant" place to live and so named it that.

As the crow flies, Dulce is only about five miles south of the Colorado border, and the Archuleta Mesa rises north of Dulce about halfway in between. The accompanying Google map above is a hybrid graphic with the highway overlaid on top of the topographical satellite image. Dulce is easy to find because it is located exactly at the V-shaped turn of the highway. I have placed a black dot on the graphic at the location of Dulce. There is a dirt road that runs from Dulce to the top of Archuleta Mesa, just above the shadowy ridges north of Dulce. As I recall, one of the local Apache men said that it was a distance of about six miles, since this dirt road would be going uphill and curving around at the same time, making it longer than the straight-line distance.

As soon as I decided to tour New Mexico, I knew at once that I would have to visit Dulce, since I have heard so much about it. My old acquaintance Carrie and I arrived in Dulce around mid-afternoon, following an easy drive through the mountains from Taos and across the Continental Divide, which near Dulce reaches an elevation of about 11,000 feet. Most of the scenery along the route from Tierra Amarilla to Chama and Dulce looked about the same as the scenery elsewhere, so I did not waste any time stopping for additional photographs (others of which can be seen at my website).

Approaching the town, we noticed a billboard advertising the Best Western Inn and Restaurant at Dulce. I suggested that we first stop there and make some inquiries. Dulce's population is only about 2,700. To a certain extent, it is merely a "wide spot" in the road, although it does host the Jicarilla Apache Headquarters and the Apache Nugget Casino, which according to the Official New Mexico State Atlas is located about halfway between Dulce and Archuleta. As with other Indian casinos, this Apache Casino must be drawing people into Dulce to gamble, even from southern Colorado. Gamblers would think nothing of driving only an hour to the nearest casino. Nevertheless, for all practical purposes, Dulce is way out in the middle of nowhere; and I had no idea of what to expect to hear about the Alien Base. Near the Best Western Inn there was also a sign indicating that visitor information could be obtained there, so we turned in and parked.

Inside the motel we were greeted by a tall, handsome Apache man.

"May I help you?" he inquired with a smile.

"We're looking for the information service," I said.

"Well, here it is!" he grinned broadly and gestured around at the small motel lobby. "This is it. What were you wanting to know?"

"Well, actually, we came here to see if we could find out anything about the reptilian Alien Base underneath the Archuleta Mesa. Do you know anything about it?"

At this point, he broke out laughing. "You are just in time! There's a conversation going on in the restaurant right now about this Alien Base. Were you planning to eat?"

"No, but we'd like to have some coffee," I replied.

"Follow me," he said and escorted us into the restaurant. Two other tables were occupied, one by two Apache women and the other by an Anglo man who was having coffee and a snack whilst conversing with a second Apache man, who was standing beside his table and who, it turned out, also worked for the motel and restaurant.

The man who showed us in promptly introduced us to the second Apache, and we introduced ourselves to the Anglo who said that he had just driven to Dulce from Duncan, Oklahoma, which is north of Fort Worth, Texas. Everyone exchanged the usual pleasantries.

Then I said, "So, y'all are talking about the Alien Base?"

The Oklahoman told us that he'd read all about Dulce on the Internet and had come to see it for himself, as was the case with Carrie and me. Then the second Apache man, whose name I have forgotten, began to expound at length upon this Alien Base, and I paraphrase him here.

"Let me tell you a story," he began. "It is rather long, but you'll find it interesting, I think. In our Indian traditions we have a custom that we follow for boys who reach one year of age. They say that if you shave a boy's head on his first birthday, when he gets older, his hair will grow long and thick. After I got married and had a son, on his first birthday we went to visit my grandmother, so that she could shave his head. This was the first time we'd had a good look at his skull without any hair, and suddenly I saw that both of his ears were pierced at the top. We had never noticed this before, and we were quite perplexed. But at the time I didn't really think much about it.

"Then my wife had a second son, and on his first birthday, we drove to my grandmother's again so she could shave his head. After she finished, we found two tiny holes or dents in his scalp, right on top of his head." He pointed at the top of his own head. "Now I was starting to wonder about these things, because clearly this is not normal. A couple of years later, my wife had a third son, and again we drove to my grandmother's for the head-shaving ceremony. This third boy had a hole in the side of his right ear, about in the middle. I didn't know what to think about it.

"Around 8 o'clock we left my grandmother's to drive back to Dulce. Usually this ride takes about an hour and 45 minutes. About 9 o'clock, we were still about 45 minutes from home, when my wife and I noticed a strange light or object flying low across the horizon. I pulled over beside the road to check it out. We sat there for a while, and we saw one UFO after another, five in all, I think, one following the other, all going in the same direction, like a caravan. The three boys were getting cranky in the back seat, and we needed to get home. It was dark, and I didn't have a watch. But when we walked in the door at home and looked at the clock, it was 11:15. I said to my wife, 'Why did it take us so long to get back? What happened?' But she didn't know either. Somehow we had 'lost' an hour, and we were certain that we hadn't stopped that long to watch the UFOs. It was very odd."

"Hmm," I said. "Do you think that you were abducted or something?"

He shrugged and nervously laughed. "I don't know, but it was definitely very strange."

"And," I asked him further, "do you think that those 'holes' in your boys' heads have something to do with the Alien Base?"

Again he said that he didn't know, but there was no reason at all for him to have been fabricating such a peculiar tale.

He then turned to the subject of cattle mutilations. He said that there are regular cattle mutilations around Dulce, and he pointed to one of the Indian women who were sitting at another table. "One of her cattle got mutilated last spring." The woman smiled and agreed that it had, and she said that the incisions were quite precise, not like a wild animal or human would make, that they were "laser-like" incisions, similar to the precision cuts that were made during the near epidemic of cattle mutilations throughout the Southwest in the 1970s. Carrie remarked to me later that she believed what this woman said because Indian people have no reason to concoct such fictions, that they are too down-to-earth and levelheaded to play those sorts of games.

"I'll tell you," said the loquacious Apache man, "everybody in Dulce has a story to tell about what they've seen around Archuleta Mesa. There are secret doors in the mesa. UFOs have been seen flying into these doors. But when somebody goes up to try to find these doors, they have been camouflaged again and can't be located. During the 1950s, or maybe 60s, a U.S. Army convoy went up to look for these doors, and they disappeared inside the mountain and were never seen again."

He paused for a moment and continued. "I was searching for Dulce information on the Internet, and I found a website about a Bigfoot Conference that was held in Dulce a couple of years ago. There were all these men in camouflage fatigues. They camped out near a small lake south of town."

"Did they spot a Bigfoot? I've never heard of any Bigfoots in these parts before."

"I don't know what they did," he said. "I've never seen a Bigfoot, and I've lived here all my life."

He was about 35 years old.

"And you know something else?" he asked. "People in Dulce are going deaf."

"Going deaf? What do you mean?"

"I mean, they can come into this restaurant, and the music is playing in the background, and I'll say to someone, I really like that song, and the person will ask me, 'What song?' 'Can't you hear the music?' I ask them." He pointed at a speaker, over which radio station music was playing quietly in the background. "They say, 'No, I can't hear any music.' This has happened many times."

"Why?" I mused. "Do you think that some noise like the Taos Hum is causing them to lose their hearing?"

He shrugged. "Who knows? But people are going deaf."

Everybody in the restaurant seemed most sincere. I generally know when people are trying to put something over on me, and I didn't get that feeling here. Neither did Carrie nor the man from Oklahoma. The three of us got up to leave and pay our bills. We thanked everybody for their stories and information and walked outside.

The Archuleta Mesa looms large on the northern horizon of Dulce. I took my photo from the Best Western parking lot. On the opposite side of the mesa there is a small river, but north of that there is nothing except national forest and Indian reservations well into southern Colorado. There is no telling what might be happening on the northern side of the mesa out of view of the citizens of Dulce. There are no other roads or towns for miles around. This is an ideally isolated location for such an Alien Base, still close enough to a small bastion of civilization for convenience but far enough off the beaten path to be ignored by most tourists and others who might pass through Dulce unaware of anything.

Atop Archuleta Mesa is a cluster of antennae. We were told that these antennae are used by local television stations and other communications businesses to transmit images and Internet data. And I personally have no doubt about that. On the other hand, if the reptilian aliens who are conducting all of these high-tech experiments inside the mesa are so advanced, then it would certainly be no problem for them to tap into these transmission antennae and, if nothing else, watch some local television programs. And don't you know that they'd get a sidesplitting chuckle out of that?!

Carrie and I said goodbye to the man from Oklahoma, and I gave him one of my calling cards with my website URLs. Then Carrie and I got into her car. Just as we were pulling out of the Best Western parking lot, the boombox in the back seat came on, all by itself.

"Did you turn that on?" I asked Carrie.

"No. Did you?" she said.

"No." I leaned over into the back seat and turned it off. "I wonder what caused that to happen," I mused to Carrie.

"Yes, I wonder." She glanced quizzically at me and then drove out onto the highway back to Chama.

Chapter 17

Introduction To Galactic Mathematics

*

As time goes by and as we come across more and more information concerning the "occult sciences", we find a sort of inconsistency in numbers, especially when dealing with the period of time known as one cycle of the "Precession of the Equinoxes". The Precession of the Equinoxes is the number of years that it takes the Earth to pass through one complete cycle of the Constellations of the Zodiac. Our spring equinox is now aligned with the Constellation of Pisces, about 10° into that constellation, "precessing" or "preceding" in a "backward" direction towards the end of the Constellation of Aquarius. It takes this planet 72 years to pass through 1° of the Zodiac, or 25,920 years to complete one full cycle of 360°.

Often one sees other numbers: the Mayas recorded that the cycle lasted for exactly 26,000 years; other people casually round the number off to 25,000 years. The Mayas were only 80 years off the mark because their celestial counting system revolved around the number 52 (or 4 X 13, or 2 X 26). This special variation of the Precession Cycle will not be dealt with here.

The geometry of the Great Pyramid of Egypt is based upon the assumption that the Earth Year lasts for exactly 365.25 days. Yet our astronomers say, and can certainly demonstrate scientifically, that our Sidereal Year (the time that is required for the Earth to complete one precise revolution around the Sun) supposedly consists of exactly 365.2563599 days and that our Tropical Year (the interval between two consecutive returns of the Sun to the point of the spring equinox) consists of exactly 365.2421990 days.

But let's change our angle of perception just a bit. One cycle of 25,920 years of exactly 365.25 days per year gives us a total of 9,467,280 days. Thus, we could say that one cycle of the Precession of the Equinoxes lasts for almost nine and a half million days.

9,467,280 ÷ 365.2563599 = 25,919.54867 years

9,467,280 ÷ 365.2421990 = 25,920.55361 years

25,920.55361 - 25,219.54867 = 1.004494 years

The difference in the number of Tropical Years in a period of 9,467,280 days and the number of Sidereal Years in that same period is only slightly greater than exactly 1 year, to wit, only 43 hours 18 minutes longer. Question: what would be the average lengths of these years over a period of 9,467,280 days if the difference were *exactly* one year? Here's the math that determines these average lengths.

25,920 X 365.2563599 = 9,467,444.848 days

25,920 X 365.2421990 = 9,467,077.798 days

9,467,444.848 - 9,467,077.798 = 367.05 days

9,467,444.848 - 9,467,280 = 164.848 days

9,467,280 - 9,467,077.798 = 202.202 days

164.848 ÷ 202.202 = 0.815263943 ratio

202.202 ÷ 164.848 = 1.226596622 reciprocal

164.848 ÷ 365.25 = 0.451329226

202.202 ÷ 365.25 = 0.553598904

0.451329226 + 0.553598904 = 1.004928130 years

Assumption: over a period of time lasting 25,920 years of exactly 365.25 days each, these two ratios would even out at 0.45 and 0.55 for a total of 1 exact year.

365.25 X 0.45 = 164.3625 days

365.25 X 0.55 = 200.8875 days

164.3625 + 200.8875 = 365.25 days

164.3625 ÷ 200.8875 = 0.81818181 ratio

200.8875 ÷ 164.3625 = 1.22222222 reciprocal

9,467,280 + 164.3625 = 9,467,444.3625 days

9,467,280 - 200.8875 = 9,467,079.1125 days

9,467,280.3625 ÷ 25,920 = 365.2563411 days

9,467,079.1125 ÷ 25,920 = 365.2422497 days

9,467,444.3625 - 9,467,079.1125 = 365.250000 days

When compared to a precise value for the year of exactly 365.25 days, the number of days in excess of 9,467,280 days is in a relationship to the number of days fewer than 9,467,280 days of 45:55, or 9:11. Thus, over a period of 25,920 years of exactly 365.25 days (*i.e.*, axial rotations) each, the correct average lengths for the Sidereal and Tropical Years would be the following:

Sidereal Year = 365.2563411 days

Tropical Year = 365.2422497 days

The differences between these two derived periods and the values currently put forth by astronomers involve only matters of seconds per year, a discrepancy which should not be considered too important over such a long period of time. Even our own modern years have occasional atomic "leap seconds" inserted into them for greater accuracy of measurement.

Because the difference between these two derived values for the Sidereal and Tropical Years is equal to 365.25 divided by 25,920; and because the difference between 9,467,444.325 days and 9,467,079.1125 days is precisely 365.250000 days, we can therefore conclude, just as the Egyptian astronomers told us with the Great Pyramid, that the correct average length of one Earth Year over a period of one cycle of the Precession of the Equinoxes is unequivocally and exactly 365.25 days!

According to the Egyptians, this exact year of 365.25 days coincided with the Star Sirius or "Sothis", as they referred to it. They used this star to calculate the exact start of each new year, so as not to get out of synch with the Precession of the Equinoxes. The Egyptian Sothis Period, as it was called, supposedly lasted for 1,461 years. What is the significance of this number?

First of all and most obviously, 1,461 is the number of days in 4 full years if the Pyramid Year of exactly 365.25 days is used. 365.25 X 4 = 1,461 (and 1,461 ÷ 3 = 487).

Secondly and more importantly, 1,461 is a reference to a cycle of years. As such, it must somehow be reconciled with the Precession Cycle of 25,920 years. But the number 25,920 is not evenly divisible by 1,461. The nearest, closest number than can be evenly divided into 25,920 is 1,440.

25,920 ÷ 1,440 = 18

1,461 - 1,440 = 21

18 X 1,461 = 26,298

26,298 - 25,920 = 378 excess years (or 18 X 21)

However, one orbit by the Planet Earth around the Sun requires 360°, as well as 365.25 axial rotations. Four years equal four orbits. 4 orbits X 360° = 1,440°

Significantly, the difference between the number of axial rotations in 4 years and the number of orbital degrees in 4 orbits is 21, which is also the difference between the Egyptian Sothis Period and the nearest equivalent number of years that can be reconciled with Precession.

Therein lies the elusive key to the Egyptian Sothis Period.

4 years = 1,461 axial rotations

4 orbits = 1,440 degrees

Eventually, after the passage of a number of years, enough increments of 21 days will accumulate in excess of the number of degrees to, in effect, create the possibility for a "statistical adjustment" of the cycle to reflect the fact that the number of days in a given even cycle of years would exactly total the number of degrees in a given even cycle of orbits. Such a point in Time and Space would theoretically allow the calendar to be modified to show exactly 360 days per year, as the Egyptians and Sumerians once recorded, and as Dr. Velikovsky has so ably discussed in Part II of *Worlds In Collision.*

365.25 - 360 = 5.25 days

5.25 X 72 years = 378 days

378 - 360 = 18

72 years X 365.25 days = 26,298 days

72 orbits X 360° = 25,920°

26,298 - 25,920 = 378

After the passage of 72 years of 365.25 days each, or after the passage of 72 orbits of 360° each, enough days would have accumulated in excess of 72 X 360 to allow the calendar to be modified to show the passage of 73 "Orbit-Years" of 360 days each, plus 18 extra days. And keep in mind that, as was stated above, 72 Earth Years are required for the Precession of the Equinoxes to "precede" by one degree. Then, after the passage of 20 such periods of 73 Orbit-Years, an additional 360 days would have further accumulated, requiring that an additional Orbit-Year be added at the end of the twentieth cycle.

360 ÷ 18 = 20

73 X 20 = 1,460

1,440 years X 365.25 days = 525,960 days

1,461 orbits X 360° = 525,960°

20(72 X 365.25) = 20(73 X 360) + (20 X 18), or 525,960 = 525,960

1,440 X 365.25 = (1,460 X 360) + 360

1,440 X 365.25 = 1,461 X 360!

If we were to refer to the Precession of the Equinoxes as "One Galactic Hour", then 1 Galactic Hour would contain 18 Egyptian Sothis Periods of Orbit-Years. By the end of 1 Galactic Hour there would be a total statistical correction by Orbit-Years ("systemic leap years", so to speak) to reflect the numerical difference between 26,298 orbits of 360° each and 25,920 years of 365.25 days each.

25,920 years X 365.25 days = 9,467,280 days

26,298 orbits X 360° = 9,467,280°

25,920 years = 26,298 orbits

378 = 360 + 18

1,461 - 1,440 = 20 + 1

(20 + 1) X 18 = 378

378 + 25,920 = 26,298

Square Root of 378 = 19.4422222222
And, 19.442222 ÷ 4 = 4.860555!

The Square Root of the numerical difference between 26,298 orbits and 25,920 years is 19.4422222222, a number that is in direct sequence with the excess amount of Galactic or Cosmic Precession in 25, 920 years, which is 1.944°. This can be shown as follows:

The annual movement of the Precession of the Equinoxes involves a rate of Zodiac passage of exactly 50.27″ per year. One complete 360° movement is equivalent to a movement of 1,296,000 seconds.

360° X 60′ X 60″ = 1,296,000 seconds

1,296,000" ÷ 50.27″ = 25,780.783 years

If we now multiply an annual rate of Precession of 50.27″ by 25,920 years, we get a movement by the Planet Earth of 361.944° in 25,920 years.

At first glance, this excess movement of 1.944° would seem to invalidate the earlier calculations involving the Egyptian Sothis Period — but don't forget the numerical similarity of 1.944° with 19.4422222222, the Square Root of the excess within the Sothis Period. Therefore, it is the conclusion here that just as the Earth itself has to "precede" by 50.27″ every 1 Earth Year, so also does this entire Solar System have to "precede" with respect to the Cosmos by 1.944° every 1 Galactic Hour. Eventually, just as with the Sothis Period, this excess Cosmic Precession which occurs at the end of every Galactic Hour would total an additional even 360°. The following calculations are performed in order to show at what point in Time and Space this additional accrual of 360° would first be reached.

25,920 X 50.27″ = 1,302,998.4″

1,302,998.4″ ÷ 3600″ (or 1°) = 361.944°

25,920 years = 361.944°

1.944° excess ÷ 25,920 years = 0.000075° per year

360° ÷ 0.000075° excess per year = 4,800,000 years

Thus, after the passage of 4,800,000 Earth Years, there is again enough excess accrual to allow the insertion of an additional Earth Orbit-Year as a systemic leap year. This greater cycle may therefore be called a "Great Cosmic Precession Cycle". It should be noted in passing, however, that the number 4,800,000 itself is not evenly divisible by 25,920; so this Cosmic Cycle is also an intermediary cycle. Both cycles finally resolve evenly at year number 129,600,000, the lowest number into which they can both be evenly divided. This might be termed 2 "Galactic Seasons", each with a duration of 64,800,000 Earth Years (or 25,920 X 2,500). It takes our Solar System approximately 220,000,000 Earth Years to revolve around the Galactic Center, during one "Galactic Orbit-Year". Since 220 million is about twice 129 million, we could say that one Galactic Orbit-Year contains 4 Galactic Seasons, and I strongly suggest that this time-period is exactly 259,200,000 Earth Years.

129,600,000 ÷ 25,920 = 5,000

129,600,000 ÷ 4,800,000 = 27

Note that 25,920 divided by 2 equals 12,960. These are definitely integrated "magical numbers"! Just think about it, readers — a lot of cosmic happenings can transpire 1 Galactic Orbit-Year!

The number 25,920 is an almost sublime number in many respects. It is evenly divisible by all of the following numbers: 1, 2, 3, 4, 5, 6, 8, 9, 10, 12, 15, 16, 18, 20, 24, 27, 30, 32, 36, 40, 48, 54, 60, 64, 72, 80, 90, 108, 120, 144, and many other higher numbers. Notice what is absent from this list: 7, 11, 13, 14, 17, 19, and 21. In fact, none of the cycles involving Precession is evenly divisible by any of those latter numbers. For one of these cycles to be evenly divisible by any of these "odd numbers", it must first be multiplied by that number! For these numbers to come into play in terms of Galactic Mathematics, enormously long periods of time must be considered and from our short-lived, Earth-bound perception really have no immediate relevance. But curious things exist in our history to alert us to these seemingly "eternal" periods. The Mayas have a cycle called the "Atautun" which lasts for 64,000,000 years. The Hindus have a cycle called the "100 Years of Brahma". It lasts for a whopping 311,040,000,000,000 (311.04 trillion) Earth Years, after which time the Universe itself dies and is reborn as a totally new Creation. In the mathematics developed here one "Galactic Night or Day" would equal 311,040 Earth Years (or the equivalent of 12 Galactic Hours of 25,920 years each)! In fact, all of the Hindu Ages and whatnot fit perfectly into this mathematical system, indicating a highly skilled mathematical civilization in æons long forgotten.

This essay is titled an "Introduction" to Galactic Mathematics because here only the Planet Earth is considered. This entire mathematical structure can be applied to all the other planets in this Solar System, as well as the Sun, the Moon and the Planet X Nibiru. 2 + 2 = 4, even in the Pleiades. These additional calculations may be made available at some future time, but suffice it to say here that Nibiru's 3,600-year orbit is also not evenly divisible into 25,920 years. It takes 5 Galactic Hours of 25,920 years each before Nibiru and Earth/Tiamat's combined cycle can complete itself.

5 X 25,920 = 129,600 Earth Years = 36 Nibiruan Years ("Shars")

Notice how all of these numbers leap in and out of one another, overlapping in some of the strangest places, connecting history and mythology to science and mathematics!

For the record, if 1 Nibiruan Year lasts for 3,600 Earth Years, then 1 Nibiruan "Orbit-Day" would last for 10 Earth Years.

3,600 ÷ 360 = 10

Thus:

10 X 12 = 120 Earth Months
120 ÷ 24 = 5 Earth Months per 1 Nibiruan Hour
5 X 30 = 150 Earth Days
150 ÷ 60 = 2.5 Earth Days per 1 Nibiruan Minute
2.5 X 24 = 60 Earth Hours
60 ÷ 60 = 1 Earth Hour per 1 Nibiruan Second

Galactic Mathematics is an almost flawless, perfect numbering system. It shows the structural relationships of numbers, not just their linear progression, as most people today think of and use numbers in their linear decimal system. The existence of Galactic Mathematics alone proves that everything in the Universe is connected.

So, what is the use of all this knowledge, because none of us will live for even 26,000 years? In normal times, probably not too much, except to know it as an occult oddity (separate completely from "numerology", it must be added); but ours are **not** normal times. Mayan scholars — linguists, archaeologists, anthropologists, astronomers, everyone! — all agree that this particular Mayan cycle will terminate on the Winter Solstice of 2012 CE. And by deduction, this date can be tied into the cataclysm years set forth by Dr. Immanuel Velikovsky, as demonstrated in Chapter 18.

We all anxiously feel that we're on the brink of a "new age" of cataclysmic cosmic enlightenment. The Hindu Iron Age is about to end and the sequential Golden Age will be reborn. This knowledge is as timeless as the Earth herself. A periodic cleansing of the Earth is imminent, and we who know that it is coming are indeed privy to some dynamite information.

According to the book *Buddhism In Translations* by Henry Clarke Warren (New York, 1972) in Section 69 titled "World-Cycles", the Buddhist religion contains "myths" of three Heavens: the lowest, in which we reside, is the "Heaven of the Radiant Gods"; higher up the scale is the "Heaven of the Completely Lustrous Gods"; and at the top echelon is the "Heaven of the Richly Rewarded Gods".

The Buddhists have a divine sequence of 64 great world-destructions, each composed of 4 "immensities", much like the four Yugas or "ages" of Hinduism. The first 7 destructions are by fire and affect only our domain of the Heaven of the Radiant Gods. The eighth destruction is by water, cleansing everything from the Heaven of the Completely Lustrous Gods on down. Then the 7 by fire repeat to be followed by a 16th (or second 8th) by water, as above. This cycle repeats 63 times; but the 64th destruction (the 8th 8) is by wind, demolishing even the Heaven of the Richly Rewarded Gods. Then the cycle of 64 starts all over again. This is most certainly what is meant by that passage from the Hindu *Mahabharata* in which it is stated: "The time for the purification of the worlds has now arrived. The period dreadful for the Universe, both moving and fixed, has come." See the quoted material from *Buddhism In Translations* at the end of this chapter for details about these cycles.

One of these great cycles is about to end, on or before the year 2012 CE. In all likelihood, we are approaching the simultaneous ending of the current Hindu "Kali Yuga" of 432,000 years and the Mayan "Atautun" of 64,000,000. There was also a "Heracleitus Cycle" in Ancient Greece (see Appendix A), lasting for 10,800 years, the length of three Nibiruan "shars". It, too, must be ending. And as far as where we are in the Buddhist cycle of 64 (NB: akin to the Mayan number!), it wouldn't be too far off the mark to suggest that we're possibly on the brink of a Buddhist "destruction by wind".

Finally, isn't it interesting indeed that so many ancient peoples, especially the Hindus, used numbering sequences for their cosmic periods that can now be shown to form the essence of Galactic Mathematics, not to mention have also been incorporated into the programming of modern computer equipment? Those peoples were much, much smarter than we give them credit for!

*

As this book was in preparation, the following information came to my attention in late 2005. I am including it here because it seems pertinent to the foregoing discussion. I quote it verbatim, but I have taken the liberty of re-editing some of the original, imprecise English translation.

*

http://www.aztlan.net/rumblings_center_galaxy.htm

"Strange Rumblings At The Center Of Our Galaxy"
By Hector Carreon, *La Voz de Aztlan*

"There are more things in heaven and earth, Horatio, than are dreamt of in your philosophy." William Shakespeare in *Hamlet*

Radio Bursts From The Galactic Center (Hunab Ku)

Los Angeles, California, 18 October 2005 — (ACN) In March of this year, physics professor and astronomer Scott Hyman of Sweet Briar College made a startling announcement. Dr. Hyman and his colleagues wrote in the science journal *Nature* that they had detected something very extraordinary at the center of our Milky Way Galaxy. While analyzing low-frequency radio-wave images taken by the Very Large Array Telescope at Socorro, New Mexico, in 2002, Dr. Hyman and his his fellow researchers detected an intermittent signal that seemed intelligently directed. The signal consisted of five highly energetic radio emissions of equal brightness that lasted 10 minutes each and appeared every 77 minutes over a 7-hour period from September 30 to October 1. The discovery has left Professor Scott Hyman and the entire astronomy community "scratching their heads".

What are these strange rumblings at the center of our Galaxy? Modern-day astrophysicists and astronomers, with their sophisticated scientific instruments, are not the only humans that have contemplated this very same question. In fact, the ancient Mayas, the superb astronomers and mathematicians that they were, also mused on what may exist at the center of our Galaxy. The Mayas knew exactly where this center was located in the sky and had a glyph representing it named Hunab Ku by Dr. José Arguelles of Princeton University, but known to the Mayas as "The Galactic Butterfly". Even more, their entire cosmology and extremely accurate calendars were based on the existence and location of Hunab Ku, and they believed that the future of mankind ultimately depends on what occurs there.

Hunab Ku was, to the Mayas, the supreme God and ultimate Creator. It represented the gateway to other Galaxies beyond our Sun as well as all of the Consciousness that has ever existed in this Galaxy. Hunab Ku, according to the Mayas, is also the Consciousness which organized all matter, from a "whirling disk", into stars, planets and solar systems. Hunab Ku is the "Mother Womb" which is constantly giving birth to new stars, and it gave birth to our own Sun and Planet Earth. They also believed that the "Creator" directs everything that happens in our Galaxy from its center through the emanation of periodic "Consciousness Energy" bursts. Today, modern astronomers have verified that at the center of our Galaxy is a "whirling disk" with a "Black Hole" at its center that is both swallowing and giving birth to stars. Could the strange rumblings observed by Dr. Scott Hyman and his associates earlier this year at the center of our Galaxy be connected with what the Mayas believed?

There appears to be a convergence between what the Mayas believed and understood and what modern scientists are discovering about our Galaxy today. One big difference is how modern physicists view and how the ancient Mayas viewed "Time". Today, physicists view Time as being linear. Time according to modern precepts flows in a straight line. Time can slow down, as per Albert Einstein, but it always flows in one direction. In contrast, the Mayas believed essentially that Time flows in a circle. There is a beginning and an end to things but there is a renewal at the end of the Time cycle. There is a "periodicity" to all manifested phenomena. The Mayan highly accurate Long Count Calendar is based on this precept. The Mayas also believed that Time originates out of the Hunab Ku and is controlled by it.

It is very difficult to dismiss the Mayan Cosmology because they left extraordinary evidence that their Time-keeping methodologies give extremely accurate results. Their pyramids like the one at Chichén Itzá in Mexico are precisely aligned to the the yearly Solstices and the Equinoxes that are caused by the precession of the Earth's axis as it orbits the Sun. The Pyramid of Kukulcan at Chichén Itzá is essentially a superb and highly accurate Time-keeping device that never fails to mark the yearly "Precession of the Equinoxes".

The ancient Mayan astronomers accurately predicted, over 1,500 years ago, the exact alignment of the Earth, the Sun, the star cluster Pleiades and the center of our Galaxy that will take place at the end of the present long cycle in the Gregorian year 2012. On the Mayan Long Calendar the day designated as 4 Ahau 3 Kankin (13.0.0.0.0) falls on December 21, 2012, and this day will mark "El Fin de los Tiempos" ["The End of Times", RS] or the end of the long cycle at which time humanity will experience a new beginning. On this day, according to the Mayas and subsequent Meso-American civilizations, the return of Kukulcan (Quetzalcoatl) will take place.

On December 21, 2012, which coincides with the Winter Solstice, as the Sun sets west of Chichén Itzá, a pattern of shadow and light will project the Plumed Serpent (Quetzalcoatl) descending on the stairs of the pyramid that has a large head of a serpent sculpted in rock at the base. This occurs twice in Chichén Itzá every year, but on the Winter Solstice of December 21, 2012, something very special will happen. As the Sun sets in the early afternoon, the shadow of the pyramid's northwest edge will project a moving pattern of light that joins and illuminates the sculpted serpent head at the base of the stairway. Within a thirty-four minute period, the serpent, formed by this play of light and shadow, will appear to descend to the earth, as the sun leaves each stair, going from the top to the bottom. This combined effect creates the visual appearance of the body of the serpent descending the pyramid. In addition, on December 21, 2012, the tail of the serpent projected up from the top of the pyramid will be pointing precisely to the star cluster Pleiades. The pyramids at Teotihuacán which means "City of the Gods", constructed just north of Mexico City by a Meso-American civilization that preceded the Mayas, are also aligned to the Pleiades as are the Egyptian pyramids at Giza. This is not all: on December 21, 2012, at around 11:11 Universal Time, there will be a precise alignment of our Sun with the Galactic center (Hunab Ku).

Exactly what will happen on December 21, 2012, still remains a mystery, but it seems that we are receiving stronger and stronger clues. The Mayas certainly placed great significance on this date. It is still 7 years away, but our World is already experiencing unprecedented, extreme and rapid changes. The number and magnitude of "natural" catastrophes appear to be increasing exponentially. Scientists have detected extreme and erratic behavior in our Sun that in turn is having strong effects on our Earth's atmosphere and measurable effects on the other planets in our solar system as well. Could the strange rumblings detected by Dr. Scott Hyman at the center of our Galaxy be a cause of the weather and other changes we are experiencing here on Earth today? Will these energetic bursts from Hunab Ku increase in number and intensity as we approach the year 2012? The Mayas would answer "Yes".

On December 12, 2012, the Mayas said that mankind will enter and begin a new Era of Heightened Consciousness. Perhaps, a colossal emission of a yet unknown form of energy will burst out of our Galactic Center on December 21, 2012, that will change the very physics of our World, a new physics that will last until the next cycle.

*

Buddhism In Translations
By Henry Clarke Warren (New York, 1972)
Original Copyright 1896 By Harvard University Press

WORLD-CYCLES (Excerpts, Chapter IV, Section 69)
Translated from the *Visuddhi-Magga* (Chapter XIII)

Call it to mind — Can remember by following either the succession of the groups, or the sequence of births and deaths. For there are six classes of persons who can call to mind former states of existence: members of other sects, ordinary disciples, great disciples, chief disciples, Private Buddhas, and Buddhas. Now member of other sects can call to mind former states of existence for forty world-cycles, and no more.

FOOTNOTE: [*Samyutta-Nikaya*, xv. 5/6] — "It is as if, O priest, there were a mountain consisting of a great rock, a league in length, a league in width, a league in height, without break, cleft, or hollow, and every hundred years a man were to come and rub it once with a silken garment; that mountain consisting of a great rock, O priests, would more quickly wear away and come to an end than a world-cycle. O priest, this is the length of a world-cycle. And many such cycles, O priest, have rolled by, and many hundreds of cycles, and many thousands of cycles, and many hundreds of thousands of cycles. And why do I say so? Because, O priest, this round of existence is without known starting-point, and of beings who course and roll around from birth to birth, blinded by ignorance, and fettered by desire, there is no beginning discernible. Such is the length of time, O priest, during which misery and calamity have endured, and the cemeteries have been replenished; insomuch, O priest, that there is every reason to feel disgust and aversion for all the constituents of being, and to free oneself from them."

And why? On account of the weakness of their wisdom. For their wisdom is weak, as they are unable to define name and form. Ordinary disciples can call to mind former states of existence for one hundred or even one thousand world-cycles, on account of the strength of their wisdom. The eighty great disciples can call to mind former states of existence for one hundred thousand world-cycles; the two chief disciples, for one immensity and one hundred thousand world-cycles; Private Buddhas, for two immensities and one hundred thousand world-cycles, for such is the limit of their earnest wish. But The Buddhas have their power unlimited. ...

As respects, however, *many destructions of a world-cycle etc.*, when a world-cycle is on the wane, that is known as the destruction of a world-cycle; when it is on the increase, that is known as the renovation of a world-cycle. Here destruction includes the continuance of destruction, from being its beginning, and renovation includes the continuance of renovation. Accordingly the four immensities of the following quotation are all included: "There are four immensities, O priests, to a world-cycle. And what are the four? The destruction, continuance of destruction, renovation, and continuance of renovation."

[**COMMENT:** There is conflicting information from around the world as to how long a "world cycle" might last, or how long is the time-period between one great destruction, such as Atlantis, and the next. One *shar* of 3,600 Earth Years is a long time, even for us today. It would have *seemed* even longer to the ancient peoples. Undoubtedly these Buddhist scholars, like the Egyptians, Greeks, Mayas, Finns and others, realized that these "world cycles" recur **after great intervals of time**, and were merely providing their own version of the time-periods involved in these cycles. If one "world cycle" contains four "immensities", and if one "world cycle" lasts for one *shar*, then one "immensity" would last for 900 years. Thus, the previous "immensity of destruction" lasted 1587-688 BCE; the previous "immensity of continuation of destruction", 687 BCE - 212 CE; the previous "immensity of renovation", 213-1112 CE; and the current "immensity of continuation of renovation", 1113-2012 CE. That, I contend, is how one should interpret this information in terms of the time-periods which are involved. RS]

Now there are three destructions: the destruction by water, the destruction by fire, the destruction by wind. And there are three boundaries: the Heaven of the Radiant Gods, the Heaven of the Completely Lustrous Gods, the Heaven of the Richly Rewarded Gods. When a world-cycle is destroyed by fire, it is consumed by fire from the Heaven of the Radiant Gods down. When it is destroyed by water, it is engulfed by water from the Heaven of the Completely Lustrous Gods down. When it is destroyed by wind, it is demolished by wind from the Heaven of the Richly Rewarded Gods down. In lateral expansion it always perishes to the extent of a Buddha's domain. ...

[COMMENT: One can take this literally and infer that after several cycles of "shars", at the end of which the Earth is always destroyed or "purified", there occurs a "greater shar" when entire galaxies are destroyed or "purified"; and at the end of several cycles of "greater shars", there occurs the "greatest shar" and the entire Universe is destroyed or "purified". Or one can interpret this to mean that mankind's memory recalls events connected with destructions by fire, water and wind, occurring simultaneously at the end of *every shar*, because certain areas of the planet would always be flooded by melting polar ice and hurricanes, others burned by erupting volcanic lava and forest fires, and still others torn apart by tornadoes and earthquakes. Personally I favor the latter interpretation. RS]

Now the perishing and the existing of a world-cycle are after the following manner:

When a world-cycle perishes by fire, there arises in the beginning a cycle-destroying great cloud, and a great rain falls throughout one hundred thousand times ten million worlds. The people are delighted and overjoyed, and bring forth seed of all kinds and sow; but when the crops have grown just large enough for cow-fodder, the clouds keep up a braying noise, but do not allow a drop to fall; all rain is utterly cut off. Concerning which the following has been said by The Blessed One:

"There comes a time, O priests, when, for many years, for many hundreds of years, for many thousands of years, for many hundreds of thousands of years, the god does not rain."

Those creatures who depend on rain die, and are reborn in the Brahma-world; likewise the divinities who live on flowers and fruits. When thus a long time has elapsed, here and there the ponds of water dry up. Then, one by one, the fishes and turtles also die and are reborn in the Brahma-world; likewise the inhabitants of the hells. But some say the inhabitants of the hells perish with the appearing of the seventh sun.

But it may be said: "Without the trances, there is no being born into the Brahma-world. Yet some of these beings were overcome by famine, and some were incapable of attaining the trances. How could they be born into that world?" Because of their having attained the trances in the lower heavens.

For when it is known that after the lapse of a hundred thousand years the cycle is to be renewed, the gods called *Loka-byuhas*, inhabitants of a heaven of sensual pleasure, wander about through the world, with hair let down and flying in the wind, weeping and wiping away their tears with their hands, and with their clothes red and in great disorder. And thus they make announcement:

"Sirs, after the lapse of a hundred thousand years the cycle is to be renewed: this world will be destroyed; also the mighty ocean will dry up; and this broad earth, and Sineru, the monarch of the mountains, will be burnt up and destroyed, — up to the Brahma-world will the destruction of the world extend. Therefore, sirs, cultivate friendliness; cultivate compassion, joy, and indifference; wait on your mothers; wait on your fathers; and honor your elders among your kinsfolk."

[**COMMENT:** The above syllable *Loka* is suspiciously similar to the Teutonic word for their mythological god *Loki*. Linguistically these words are identical. Loki was another name for the Teutonic god Midgard, whose name crops up in *Ragnarok* as "Midgard-Serpent". This is the same god as the Nibiruan Duke Nergal. See Chapter 12. If Loki and the other gods inhabit a "heaven of sensual pleasure", then this heaven is the tethered/anchored Planet X Nibiru beyond the north. The word *Sineru* can be compared to the word *Meru*, which was the name of the "Sacred Mountain" of the gods. Mount Sineru, Mount Meru, Mount Olympus, Mount Zion. It may not be that Planet X Nibiru is also destroyed "by fire" at the end of certain world-cycles but rather that it only *appears to be destroyed* from the standpoint of Earth observers, who will see *their own destruction* originating from this "heaven"! RS]

When the people and the terrestrial deities hear these words, they, for the most part, become agitated, and their minds soften towards each other, and they cultivate friendliness, and do other meritorious deeds, and are reborn in the world of the gods. There they have heavenly ambrosia for food, and induce the trances by means of the air-kasina. Others, however, are born into the world of the gods by the alternation of the rewards of their good and evil deeds. For there is no being in the round of rebirth but has an alternation of the rewards of his good and evil deeds. Thus do they attain the trances in the world of the gods; and having there attained the trances, all are reborn in the Brahma-world.

[COMMENT: This reference to "ambrosia" is quite intriguing. It is the contention of Dr. Immanuel Velikovsky in *Worlds In Collision* that the ancient "manna" of the Israelites, the "ambrosia" of the Greeks, and the "morning honey-dew" of others all referred to a hydrocarbon precipitate that fell from the atmosphere early each morning. It was gathered and stored for food during that period following Planet X's previous arrival when the land was infertile as a result of the accompanying destruction. Ancient Buddhist historians might have been simply confused about who exactly is given this ambrosia. As for the word *air-kasina*, I do not know what it means, and it is not explained in the text. RS]

When now a long period has elapsed from the cessation of the rains, a second sun appears. Here is to be supplied in full what was said by The Blessed One in the Discourse on the Seven Suns, beginning with the words, "There comes, O priests, a time."

When this second sun has appeared, there is no distinction of day and night; each sun rises when the other sets, and an incessant heat beats upon the world. And whereas the ordinary sun is inhabited by its divinity, no such being is to be found in the cycle-destroying sun. When the ordinary sun shines, clouds and patches of mist fly about in the air. But when the cycle-destroying sun shines, the sky is free from mists and clouds, and as spotless as a mirror, and the water in all streams dries up, except in the case of the five great rivers.

[COMMENT: Of course, there is no concrete way to comprehend this concept of "the seven suns" without witnessing it for ourselves. However, this idea that there is no distinction of day and night after the "second sun" appears certainly refers to the "Night Sun", that is, Planet X Nibiru at "Hyperborea" as Yggdrasill. See Chapters 6 and 7 of *Slow-Motion*

Doomsday for additional information. Perhaps some of these succeeding "suns" are simply other planetoids and moonlets appearing in the North and dangling from Planet X, as is shown in illustrations on the cover of my previous book, the Voynich Manuscript and the Korean Crown. See Illustrations 26 and 32 of *Slow-Motion Doomsday.* RS]

After the lapse of another long period, a third sun appears, and the great rivers dry up. After the lapse of another long period, a fourth sun appears, and the sources of the great rivers in the Himalaya Mountains dry up, namely, the seven great lakes, Sihapapatana, Hamsapapatana, Kannamundaka, Rathakaradaha, Anotattadaha, Chaddantadaba, Kunaladaha. After the lapse of another long period, a fifth sun appears, and the mighty ocean gradually dries up, so that not enough water remains to moisten the tip of one's finger. After the lapse of another long period, a sixth sun appears, and the whole world becomes filled with smoke, and saturated with the greasiness of that smoke, and not only this world but a hundred thousand times ten million worlds. After the lapse of another long period, a seventh sun appears, and the whole world breaks into flames; and just as this one, so also a hundred thousand times ten million worlds.

All the peaks of Mount Sineru, even those which are hundreds of leagues in height, crumble and disappear in the sky. The flames of fire rise up and envelop the Heaven of the Four Great Kings. Having there burnt up all the mansions of gold, of jewels, and of precious stones, they envelop the Heaven of the Thirty-three. In the same manner they envelop all the heavens to which access is given by the first trance. Having thus burnt up three of the Brahma-heavens, they come to a stop on reaching the Heaven of the Radiant Gods. This fire does not go out as long as anything remains; but after everything has disappeared, it goes out, leaving no ashes, like a fire of clarified butter or sesame oil. The upper regions of space become one with those below, and wholly dark.

Now after the lapse of another long period, a great cloud arises. And first it rains with a very fine rain, and then the rain pours down in streams which gradually increase from the thickness of a water-lily stalk to that of a staff, of a club, of the trunk of a palmyra-tree. And when this cloud has filled every burnt place throughout a hundred thousand times ten million worlds, it disappears. And then a wind arises, below and on the sides of the water, and rolls it into one mass which is round like a drop on the leaf of a lotus. But how can it press such an immense volume of water into one mass? Because the water offers openings here and there for the wind. After the water has thus been massed together by the wind, it dwindles away, and by degrees descends to a lower level. As the water descends, the Brahma-heavens reappear in their places, and also the four upper heavens of sensual pleasure. When it has descended to its original level on the surface of the earth, mighty winds arise, and they hold the water helplessly in check, as if in a covered vessel. This water is sweet, and as it wastes away, the earth which arises out of it is full of sap, and has a beautiful color, and a fine taste and smell, like the skimmings on the top of thick rice-gruel.

Then beings, who have been living in the Heaven of the Radiant Gods, leave that existence, either on account of having completed their term of life, or on account of the exhaustion of their merit, and are reborn here on earth. They shine with their own light and wander through space. Thereupon, as described in the Discourse on Primitive Ages, they taste that savory earth, are overcome with desire, and fall to eating it ravenously. Then they cease to shine with their own light, and find themselves in darkness. When they perceive this darkness, they become afraid. Thereupon, the sun's disk appears, full fifty leagues in extent, banishing their fears and producing a sense of divine presence. On seeing it, they are delighted and overjoyed, saying, "Now we have light; and whereas it has banished our fears and produced a sense of divine presence [*sura-bhava*], therefore let it be called *suriya* [the sun]." Hence they named it *suriya.*

[COMMENT: Is "suriya" another name for Planet X Nibiru, the glorious Night Sun? RS]

After the sun has given light throughout the day, it sets. Then they are alarmed again, saying, "The light which we had has perished." Then they think: "It would be well if we had some other light." Thereupon, as if divining their thoughts the disk of the moon appears, forty-nine leagues in extent. On seeing it, they are still more delighted and overjoyed, and say, "As if divining our wish [*chanda*], has it arisen: therefore it is *canda* [the moon]." And therefore they named it *canda.* When thus this sun and moon have appeared, the constellations and the stars arise. From that time on, night and day succeed each other, and in due course the months and half-months, seasons and years.

Moreover, on the same day with the sun and the moon, Mount Sineru, the mountains which encircle the world, and the Himalaya Mountains reappear. These all appear simultaneously on the day of the full moon of the month Phagguna. And how? Just as when panickseed porridge is cooking, suddenly bubbles appear and form little hummocks in some places, and leave other places as depressions, while others still are flat; even so the mountains correspond to the little hummocks, and the oceans to the depressions, and the continents to the flat places.

Now after these beings have begun to eat the savory earth, by degrees some become handsome and some ugly. Then the handsome despise the ugly, and as the result of this despising, the savoriness of the earth disappears, and the bitter *pappataka* plant grows up. In the same manner that also disappears, and the *padalata* plant grows up. In the same way that also disappears, and rice grows up without any need of cultivation, free from all husk and red granules, and exposing the sweet-scented naked rice-grain. Then pots appear for the rice, and they place the rice in the pots, and place these pots on the tops of stones. And flames of fire spring up of their own accord, and cook the rice, and it becomes rice-porridge resembling the jasmine flower, and needing the addition of no broth or condiments, but having any desired flavor.

Now when these beings eat this material food, the excrements are formed within them, and in order that they may relieve themselves, openings appear in their bodies, and the virility of the man, and the femininity of the woman. Then the woman begins to meditate excessively on the man, and the man on the woman, and as a result of this excessive meditation, the fever of lust springs up, and they have carnal connection. And being tormented by the reproofs of the wise for their low conduct, they build houses, for its concealment. And having begun to dwell in houses, after a while they follow the example of some lazy one among themselves, and store up food. From that time on the red granules and husks envelop the rice-grains, and wherever a crop has been mown down, it does not spring up again. Then these beings come together, and groan aloud, saying, "Alas! wickedness has sprung up among men; for surely we formerly were made of mind." The full account of this is to be supplied with the Discourse on Primitive Ages. ...

[COMMENT: This Discourse on Primitive Ages will not be included here. RS]

Now from the cycle-destroying great cloud to the termination of the conflagration constitutes one immensity, and is called the period of destruction. And from the cycle-destroying conflagration to the salutary great rains filling one hundred thousand times ten million worlds is the second immensity, and is called the continuance of destruction. From the salutary great rains to the appearing of the sun and moon is the third immensity, and is called the period of renovation. From the appearing of the sun and moon to the cycle-destroying great cloud is the fourth immensity, and is called the continuance of renovation. These four immensities form one great world-cycle.

[COMMENT: In terms of a single *shar*, as mentioned above, this description of the four "immensities" simply does not make any sense. It is quite impossible to understand this intended meaning without additional information. RS]

This, then, is the order of events in a world-cycle when it perishes by fire.

But when a world-cycle perishes by water, it perishes in the manner above described, where it was said, "There arises in the beginning a cycle-destroyed great cloud." But there are the following points of difference: — Instead of the second sun, there arises a cycle-destroying great cloud of salt water. At first it rains with a very fine rain which gradually increases to great torrents which fill one hundred thousand times ten million worlds, and the mountain-peaks of the earth become flooded with saltish water, and hidden from view. And the water is buoyed up on all sides by the wind, and rises upward from the earth until it engulfs the heavens to which access is given by the second trance. Having there flooded three of the Brahman-heavens, it comes to a stop at the Heaven of the Completely Lustrous Gods, and it does not settle as long as anything remains, but everything becomes impregnated with water, and then suddenly settles and disappears. And the upper regions of space become one with those below, and wholly dark. This is all as described above; only in this case the world begins to appear again at the Heaven of the Radiant Gods, and beings leave the Heaven of the Completely Lustrous Gods, and are reborn in the Heaven of the Radiant Gods, or in a lower heaven.

Now from the cycle-destroying great cloud to the termination of the cycle-destroying rain is one immensity; from the termination of the rain to the salutary great rains is the second immensity; from the salutary great rains to the appearance of the sun and moon is the third immensity; and from the appearing of the sun and moon to the cycle-destroying great cloud is the fourth immensity. These four immensities form one great world-cycle.

When a world-cycle is destroyed by wind, it perishes in the manner above described, where it was said, "There arises in the beginning a cycle-destroying great cloud." But there are the following points of difference: — Instead of the second sun, there arises a wind to destroy the world cycle. At first it raises a fine dust, and then coarse dust, and then find sand, and then coarse sand, and then grit, stones, *etc.*, up to boulders as large as the peak of a pagoda, and mighty trees on the hill-tops. These mount from the earth to the zenith, and do not fall again, but are there blown to powder and annihilated. And then by degrees the wind arises from underneath the earth, and turns the ground upside down, and throws it into the sky, and areas of one hundred leagues in extent, two hundred, three hundred, five hundred leagues in extent, crack, and are thrown upwards by the force of the wind, and are blown to powder in the sky and annihilated. And the wind throws up also into the sky the mountains which encircle the earth, and Mount Sineru. These meet together, and are ground to powder and destroyed.

Thus are destroyed all the mansions on earth, and in the skies, also the six heavens of sensual desire, and a hundred thousand times ten million worlds. **Worlds clash with worlds,** Himalaya Mountains with Himalaya Mountains, and Mount Sinerus with Mount Sinerus, until they have ground each other to powder and have perished. From the earth upward does the wind prevail, until it has embraced all the heavens to which access is given by the third trance. Having there destroyed three of the Brahma-heavens, it comes to a stop at the Heaven of the Richly Rewarded Gods. When it has thus destroyed everything, it perishes. And the upper regions of space become one with those below, and wholly dark. All this is as described above. But now it is the Heaven of the Completely Lustrous Gods which first appears, and beings leave the Heaven of the Richly Rewarded Gods, and are reborn in the Heaven of the Completely Lustrous Gods, or in some lower heaven.

Now from the cycle-destroying great cloud to the termination of the cycle-destroying wind is one immensity; from the termination of the wind to the salutary great [rain] cloud is the second immensity; from the salutary great [rain] cloud to the appearing of the sun and moon is the third immensity; and from the appearing of the sun and moon to the cycle-destroying great cloud is the fourth immensity. These four immensities form one great world-cycle.

This is the order of events in a world-cycle when it perishes by wind.

Why does the world perish in these particular ways? It is on account of the special wickedness that may be at bottom. For it is in accordance with the wickedness preponderating that the world perishes. When passion preponderates, it perishes by fire; when hatred, it perishes by water. — But some say that when hatred preponderates, it perishes by fire, and that when passion preponderates, it perishes by water. — When infatuation preponderates, it perishes by wind.

Now the world, in perishing, perishes seven times in succession by fire, and the eighth time by water; and then again seven times by fire, and the eighth time by water. Thus the world perishes each eighth time by water, until it has perished seven times by water, and then seven more times by fire. Thus have sixty-three world-cycles elapsed. Then the perishing by water is omitted, and wind takes its turn in demolishing the world; and when the Completely Lustrous Gods have reached their full term of existence of sixty-four world-cycles, their heaven also is destroyed.

Now it is of such world-cycles that a priest who can call to mind former existences and former world-cycles, can call to mind many destructions of a world-cycle, and many renovations of a world-cycle, and many destructions and renovations of a world-cycle.

And after what manner?

"I lived in such a place," etc.

*

In these last paragraphs there exists a contradiction, and one wonders if these Buddhist historians even realized it. If there are seven destructions by fire and an eighth by water, and so on until the final destruction by wind, then this process is quite specific and systematic. Human emotions would never be so predictable. Logically thinking, if emotions play a part in these destructions, then emotions would be more "random" than described here. And even when these emotions were mentioned above, there was a discrepancy regarding the particular emotions that result in destructions by fire and water.

The Himalaya Mountains are in the direct path of 2-3 North Polar Axis Zones. Thus in India there would be numerous "myths" and "old wives' tales" about the various "causes" for these destructions. Old legends about emotional excess would have been passed down, as well as old legends about the specific causes (the differences!) between one shift of the Polar Axis and the next. So I don't really think that any emotions *per se* have anything at all to do with these events.

However, assuming that these cycles have some order to them, and that after a period of "shars" a "greater cycle" occurs, then let's apply some mathematics.

3,600 Earth Years X 8 World-Cycles = 28,800 Earth Years =
1 "Greater Shar" (8 "Shars")

Thus, for 7 "shars" this destruction "by fire" would be localized to our Solar System. At the end of the 8th "shar", there would be a destruction "by water" which would extend even to the Planet X Nibiru, and perhaps beyond into the whole Galaxy itself. This cycle of 8 "shars" repeats 8 times, until the 64th destruction is "by wind".

28,800 Earth Years X 8 Greater Cycles = 230,040 Earth Years =
1 "Greatest Shar" (64 "Shars")

It would seem impossible for humanity to organize itself sufficiently to keep accurate records for a whole period of 230,040 Earth Years. We are incapable of keeping accurate records for only *one shar*, let alone 64 of them!

This sub-cycle of 8 destructions (7 fire + 1 water) could correspond to a series of 8 Polar Axial Zones, and to the 8 cardinal angles of The Great Pyramid of Egypt. Either a corner or a face of The Great Pyramid would always point due-north after *all* Polar Axial Shifts. I would venture a guess that when the Sphinx again points due-north, there will be a destruction "by water", or another 8th in the series. The Sphinx now points due-east. After the North Pole shifts to the area around Siberia's Lake Baikal in or about 2012 CE, then the Sphinx will point due-northeast. Then after this next *shar* ends and the Poles shift again, the Sphinx will move to point due-north again. **THAT NEXT SHIFT** could be a destruction "by water" or even "by wind"! That Polar Axis Shift will occur in 2012 CE + 3,600 Earth Years = 5612 CE. That is a long, long way away, don't you think? It is almost impossible to imagine it.

Assuming that all of this may be valid, then the 2012 Polar Axis Shift will be the 7th in the series and thus be a destruction "by fire", one of the lesser destructions. Where it would fit into the whole scheme of 64 world-cycles is impossible to know. If the destruction in 5612 CE is "by wind", then in 5613 CE a brand-new cycle of 64 will begin. If the 5612 destruction is only "by water" again, then humans will have to wait another 28,800 Earth Years from 5612 CE to see what the subsequent eighth-level destruction will entail, only water or finally wind. I hope that readers can appreciate the enormous problem that we face in keeping such long-term records.

And all of these Buddhist "details" might not even matter in the first place, because this entire ancient document could be wholly historical speculation. But it sure does provide an enjoyable way to pass a few hours in quiet meditation upon "the meaning of life"!

Chapter 18

15 June 762 BCE
A Mathematical Analysis Of Ancient History

*

This treatise was originally completed in the summer of 1979. At that time a copy was sent to Dr. Immanuel Velikovsky in New Jersey. However, Dr. Velikovsky fell gravely ill that autumn and died in November, so I do not know if he ever saw or read it. Eventually in August 1994, Charles Ginenthal of Forest Hills, New York, published the treatise in *The Velikovskian* Journal (Volume II, Number 3). The following version, which I am transcribing for inclusion in *Osiris, Isis & Planet X*, has been edited and updated in places, and an "Epilogue" of chronology prior to 1000 BCE has been appended. Footnote numbers are enclosed in {braces}, with the footnotes listed at the end.

Also, to comprehend fully what is being discussed, one should have already read or at least be knowledgeable about the following books by Dr. Velikovsky: *Ages In Chaos*, *Oedipus & Akhnaton*, *Ramses II & His Time*, and *Peoples Of The Sea*. This treatise is based upon the assumption that the reader is familiar with the *Ages In Chaos* series.

*

INTRODUCTION

The core proposition to be discussed in this chronological treatise is that Dr. Immanuel Velikovsky's date of 26 February 747 BCE for the cataclysm of the ancient 8th century is not correct. To be precise, this cosmic and geological upheaval took place on 15 June 762 BCE, the traditional date of the so-called "Great Eclipse".

Although the notation "BCE" (Before Common Era) is used with the title of this treatise, all dates mentioned are derived from the astronomical base date of -762. If any single date may seem to be "off" by a year or so from other related material on this subject, that is to be expected every once in a while. It is the total structure of the combined synchronicities that is of primary importance here. The revised history of this period cannot be perfected without first having an historical skeleton to hold it together.

The date presented by Dr. Velikovsky{1} is based upon a formidable compilation of calendrical and cultural data, not the least of which — in order of importance — is the celebrated Era of Nabonassar at Babylon, an era that is well established and accurately dated in a number of reliable sources. That era, in and of itself, conforms to what will be outlined here. What is in error is Dr. Velikovsky's interpretation of that era. It did not start as a direct, contemporaneous consequence of the chaos caused by the cataclysm. Rather, the Era of Nabonassar started 15 years *after* the cataclysm.

As Dr. Velikovsky has pointed out, the difference between 747 and 687 is 60 years. This is equivalent to the four 15-year "Martian Cycles" which began in 747, 732, 717 and 702.{2} Significantly, therefore, the date of 762 is exactly 15 years, or one *extra* "Martian Cycle", back from 747. So even my modification of Dr. Velikovsky's date does not contradict his general astronomical formula.

According to George Rawlinson's history of Assyria{3} and Theodor Ritter von Oppolzer's *Canon Of Eclipses*{4}, the "Great Eclipse" that affected most of the countries of the Middle East occurred on 15 June 762 BCE. To be more specific, the darkest path of this total solar eclipse passed across the following present geographical areas: southern Morocco, southern Algeria, Tunisia, southern Greece, Cyprus, Turkey, Iran, Russia, China and the Philippines. All other countries within at least a 300-mile-wide band would also have been in the path of totality — including Egypt, Israel,

Jordan, Syria, Iraq, Ukraine, Italy and Tibet. However, according to the *Canon*, in 747 there was no solar eclipse over any part of the Middle East. Dr. Velikovsky brushes this matter aside by saying the following:

"Joel, Micah and Amos warned in similar terms of 'a day of thick darkness' and 'the day dark with night'. Astronomers, who thought that all this refers to a common eclipse of the Sun, wondered: 'From -763 down to the destruction of the First Temple in -586, no total eclipse was visible in Palestine.' They took it for granted that the Earth revolves along exactly the same orbit and on a slowly rotating axis, and so they questioned: Why did the prophets speak of eclipses when there were none? However, other descriptions of the world catastrophe in these prophets do not accord with the effects of an ordinary eclipse, either."{5}

Dr. Velikovsky based his idea on the premise that Oppolzer's *Canon* is valid only for eclipses that *followed* the final cataclysm because, beforehand, the mechanics of our solar system were completely different from today; thus, all data projections beyond 687 or, more likely, 747 are highly suspect, if not totally inaccurate. It is the contention here, though, that any "cutoff date" for retrospective calculations should be 15 June 762. This "Great Eclipse" marks the point from which our current system of mathematical astronomy should commence.

The date of 15 June 762 BCE is the only opportunity we have during this general time-period to link up the historical record with the astronomical one. And because the combined, synchronized chronologies of the ancient Middle East only fall into place using this traditional date, we must not fail to seize this opportunity in time. Regarding the date of 23 March 687 for the second cataclysm, no major changes need to be made (although the exact month and day of that cataclysm probably need a little more refinement — not a matter that will be addressed in this treatise). That is not to understate the severity of the second event. It merely means that the second event, unlike the first, did not alter the structure of timekeeping on this planet.

The purpose of this treatise is to demonstrate, by means of an integrated, international, ancient chronology, that Dr. Immanuel Velikovsky's date for this cataclysm should be changed. For many years, those who have been expounding on various aspects of Dr. Velikovsky's ideas have been concentrating primarily on the process of historical revision; but to date, no one has come up with a year-by-year, synchronized, historical timetable.

So there continues to be a considerable variety of opinion concerning the specific sequence of events between the end of Egyptian Dynasty 18 and the beginning of Dynasty 19 (or 26). The reason for this disagreement stems directly from the inaccuracy of the cataclysm date of 747, not from any intrinsic failings in the historical reconstruction *per se*. It is absolutely mathematically impossible to reconstruct the history of the region if a date of 747 is adhered to; and if the prophets predicted eclipses when there seemed to have been none, it was not that their predictions were incorrect but that they, as the predictors, are slightly out of synch in history and, therefore, in astronomy.

This treatise is geared more towards mathematics and logic than towards history. Two plus two always equals four. The mathematics of the years (the reigns of the kings, the ages of people, the passing of celestial events, and so forth) must be worked out completely before the events of the years can be satisfactorily correlated. Thus, in this treatise, I have sought to collect — in a single source — most, if not all, of the primary mathematical cross-links pertinent to the reconstructed period and to piece them together. The conclusions that result do not always agree with the ideas of Dr. Immanuel Velikovsky, but, in combination, these conclusions mathematically demonstrate beyond a doubt that Dr. Velikovsky's basic principles are correct. To do this, though, requires that all of the histories of the region be analyzed and synchronized simultaneously. Until now, specialists in one field or another have tackled only a portion of this ancient history, always leaving the impression that what they say is accurate, but never totally quenching the thirst for answers to all the remaining, undiscussed, correlative questions.

In the development of this "synchronology", I followed a technique which generally seems to be ignored or at least downplayed by other historians. According to Rawlinson, the Assyrians and Babylonians counted a year only *after* its passage.{6} If a king ascended to the throne in, say, 700, his "first year" would be 699 and his "fifth year" would be 695. The exact number of years stated is always subtracted from the starting date, as we in the modern-day world commonly count our years (except that we add to an AD date). On the other hand, the Egyptians and the Hebrews counted a year from its beginning, not its end. Thus, if an Egyptian or a Hebrew king began to rule in 700, his "first year" would be 700 and his "fifth year" would be 700-4, not 5, or 696. This is similar to the way that we today count our centuries. The 1900s were called "The 20th Century". I have striven to adhere to this mathematical principle as closely as possible.

What few exceptions exist are readily apparent from a close study of the text and need not be discussed. For Persia, Media and Lydia, the Assyro-Babylonian system of counting was generally employed. Ethiopian dates were always computed according to the Judeo-Egyptian system. The fact that this dual method of counting years is built into the synchronology from the outset further strengthens its already tightly women framework.

This treatise was limited in its details to the span of time between the accession of Egyptian King Amenhotep III The Magnificent in 905 BCE and the Persian Conquest of Egypt in 504 BCE. The mathematical analyses of the auxiliary historical periods before 905 and after 504 adhere to all the major tenets of Dr. Velikovsky in his *Ages In Chaos* series. In the former period, all synchronisms can be developed with accuracy and precision as far back as 1588 BCE, which I feel is the exact time of the Exodus, the "Hyksos" Invasion of Egypt and the Santorini Cataclysm, not to mention the previous Crossover Sequence of Planet X Nibiru. Appended, but not discussed or detailed in this treatise, is a synchronized chronology of these earlier histories of Egypt, Assyria, Babylonia and Israel, back to the year 1588, exactly one "shar" ago from the Mayan End-Time Date of 21 December 2012.

It is my contention that these cosmic events are not *random* occurrences. They are as systematic as the passing of the seasons during the course of an earthly year. Sudden events *do* regularly occur in Space. The Poles do *not* shift at random. There is a structure to all of this, and the only way we are going to unlock its time secrets is by agreeing on dates when we know that events did indeed occur, such as 1588, 1587, 762 and 687 BCE. In the not-too-distant future, Planet X Nibiru will return again. It behooves us now to determine a sequence for this, lest each and every time something "cosmic" happens to the Earth, we again revert back to an age of barbarism and ignorance, attributing the event to the work of an unseen "god". It is imperative that we focus on the larger, scientific picture; and, in this regard, I would welcome correspondence from anyone with similar beliefs.

In the latter period there are no insurmountable problems, with one possible exception: Because the Persian Conquest of Egypt is brought down by 21 years from its normal time, the reigns of Xerxes I and Artaxerxes I should be adjusted to approximately 6 years and 34 years, respectively. This, in turn, forces a short series of duplicate years in the history of Greece. At least five full Olympiads must be removed and recombined with five duplicate Olympiads before or during the Battles of Thermopylae, Salamis and the Eurymedon. Following this, from about the year 449 on down, no more mathematical adjustments need to be made. Egyptian Dynasty 20 poses no problems of synchronization and "ghost correlation", provided that the revolutionary leaders Inaros and the first Amyrtaeus are counted among the last nine Ramssides.

But enough of this Introduction. Let the cold, unbiased mathematics speak for itself.

*

ISRAEL & JUDAH

1. In Table 1, at the end of this section, the period of years from Asa and Ahab to Queen Athaliah and Jehu (from 897 to 855 altogether) has been structured in such a way as to satisfy the requirements for the revised Amarna Period, as discussed in *Ages In Chaos*, Chapter 7. The reign of Ahab (897-868) is the key adjustment to this period. It must be lengthened from its usual 22 years to 29 years,{7} so that it will overlap the sixth year of Shalmaneser III in Assyria (872) and the first year or two of Akhnaton in Egypt (870-868).

2. Uzziah is said to have begun his reign in the 27th year of Jeroboam II.{8} Jeroboam II started in 802; 802 - 26 = 766. If this report is accurate, then all of my subsequent synchronizations collapse *in toto*. Thus, rather than everything else's being wrong, probably only this report about a 27th year is in error; if it had been reported as the 17th year, then everything else would synchronize. Thus, here Uzziah starts in 786, the 17th, not 27th, year of Jeroboam II.

3. The Vision of Amos occurred two years before "The Great Earthquake", which took place during the reigns of Uzziah and Jeroboam II.{9} Thus, the vision was in 764. Even after making all of the required adjustments for the period from 897 to 855, as per *Ages In Chaos*, these two chronologies were once more in perfect internal synchronization with Amaziah and Jehoash. Only then was the modification regarding Uzziah implemented. However, with or without this modification, the year 747 does not and cannot fall into the reign of Jeroboam II (802-760), whereas the year 762 falls into both reigns. This is another reason for the validity of the earlier date of 762.

4. The reign of Tiglath-Pileser III in Assyria *must* overlap the reigns of Menahem in Israel (748-737) and Ahaz in Judah (719-704). Tiglath-Pileser III's fourth to eighth years were contemporaneous with five of Menahem's years,{10} and Ahaz's first year was contemporaneous with one of Tiglath-Pileser III's years.{11} Tiglath-Pileser III's reign must, therefore, last from 744 to at least 718, which increases his reign from its usual 18 years to 26 years. This modification poses no problems elsewhere in the other chronologies.

5. Hoshea of Israel reigned for nine years from the 12th year of Ahaz.{12} Hezekiah ruled 29 years from the third year of Hoshea.{13} During the seventh year of Hoshea and the fourth year of Hezekiah, Samaria came under siege by the King of Assyria. Israel was conquered by Assyria in the ninth year of Hoshea and the sixth year of Hezekiah.{14} Sargon II recorded that he had attacked Samaria in his first year.{15} Sargon's first year in Babylonia was 701. He had ascended to the Assyrian throne 12 years earlier. He and Marduk-Baladan II of Babylonia had ascended to the throne in the same year, and, 12 years later, Sargon II conquered Babylonia.{16} Sargon II started to reign in 714, four years after Tiglath-Pileser III died. His 12th year was 702. A year later, in 701, he attacked Samaria. Here, the year 701 corresponds exactly with the seventh year of Hoshea (707 - 6 = 701) and the fourth year of Hezekiah (704 - 3 = 701).

In turn, these synchronizations permit the 14th year of Hezekiah (691) to coincide with the third year of Sennacherib in the following way: Sennacherib attacked Jerusalem in the 14th year of Hezekiah. He besieged the city for a long time. Then his army was destroyed by unknown "cosmic forces" as it prepared to do battle with the counterattacking Ethiopians who were friends of Hezekiah.{17} Sennacherib's own annals state that he attacked "King Hizkiah" of Jerusalem in his third year, a year or so before his army was destroyed.{18} Here, Sennacherib ascended to the throne in 695. His third year was 692. If Sennacherib started out for Jerusalem in 692 and reached it in 691, then the requirements of both Hebrew and Assyrian histories are satisfied. Elsewhere, this attack on Hezekiah is said to have started in Sennacherib's fourth year,{19} which would coincide exactly with 691, Hezekiah's 14th year.

6. After the destruction of the Assyrian army in 687, Hezekiah became ill. Whilst he was ill, a diplomatic delegation of well-wishers was sent to him from Marduk-Baladan II, who had been restored to the Babylonian throne.{20} The year was 686. Sennacherib reconquered Babylonia in his

tenth year, after a nine-year period of Babylonian independence,{21} and he ruled there for six years:

695 - 10 = 685
685 - 6 = 679

Thus, Marduk-Baladan II was ousted again a year or so after he dispatched the diplomats to Jerusalem.

7. From the beginning of the reign of Uzziah, in 786, through the beginning of the reign of Josiah, in 619, a period of 167 years, no internal structural modifications have been arbitrarily made in the chronologies of Israel and Judah. This period corresponds to biblical data, give or take up to six months. To change the date for the beginning of Josiah's 31-year reign would require a total restructuring of everything else before and after it. Therefore, unless the 31-year reign of Josiah can be satisfactorily synchronized with the complex contemporaneous events in Egypt and Babylonia, the entire chronology presented here collapses. In fact, this whole integrated, international, historical scheme is synchronized so exactly that, for all practical purposes, it must be accepted in its totality or not at all.

Regarding the reigns of Josiah, Ramses II/Necho II and Nebuchadnezzar II/Hattushilis III, Dr. Velikovsky has proposed the following scenario: Ramses II began to rule in 609 BCE, a year before Josiah was killed. Josiah was killed in 608 by an Egyptian archer during Ramses II's first campaign of conquest through Palestine in his second year. To replace Josiah, Ramses II chose Jehoahaz; but, three months later, Jehoahaz was replaced by Jehoiakim. For about the next three years, Egypt controlled Palestine. Then in Ramses II's fifth year, 604, he fought Nebuchadnezzar II at Kadesh-Carchemish. Egypt was defeated and Ramses II's army hastily retreated home in disgrace. They were pursued by Nebuchadnezzar II; but, when Nebuchadnezzar II reached Egypt, he was informed of dynastic troubles in Babylon. So Nebuchadnezzar II hurriedly negotiated a truce with Ramses II and returned to his capital, where in 603, during the fourth year of Jehoiakim, Nebuchadnezzar II became King of Babylonia. Then, for the next few years, Babylonia, not Egypt, controlled Syria and Palestine.{22}

In about his eighth year, however, Ramses II invaded Palestine again.{23} This time, he captured Ashkelon. The next year, Ramses II's ninth year,

Jehoiakim rebelled against Nebuchadnezzar II. Three years later, in 597, the 12th year of Ramses II, Nebuchadnezzar II sent an army to Jerusalem to depose Jehoiakim, who was replaced with Jehoiachin. Nebuchadnezzar II began to distrust Jehoiachin, so a few months later, Zedekiah was installed as king. (Jehoiachin was imprisoned at Babylon for 37 years. He was released by Nebuchadnezzar II's son, Evil-Merodach.)

Eight years later, *c*588, Zedekiah revolted.{24} Nebuchadnezzar II besieged Jerusalem again and Egypt went to the aid of Judah. This was the 21st year of Ramses II. The presence of the advancing Egyptians caused alarm amongst the Babylonians, who began to retreat. During this conflict, Ramses II and Nebuchadnezzar II/Hattushilis III decided that they had had enough of wars against each other. Soon after Nebuchadnezzar II destroyed Lachish, they met and signed the famous Egyptian-Hittite Peace Treaty.{25} This was in the ninth year of Zedekiah. The Egyptians returned home and, two years later, in 586, Nebuchadnezzar II, unopposed by Egypt, finally conquered Jerusalem.

Dr. Immanuel Velikovsky provides some compelling arguments for this historical scenario, but this scenario is out of synch in one key aspect. The death of Josiah did not occur during the second year of Ramses II. It occurred in his 18th year, which was 588 (the same numerical year that Dr. Velikovsky synchronizes with the destruction of Lachish and the peace treaty). This, in turn, means that the Egyptian King who marched to assist Zedekiah was Merneptah I/Hophra-Apries, not Ramses II/Necho II. Therefore, the following scenario is proposed as a substitute for that of Dr. Velikovsky.

Ramses II ascended to the throne as sole ruler in 605, during the reigns of Nabopolassar/Murshilis II in Babylonia and Josiah in Judah. As will be independently demonstrated further on, the fall of Ninevah and the death of Sinsharishkun occurred in 608, but Ashur-Uballit II held on at Harran until 604. In his second year, 604, Ramses II invaded Palestine on his way to assist Ashur-Uballit in his war with Nabopolassar, but Assyria and its ally, Egypt, were defeated. Assyria was conquered by Nabopolassar, but Egypt retained hegemony over Syria and Palestine. In his fifth year, 601, Ramses II marched across Palestine to do battle with Nergilissar I/ Muwattalis at Kadesh-Carchemish. Muwattalis had ascended to the throne in 602 and was assisted in his military ventures by his younger brother, Hattushilis III, who was head of the armed forces. Egypt was defeated. Ramses II returned home in disgrace. Anarchy and confusion prevailed in Palestine, but three years later, in his eighth year, 598, Ramses II did recapture Ashkelon. Then in 588, Ramses II's 18th year, Egypt invaded Babylonian-controlled Palestine once again. It was during this invasion that Josiah was killed and replaced by Jehoahaz and, three months later, by Jehoiakim. This was either at the end of the reign of Muwattalis or at the beginning of the reign of Urkhi-Teshub in Babylonia/Hatti. For the next three years, Egypt controlled the Levant.

In the 21st year of Ramses II, 585, the Egyptians met the Babylonians in battle again in the area of Kadesh-Carchemish. The Egyptians were defeated again. Ramses II retreated and was pursued by Nebuchadnezzar II. However, when Nebuchadnezzar II reached the borders of Egypt after sacking Lachish, he was called home to Babylon because of a dynastic emergency. Before he left the Sinai, Nebuchadnezzar II made peace with Ramses II. This was the famous peace treaty. Then Nebuchadnezzar II returned home, deposed Urkhi-Teshub and was crowned King of Babylonia in 584 — which was, as has been noted, the fourth year of Jehoiakim.

The reign of Nebuchadnezzar II lasted for 43 years: 584 - 43 = 541. His successor, Evil-Merodach, reigned for two to three years. Evil-Merodach released Jehoiachin from prison in the 37th year of Jehoiachin's exile. Jehoiachin was deposed in 576: 576 - 36 = 540, which corresponds to the computed second year of Evil-Merodach.

Going purely by the Hebrew chronology, if the fourth year of Jehoiakim was 584, then the fall of Jerusalem in the 11th year of Hezekiah{26} would have been in 566. It is also reported that Nebuchadnezzar II conquered Jerusalem in his own 19th year,{27} which, by the Hebrew timetable, would be 584 - 18 = 566. The two chronologies are perfectly synchronized because Nebuchadnezzar II's 19th year equals Zedekiah's 11th year.

Ramses II/Necho II died in 570. Necho II was succeeded by Psammetichus II, who reigned no more than six years. Ramses II was succeeded by Merneptah I and later by Seti II and his consort, Queen Tuosri. If Merneptah I equals Hophra-Apries,{28} then he cannot equal Psammetichus II, who immediately succeeded Necho II. Therefore, Psammetichus II probably equals Seti II. Queen Tuosri was probably one of Ramses II's wives. Their reign *preceded*, not followed, Merneptah I.

Hophra-Apries ascended to the Egyptian throne at the end of the reign of Zedekiah, because Zedekiah appealed to him for military assistance when, during his ninth year, the Babylonians attacked.{29} Thus, Hophra-Apries was on the throne in 568, which allows two years for the reign of Seti II/ Psammetichus II, just after Ramses II/Necho II.

Merneptah I fought the Libyans and their allies in his fifth year, which was 564. This battle occurred at the peak of the flood of Hebrew refugees into Egypt, which began in 565, a year or so after the fall of Jerusalem.{30} Then, in the last year of his reign, Hophra-Apries fought the Libyans (or Cyrenians) again. Egypt was defeated. Amasis became king and Hophra-Apries was killed by a mob. But, just before he was deposed and murdered, Hophra-Apries' army was defeated by Nebuchadnezzar II, who triumphantly sailed up the Nile as far as Aswan.{31} If we allow Merneptah I/Hophra-Apries a reign of 21 years, then he would have been deposed in 547. Conveniently, this year corresponds to the 37th year of Nebuchadnezzar II in Babylonia: 547 + 37 = 584.

This derivation leads, in turn, to another minor flaw in Dr. Velikovsky's process of synchronization, concerning the state visit to Egypt by Hattushilis III during the 34th year of Ramses II. Dr. Velikovsky places this event in the 37th year of Nebuchadnezzar II because of a fragmentary inscription on a Babylonian tablet referring to contact between Babylonia and Egypt in Nebuchadnezzar II's 37th year.{32} But:

609 - 33 = 576, and
608 - 37 = 571.

Dr. Velikovsky's mathematics are out of synch by five years. However, according to the chronology presented here,

605 - 33 = 572, and
584 - 37 = 547.

This chronology seems to be even more out of synch — by 25 years. Thus, which approach is correct?

The fragmentary tablet referred to above says, "In the 37th year of Nebuchadnezzar, King of Babylon — the King of Egypt came up to do battle (?) and *-es*, the King of Egypt, — and — of the city Putu-Jaman — far way regions which are in the sea — numerous which were in Egypt — arms and horses — he called to — he trusted."{33}

This tablet could just as easily refer to a state of war as to a peaceful international visit, despite Dr. Velikovsky's interpretation. In this chronology, the 37th year of Nebuchadnezzar II, 547, falls 23 years after the death of Ramses II/Necho II and corresponds to the end of the reign of Merneptah I/Hophra-Apries in Egypt. "Apries" ends in *es* just like "Ramses",{34} even though the former is a Greek name and the latter is Egyptian. Therefore, any reference to Nebuchadnezzar II's state visit to Egypt should be dated in 572, during the 34th year of Ramses II and the 12th year of Nebuchadnezzar.

It is also said that Hophra-Apries' successor, Amasis (whose name ends in *is*), suffered an early defeat by Nebuchadnezzar II but bounced back to rule for over 40 years.{35} Thus, this early defeat probably occurred in Amasis' first year, which corresponded to Hophra-Apries' last year and Nebuchadnezzar II's 37th year.

The end of a 13-year siege of Tyre by Nebuchadnezzar II resulted in his conquering Phoenicia and defeating Egypt. This siege of Tyre was undertaken because of Nebuchadnezzar II's wrath over the flight of Hebrew refugees into Egypt. As a consequence of this defeat, Amasis became King of Egypt.{36} Thus, the siege of Tyre ended in 547. Since it lasted for 13 years, it began in 560, six years after the fall of Jerusalem. This allows an intermediate period of six tense "cold war" years of Hebrew exile into Egypt between 566 and 560, which corresponds nicely to the rest of the combined chronologies.

8. The Edict of Cyrus II The Great of Persia, proclaiming the rebuilding of the Temple at Jerusalem, was issued in Cyrus II's first year.{37} Reconstruction of the Temple was completed in the sixth year of Darius I.{38} As determined independently from the combined chronologies of Persia, Babylonia and Media, the first year of Cyrus II, when the edict was issued, was 538 BCE. Darius I ascended to the throne 36 years later (28.5 + 7.0 + 0.5), in 502. His sixth year was 496 (502 - 6).

Josephus wrote that there was a 70-year interval between the Babylonian conquest of Jerusalem and the first year of Cyrus II.{39} Since the Babylonian conquest was in 566, then the difference between 566 and 538 is only 28 years. But if the accession of Cyrus II is fixed at 538, then 538 + 70 = 608, which is much too early for the Babylonian conquest. Similarly, 566 - 70 = 496, which is much too late for the accession of Cyrus II. Coincidentally, however, 566 - 70 equals the aforementioned sixth year of Darius I. Thus, Josephus must have meant that the difference between the Babylonian conquest and the completion of the Temple, *not* the Edict of Cyrus II, was 70 years. This conclusion reinforces the separately derived accession date of 502 for Darius I, since 496 + 6 = 502.

TABLE 1
CHRONOLOGY OF ISRAEL & JUDAH
*** = Throne Data (All Tables)**

	Israel	Judah
995	*Solomon	
992	Commencement of Temple Construction by King Solomon	
975	State Visit to Jerusalem by Egyptian Queen Hatshepsut Makeda Sheba	
955	Death of Solomon & Division of the Kingdom	
955	*Jeroboam I	*Rehoboam
950		Attack by Egypt, 5 Rehoboam
938	18 Jeroboam I	*Abijah
935	20 Jeroboam I	*Asa
934	*Nadab	2 Asa
932	*Baasha	3 Asa

924		Attack by Egypt, 12 Asa
909	*Elah	26 Asa
908	*Zimri	27 Asa
908	*Omri	27 Asa
897	*Ahab	38 Asa
894	4 Ahab	*Jehoshaphat
869	28 Ahab	*Jehoram
868	*Ahaziah	2 Jehoram
866	*Jehoram	4 Jehoram
862	5 Jehoram	*Ahaziah
861	*Jehu	*Queen Athaliah
855	7 Jehu	*Joash
833	*Jehoahaz	23 Joash
818	*Jehoash	37 Joash
816	2 Jehoash	*Amaziah
802	*Jeroboam II	15 Amaziah
786	17 Jeroboam II	*Uzziah

764		Vision of Amos
762	June 15, "The Great Earthquake"	
760	*11-Yr. Interregnum	27 Uzziah
749	*Zechariah	38 Uzziah
749	*Shallum	38 Uzziah
748	*Menahem	39 Uzziah
740	War with Assyria, 4 Tiglath-Pileser III	
737	*Pekahiah	50 Uzziah
735	*Pekah	52 Uzziah
734	2 Pekah	*Jotham
719	17 Pekah	*Ahaz
719		Conflict with Assyria, 25 Tiglath-Pileser III
714	*8-9 Yr. Interregnum	4 Ahaz
707	*Hoshea	12 Ahaz
704	3 Hoshea	*Hezekiah
701	Assyrian Siege of Samaria, 7 Hoshea	4 Hezekiah

699	*Conquest by Assyria, 9 Hoshea	6 Hezekiah
691		Attack by Assyria, 14 Hezekiah, 4 Sennacherib
688		Vision of Isaiah
687		Destruction of the Assyrian Army, 18 Hezekiah, 8 Sennacherib
686		Visit by Babylonian Diplomats Sent by Marduk-Baladan II
675		*Manasseh
621		*Amon
619		*Josiah
604		End of the Assyrian Empire, 16 Josiah, 2 Ramses II
601		First Battle of Kadesh-Carchemish, 19 Josiah, 5 Ramses II
588		Death of Josiah, 18 Ramses II

588	*Jehoahaz
587	*Jehoiakim
585	Second Battle of Kadesh-Carchemish, Egyptian-Babylonian Peace Treaty, 3 Jehoiakim
584	Accession of Nebuchadnezzar II, Median-Lydian Peace Treaty, 4 Jehoiakim
576	*Jehoiachin
576	*Jehoiachin's Arrest & Imprisonment in Babylon
576	*Zedekiah
572	Vision of Ezekiel, 5 Jehoiachin's Exile in Babylon
567	Military Assistance from Egypt, 9-10 Zedekiah, 2 Merneptah I

566	*Babylonian Conquest 11 Zedekiah, 19 Nebuchadnezzar II
540	Jehoiachin's Release from Prison after 36 Years
538	Edict of King Cyrus II of Persia
496	Temple Restoration, Jerusalem, 6 Darius I

*

PERSIA

1. The complex chronology of Persia (and its subsequent relationships with Greece and Egypt) after the sixth year of Darius I, 496, will not be analyzed here; however, a suggested outline of Persian history, from 575 to 331 (Table 2), is provided for auxiliary reference purposes at the end of this section. Suffice it to say that *the traditional Persian chronology must be shortened by 20 years* after the death of Darius I. This automatically forces a similar shortening, or duplicating, in the history of Greece between 496 and 449, whether this or another as yet developed scenario is eventually adopted. Greek "traditional historians" will not want to tamper with the sequence of the Olympiads, but unfortunately it *must* be done. I certainly hope that Dr. Velikovsky understood the ultimate necessity for this before he died.

2. The sixth year of Darius I was 496: 496 + 6 = 502. Darius I ascended to the throne in 502. He was preceded by the Pseudo-Smerdis and the Magi, who ruled less than one year.{40} Thus, the Pseudo-Smerdis clique ruled in 502.

This Magi government was preceded by Cambyses II, who ruled for seven years{41} and who conquered Egypt around his fifth year.{42}

Thus, Cambyses II reigned from 509 to 502 and conquered Egypt *c*504. Cambyses II's conquest of Egypt ended the one-year reign of Psammetichus III (or Merneptah II/Siptah). From foregoing Hebrew synchronisms, it was determined that the predecessor of Psammetichus III, Amasis (or Amenmeses), began to rule in 547, the 37th year of Nebuchadnezzar II. Thus, Amasis reigned for 42 years, from 547 to 505. Since Amasis' reign is generally reported as lasting for 43 or 44 years, this derived period of 42 years closely corresponds to traditional history.

3. Cyrus II The Great was the father and predecessor of Cambyses II, and he ruled for 29 years.{43} The accession of Cyrus II corresponded to the first or second year of Nergilissar II in Babylonia.{44} Nergilissar II, whose reign is determined from the death of Nebuchadnezzar II in 541, as already discussed, ruled for about three years during 538-535. By first working forwards through the Hebrew and related chronologies to the estimated sixth year of Darius I and then backtracking through the Persian chronology to an accession date of 509 for Cambyses II, we can now double-check the validity of these two dates by adding 29 to 509 to see if the result synchronizes with the reign of Nergilissar II: 509 + 29 = 538. It does. Therefore, Cyrus II did ascend to the throne in 538 and the sixth year of Darius I was indeed (566 - 70 =) 496 BCE.

The accession of Cyrus II occurred simultaneously with the Persian conquest of Media, during which his father, Cambyses I, was killed.{45} Cyrus II conquered Lydia in his 13th year:{46} 538 - 13 = 525 for the conquest of Lydia. Cyrus II conquered Babylonia in his 19th year:{47} 538 - 19 = 519 for the conquest of Babylonia. Now, since the reign of Nergilissar II ended in 535 and that of co-rulers Nabonidus and Belshazzar ended in 519, with the Persian conquest, the intervening 16 years are the correct length of the regnal period of Nabonidus and Belshazzar, the last rulers of Babylonia.

4. Josephus reported that there was an interval of 253 years between the first year of Cyrus II and the death of Alexander III The Great.{48} The date of Alexander III's death is universally undisputed: "He was attacked by a fever, probably brought on by his recent exertions in the marshy districts around Babylon, and aggravated by the quantity of wine he had drunk at a banquet given to his principal officers. He died after an illness of 11 days, in the month of May or June [323 BC]. He died at the age of 32, after a reign of 12 years and eight months."{49}

This reference to 323 BC indicates the historical year 323, which would be the equivalent of scientific -322, referred to here as 322 BCE. This date cannot be altered.

A date of 575 (322 + 253) for the first year of Cyrus II, as reported by Josephus, is clearly impossible.

The date of 285 (538 - 253) for the death of Alexander is also impossible.

However, since Cyrus II was definitely preceded by his father, Cambyses I, the father's reign could have logically and easily lasted for 37 years, from 575 to 538. Because 575 corresponds to 322 + 253, Josephus' 253-year period may actually refer to the interval between Alexander III's death and the dynastic origin of the Persian Empire during the reign of Cambyses I, in 575, rather than to the accession of Cyrus II in 538. Thus here, Cambyses I has been assigned an accession date of 575 BCE.

According to George Rawlinson{50}, the Persian Empire grew out of an already existing monarchy. Darius I recorded that he had been preceded by eight former kings, including Cambyses I, Cyrus II, Cambyses II and the Pseudo-Smerdis. Thus, four other kings antedated Cambyses I. The first of these was Achaemenes. He was followed by his son, Teispes, and then by another conjectural Cambyses and by Cyrus I, who was said to have been the grandfather of Cyrus II The Great.{51}

Allowing these four early monarchs an average reign of 20 years each and adding these 80 years to 575 BCE puts the birth of the Persian monarchy at about 655 BCE, a date which falls within the turbulent period connected with the reigns of Ashurbanipal in Assyria and Tirhakah in Egypt. This period was to have been the subject of Dr. Velikovsky's *Assyrian Conquest*, the middle or third volume in the *Ages In Chaos* series; however, this book was never published, and I feel certain that the reason had to do with Dr. Velikovsky's stubbornness of sticking with the cataclysm date of 747 rather than 762. The mathematics of this middle period simply cannot be determined correctly using a date of 747. Also, Dr. Velikovsky failed to make the identification ("ghost correlation") of Tirhakah with Horemhab in Egypt, a matter which is discussed in the later section about Egyptian chronology.

Regarding the Battle of Marathon, it is usually dated at 490 BCE. Here, however, I have brought it down by 20 years to fit it into the end of the reconstructed chronology. As has been shown, this chronology of Persia is already mathematically linked to other chronologies. It would be impossible to disturb it without disturbing all the others, but its revision conflicts with the history of Greece for 20 years. If Greek history is to remain untouched, then my mathematical analysis is invalidated. This matter is in urgent need of some scholarly investigation to determine exactly how the events of this short period of Greek history should be rearranged. The duplication of 5 Greek Olympiads with 5 subsequent Olympiads also brings down the start of the Olympiads from the traditional date of 776 BCE to 756 BCE, which in turn places it *after* 15 June 762 and closer to the Founding of Rome (753) and Babylonian Era of Nabonassar (747). The original purpose of the inauguration of the Greek Olympiads was to commemorate the end of the Trojan War, but traditionalists would place the end of the Trojan War about 400 years earlier than the Olympiads, followed by a "dark age" devoid of all history. Doesn't it seem more reasonable to date the start of the Olympiads *four* years after the Trojan War rather than *four hundred* years later, if there was indeed a direct cause-and-effect of one upon the other? I think so. And in the process this useless "dark age" is completely eliminated from the history of Greece, as Dr. Velikovsky himself proposed that it should be.

TABLE 2
CHRONOLOGY OF PERSIA

575	*Cambyses I
538	Conquest of Media & Battlefield Death of Cambyses I
538	*Cyrus II The Great
538	Edict of Cyrus II Regarding Jerusalem
525	Conquest of Lydia, 13 Cyrus II
519	Conquest of Babylonia, 19 Cyrus II
509	*Cambyses II
504	Conquest of Egypt, 2 Cambyses II

502 *The Magi & Pseudo-Smerdis

502 *Darius I

496 Restoration of Jerusalem, 5 Darius I

470? Battle of Marathon with Greece

466 *Xerxes I

461 Battles of Thermopylae, Salamis & The Eurymedon with Greece

460 *Artaxerxes I

455 Revolt of Inaros in Egypt

449 End of the Six-Year Egyptian War & Execution of Inaros

449 Battle of Cyprus & Peace of Callias

425 *Xerxes II

425 *Sogdianus

424 *Darius II

404 Restoration of Egyptian Independence & Death of Darius II

404 *Artaxerxes II

358 *Artaxerxes III

340 Reconquest of Egypt

338 *Artaxerxes IV

336 *Darius III

331 *Conquest by Alexander III The Great of Macedonia

*

MEDIA & LYDIA

1. The Kingdom of Media ended in 538, the first year of Cyrus II. According to Herodotus, the Medes had ruled Asia beyond the River Halys for 128 years, not including the Scythian Period, before they were conquered by Cyrus II.{52} The Scythian Period lasted 28 years:{53} 128 + 28 = 156 years. The reigns of the Median Kings are listed below.

TABLE 3
MEDIAN KINGS BEFORE CYRUS II

Deioces	53 Years
Phraortes	22 Years
Cyaxares & Scythians	40 Years
Astyages	35 Years

The total of the above time-periods is 150 years. Thus, Herodotus contradicts himself by six years; and so we must determine the correct sequence. (See Table 5 at the end of this section.)

Cyaxares began to rule in the 34th year of Ashurbanipal in Assyria.{55} This is computed separately as being 632 BCE. Cyaxares survived the Eclipse of Thales, on 28 May 584, by no more than three years; and during that time, he assisted Nebuchadnezzar II's expedition into Syria.{56} Nebuchadnezzar II ascended to the throne in 584; he probably mediated the Median-Lydian Peace Treaty on 28 May 584, as he was returning to Babylon from Egypt, after signing the famous pact during the 21st year of Ramses II, just before he deposed Babylonian ruler Urkhi-Teshub. Herodotus also reported that Cyaxares was the King of Media "who fought the Lydians on the occasion when the day was darkened".{57}

As an aside here, the term "Eclipse of Thales" refers to the fact that this was the first historically recorded actual prediction of a future eclipse. The Greek astronomer who predicted the eclipse was named Thales. And regarding the Eclipse of Thales *per se*, Dr. Velikovsky's date of 30

September 610 BCE{58} cannot fit the integrated mathematics of this historical scheme. Therefore, I believe that his hypothetical date is wrong and that the Eclipse of Thales should be left at its traditional date of 28 May 584.

Thus, Cyaxares' reign must stretch from 632 to at least 582, during which time the Scythians conquered Media. If Cyaxares survived the Eclipse of Thales by no more than three years, then he died *c*582. The difference between 632 and 582 is 50 years, ten years more than Herodotus' 40-year period for Cyaxares and the Scythians. If Cyaxares died in 582 and Cyrus II conquered Media in 538, then the intervening 44-year period automatically becomes the length of the reign of Astyages. This period is nine years longer than Herodotus reported.

Soon after Cyaxares ascended to the throne, he attacked Assyria. During this attack the Scythians invaded Media, so Cyaxares was forced to return home to defend his capital city. He was unsuccessful and the Scythians overran Media.{59} Cyaxares ascended to the throne in 632. If he attacked Assyria in, say, 630, then the Scythian domination of Media began around that same date. Next, we find that by the 11th-12th years of Nabopolassar in Babylonia, Cyaxares, already restored to the Median throne, formed an alliance with Babylonia against Assyria.{60} The 11th-12th years of Nabopolassar were 611-610. Therefore, the Scythian Period in Media lasted from no earlier than 630 to no later than 611. The difference is 19 years. So if Herodotus' 28-year period of Scythian domination is reduced to 18 years, everything else relating to Cyaxares would synchronize with Assyria and Babylonia. Allowing at least one year between the end of the Scythian domination of Media and the alliance between Cyaxares and Babylonia in 611, we arrive at the dates of 630-612 for the Scythian Period in Media.

After the Scythians conquered Media, they pressed on towards Egypt but were repulsed in Syria by Psammetichus I.{61} Psammetichus I ruled 659-605. The year 630 and those immediately following it synchronize perfectly with the middle of the reign of Psammetichus I, so this auxiliary Scythian correlation is valid.

2. According to Herodotus, the Kingdom of Media began from 150 to 156 years before the accession of Cyrus II in Persia: 538 + 150 = 688. This date corresponds nicely with the cataclysm that destroyed Assyrian King Sennacherib's army, as noted by Dr. Velikovsky in *Worlds In Collision.*

However, it is also stated that a revolt of the Medes occurred during the middle of the reign of Sargon II;{62} and Josephus wrote that a revolt of the Medes, against Assyrian domination, occurred just after the destruction of Sennacherib's army.{63} Sennacherib's army was destroyed in 687. The middle third of Sargon II's 19-year reign was from 708 to 701. The period of time between the revolt of the Medes under Sargon II and the actual accession of Deioces could have lasted for 22 years.{64} Adding 22 years to 686 gives the date of 708, which falls into the middle third of Sargon II's reign, or during his sixth through seventh years. If the 53 years of Deioces are subtracted from 708 BCE, then the date of 655 is obtained for the accession of Phraortes. Phraortes was killed and Cyaxares ascended to the throne in 632, the 34th year of Ashurbanipal. The difference between 655 ad 632 is 23 years, and Herodotus reported that Phraortes ruled for 22 years.

3. The Median-Lydian War began six years before the Eclipse of Thales on 28 May 584:{65} 584 + 6 = 590. Alyattes was King of Lydia at the time of the war.{66} Alyattes ruled for 57 years and his son, Croesus, ruled for 14 years before Lydia fell to Persia.{67} Cyrus II conquered Lydia in 525. Adding 14 years to 525 gives the date of 539 for the accession of Croesus. Adding the 57 years of Alyattes takes us back to 596, which was six years prior to the Median-Lydian War. Another synchronism is verified, but please note that this synchronism would not be possible if the Eclipse of Thales is pushed back to 610, as Dr. Velikovsky tried to do.

Alyattes was preceded by Sadyattes, who reigned for 12 years.{68} Sadyattes was preceded by Ardys and by Ardys' father, Gyges. Gyges was the first King of the Mermnadae Dynasty in Lydia, and this dynasty lasted for 170 years before Lydia was conquered by Persia. Gyges and Ardys supposedly ruled for 38 and 49 years, respectively.{69} Adding 170 years to 525 gives the date of 695 for the establishment of the Lydian Mermnadae Dynasty. This conveniently corresponds to the first year of Sennacherib in Assyria.

According to Herodotus{70}, the reign of Gyges commenced with the murder of the last king of the previous Lydian dynasty. This king was named Candaules, but was called "Myrsilis" by the Greeks. As will be shown in the following section on Hatti, Sennacherib — who was murdered like Candaules — equals Murshilis I, who also was murdered and whose name is suspiciously like the Greek name above. Sennacherib ascended to the throne in 695, which corresponds exactly with Herodotus' 170-year period from 695 to 525.

4. The reign of Gyges must synchronize with the reign of Psammetichus I in Egypt because Gyges sent troops to assist Psammetichus I in his wars with Assyria.{71} Psammetichus I equals Seti I The Great. Psammetichus I ruled for 54 years and was succeeded by Ramses II/Necho II in 605: 605 + 54 = 659. Gyges is said to have ruled for 38 years, starting in 695: 695 - 38 = 657. Thus, the reigns of Gyges and Psammetichus I overlap during 659-657. Seti I attacked Palestine in his first year and conquered Asia soon thereafter.{72} Seti's first two years synchronize with Gyges' final two years, so Gyges sent troops to Egypt *c*658.

5. Ashurbanipal attacked Gyges in Lydia shortly after his second attack on Egypt, during the reign of Tanutamon, who succeeded Tirhakah.{73} I shall show in the sections on Hatti and Egypt that Tirhakah died in 653, which would mark the accession of Tanutamon. In his ninth year, Psammetichus I completely defeated Assyria and conquered all of Egypt:{74} 659 - 8 = 651. The reign of Tanutamon ended in 651. Thus, the second Assyrian attack on Egypt occurred *c*652, the second year of Tanutamon and the 14th year of Ashurbanipal.

If Gyges ruled for only 38 years, from 695 to 657, he died too early to have been the King of Lydia when Ashurbanipal attacked Lydia *c*651, just after his second attack on Egypt. If, however, we reverse the lengths of the reigns of Gyges and Ardys, as reported by Herodotus, who exhibited some slight confusion elsewhere in the chronologies of Media and Lydia, then a 49-year reign for Gyges and a 38-year reign for Ardys would make Gyges die in 646, five years after Ashurbanipal's attack on Lydia in 651. Thus, in the chronology presented here, the lengths of these two reigns have been reversed.

6. Revision of the Egyptian history of Dynasty 18 dictates that the history of Mitanni be revised downwards by the same length of time, since certain Kings of Mitanni were contemporaneous with Egyptian Dynasty 18. These Mitannian, or Old Median, Kings also had contact with Assyria and Babylonia. Table 4, below, lists the 12 Old Median Kings before the country was conquered by Assyria *c*730 BCE, 22 years prior to the first revolt of the Medes in 708.

The dates given for this Old Median or Mitannian line are purely conjectural. They are provided for reference purposes only and will not be examined further, except to note that the placement of Tushratta coincides, as it should, with the Amarna Period in Egypt.

TABLE 4
OLD MEDIAN KINGS BEFORE ASSYRIAN CONQUEST

Parattarna	1000-980
Shutarna I	980-960
Saushattar I	960-930
Shattuara I	930-910
Artatama I	910-900
Shutarna II	900-880
Tushratta	880-855
Assyrian Interregnum	855-845
Mattiwaza	845-825
Shattuara II	825-800
Wasashata	800-780
Saushattar II	780-750
Artatama II	750-730
Assyrian Interregnum	730-686 {75}

TABLE 5
CHRONOLOGY OF MEDIA & LYDIA

	Media	**Lydia**
708	First Revolt of the Medes, 6 Sargon II	
708	*22-Year Interregnum	
695		*Gyges
686	*Deioces	

658		Military Alliance with Egypt, 2 Seti I
655	*Phraortes	
651		War with Assyria, 15 Ashurbanipal
646?		*Ardys
632	*Cyaxares, 34 Ashurbanipal	
630	War with Assyria	
639	*18-Year Scythian Interregnum	
612	*Cyaxares (restored)	
611	Alliance with Babylonia	
608	Attack on Ninevah	*Sadyattes
604	Victory over Assyria	
596		*Alyattes
590		Start of Six-Year War Between Media and Lydia
584		End of War & Eclipse of Thales on May 28

582	*Astyages	
539		*Croesus
538	*Conquest by Persia, 1 Cyrus II	
525		*Conquest by Persia, 13 Cyrus II

*

HATTI

According to the Velikovskian School, what are some generally accepted facts about the Hittites, facts that really need no citations?

1. The Hittites were the Neo-Babylonians.

2. Nabopolassar/Murshilis II was the father of Nergilissar I/Muwattalis and Nebuchadnezzar II/Hattushilis III.

3. Nergilissar I/Muwattalis was the father of Labash-Marduk/Urkhi-Teshub, who reigned briefly just *before* Nebuchadnezzar II/Hattushilis III.

4. Nebuchadnezzar II/Hattushilis III was followed by Evil-Merodach (his son), then by Nergilissar II (a cousin, probably, of Evil-Merodach) and finally by Nabonidus (a son of Evil-Merodach) and his co-ruler and heir, Belshazzar (a son of Nabonidus).

5. The end of Neo-Babylonia was, therefore, the end of Hatti.

6. There was an early Hittite king named Suppiluliumas who was contemporaneous with the Amarna Period in Egypt.

7. Another Suppiluliumas was a contemporary of the Ethio-Egyptian Queen Dakhatamon, the widow of King Tirhakah, who made war against

Assyria during the reigns of Esarhaddon and Ashurbanipal. The reign of this Suppiluliumas is, therefore, dependent upon the death of Tirhakah.

8. The Egyptian-Hittite Peace Treaty, dated in the 21st year of Ramses II/Necho II, was signed by Hattushilis III, son of Murshilis II, grandson of a Suppiluliumas. Therefore, if this Hattushilis III equals Nebuchadnezzar II and this Murshilis II equals Nabopolassar, then this Suppiluliumas *must be* the father of Nabopolassar.

A quick mathematical analysis of these eight points yields a fascinating solution. The reigns of the Hittites, from Muwattalis on down, are fixed by the precise synchronizations between Babylonia and the other countries already discussed. Muwattalis ascended to the throne in 602. Nabopolassar/ Murshilis II reigned for 21-22 years:{76} 602 + 22 = 624.

Now, when did Tirhakah reign in Egypt? Tirhakah murdered Shabataka and ascended to the throne of Egypt in the same year that Sennacherib was murdered and Esarhaddon became King of Assyria.{77} Fortunately in this regard, the death of Sennacherib can be calculated with absolute certainty. After the cataclysm of 687, when the Assyrian Army was destroyed at Pelusium, Sinai, Egypt, by "cosmic forces", the New Year Festivals at Babylon were canceled for 20 years — for eight years under Sennacherib and 12 years under Esarhaddon.{78} This means that Sennacherib died in (687 - 8 =) 679 BCE. Tirhakah ruled for 26 years:{79} 679 - 26 = 653. Therefore, Tirhakah died in 653 BCE. His widow, Queen Dakhatamon, wrote a letter to Suppiluliumas I, the contemporary King of Hatti. Therefore, this Suppiluliumas was alive and sitting on the "Hittite" throne in 653.

The difference between 653 (the death of Tirhakah) and 624 (the approximate accession of Nabopolassar/Murshilis II is only 29 years. If the name of Murshilis II's father was Suppiluliumas I, then it logically follows that the Suppiluliumas who corresponded with Queen Dakhatamon was the father of Nabopolassar/Murshilis II.

In 653, Ashurbanipal was King of Assyria whilst his brother, Shamash-Shum-Ukin, was King of Babylonia. Of this there can be no doubt, for these reigns are determined by the rigid date of 679 for the death of Sennacherib. So one of these two men automatically equals Suppiluliumas I, the father of Nabopolassar/Murshilis II. Which one was it? Most certainly Ashurbanipal, because both Ashurbanipal and Suppiluliumas I were dynamic, expansionist movers-and-shakers. Shamash-Shum-Ukin was a weakling who was eventually ousted by Ashurbanipal, who then became King of Babylonia and could, logically, have earned a Hittite title.

A sickly Hittite King named Arnuvantas III ruled for a short time between Suppiluliumas I and his son, Murshilis II.{80} Thus, this Arnuvantas III ruled at some point *c*624, the year of the death of Ashurbanipal/ Suppiluliumas I and (coincidentally?) the date computed above for the accession of Nabopolassar/Murshilis II. As I shall show in connection with the Assyro-Babylonian chronology, the fall of Assyria occurred in 604, the second year of Egyptian Ramses II The Great. Nabopolassar defeated Ashur-Uballit II at Harran in his 18th year.{81} The fall of Ninevah occurred four years earlier{82} in 608. Since Muwattalis took power in 602, just before the Battle of Kadesh-Carchemish in 601 in the fifth year of Ramses II, then the intervening 20-year period is the correct length of Nabopolassar's reign. Thus, the two-year period from 624 to 622 becomes the reign of Assyrian Ashur-Etel-Ilani/Arnuvantas III and satisfies this Hittite requirement.

After Ashurbanipal took control of Babylonia from his brother, Shamash-Shum-Ukin, he appointed a certain Kandalanu to govern Babylonia on his behalf. After Ashurbanipal died, Kandalanu attempted to seize power in Babylonia for himself, but he was routed by Nabopolassar, who then launched his successful revolt from Assyria.{83}

During the two-year period 624-622, Ashur-Etel-Ilani, a son of Ashurbanipal, reigned in Assyria and — at least nominally — in Babylonia. Ashur-Etel-Ilani ruled for about four years and was succeeded by his brother, Sinsharishkun.{84} Since Ashur-Etel-Ilani nominally preceded Nabopolassar in Babylonia and since both were, apparently, the sons of Ashurbanipal/Suppiluliumas I, Ashur-Etel-Ilani was probably Arnuvantas III, the sickly king with the brief reign; but, because Kandalanu also governed Babylonia, he, too, must have a Hittite designation. For the present, therefore, let him equal one of the lesser kings, such as Hantilis I.

Hattushilis III was succeeded by his son, Todkhalijas IV. Two weak Hittite kings named Arnuvantas IV and Suppiluliumas II reigned afterwards, before the Hittite Empire came to a dismal end.{85} Evil-Merodach was the son of Nebuchadnezzar II. Therefore, Evil-Merodach equals

Todkhalijas IV. The last two Babylonian monarchs were Nabonidus and his co-ruler, Belshazzar.{86} Let Nabonidus equal Arnuvantas IV and Belshazzar equal Suppiluliumas II. This leaves only Nergilissar II to synchronize with a Hittite, and he can easily be matched up with another lesser known king, like Huzziyas I. Since Hittite history is quite vague in general, it is not "unreasonable" to duplicate some of the Babylonians with Hittites who seem "out of synch" in earlier time. You readers can be the ultimate judges as to whether you accept this modified Velikovskian Historical Reconstruction.

Table 6, below, lists a generally accepted sequence for the Kings of Hatti.{87} All told, there were about 30 Hittite Kings, of which only 11 have, so far, been synchronized with the known Kings of Babylonia. But the other 19 men *did* exist, so they *must be* correlated with other valid historical personages.

TABLE 6
CORRELATION OF HITTITE & BABYLONIAN KINGS

Hittites		Babylonians
Pitkhanas of Kussara		
Anittas of Kussara		
Todkhalijas I		
Pusarrumas		
Labarnas		
Hattushilis I		
Murshilis I		
Hantilis I	=	Kandalanu
Zidantas I		
Ammunas		
Huzziyas I	=	Nergilissar II
Telipinus		
Alluvamnas		
Hantilis II		
Zidantas II		
Huzziyas II		

Todkhalijas II		
Arnuvantas I		
Hattushilis II		
Todkhalijas III		
Arnuvantas II		
Suppiluliumas I	=	Ashurbanipal
Arnuvantas III	=	Ashur-Etel-Ilani
Murshilis II	=	Nabopolassar
Muwattalis	=	Nergilissar I
Urkhi-Teshub	=	Labash-Marduk
Hattushilis III	=	Nebuchadnezzar II
Todkhalijas IV	=	Evil-Merodach
Arnuvantas IV	=	Nabonidus
Suppiluliumas II	=	Balshazzar

Because Dr. Velikovsky has identified some of these Hittites with Babylonians, it is easy to assume that the rest of them were also Babylonians — or, more specifically, Chaldeans. But there is a catch, a hitch, to this presumption. In a one-on-one match-up between the Chaldeans and the Hittites, which *must* be done for reasons of logic, counting back from Ashurbanipal through the previous 19 Kings of Babylonia, or Chaldea, we arrive at a date of only 800 BCE! Table 7, below, lists the Chaldean kings.{88}

Of these 19 kings, four were definitely Assyrians who conquered Babylonia: Tiglath-Pileser III (or Pul), Sargon II, Sennacherib and Esarhaddon. Ululai was probably an Assyrian (Shalmaneser V) and Shamash-Shum-Ukin, though not a King of Assyria himself, was the son of a King of Assyria. Therefore, we must ask ourselves this highly critical question: Was the Hittite line really a Babylonian, or Chaldean, line; and, if so, did it really begin only in 800 BCE? I think that the answer is "no" because of one primary reason: The Hittites were already a powerful nation during the Egyptian Amarna Period, from about 890 through 842, yet no Hittites remain to fill out a duplicate century in Chaldea — where at least eight kings reigned.

TABLE 7
CHALDEAN KINGS

Marduk-Baladan I
Eriba-Marduk
Naboshumishkun II
Nabonassar
Nabonadinzeri
Nabu-Shum-Ukin
Nabomukinzeri II
Pul (= Tiglath-Pileser III)
Ululai (= Shalmaneser V?)
Marduk-Baladan II
Sargon II
Sennacherib
Marduk-Zakir-Shumi II
Belipni
Ashur-Nadin-Shumi
Nergilishezib
Mushezib-Marduk
Esarhaddon
Shamash-Shum-Ukin

On the other hand, if the Hittite line were, in fact, identical to the *Assyrian* line — since at least five Assyrians must be correlated with Hittite names anyway — then a one-on-one match-up with the Assyrians would take the Hittite line back to *c*980 BCE, a century *before* the Amarna Period. This alternate one-on-one match-up even includes Shamash-Shum-Ukin, Sinsharishkun and Ashur-Uballit II. I believe that this match-up is much more logical and is bolstered by Dr. Velikovsky's own findings.

According to Dr. Velikovsky, the Suppiluliumas of the Amarna Period was probably Shalmaneser III, King of Assyria.{89} The royal annals of Boghazkoi have a close relation to the Assyrian royal annals of Sennacherib, Esarhaddon and Ashurbanipal.{90} Assyria civil law had much in common with the civil laws of Boghazkoi.{91} Hittite culture closely resembles that of the late Assyrian Empire, as well as that of the Neo-Babylonian Empire.{92} The ruined palace of Boghazkoi greatly resembles the Northwest Palace of Ninevah built by Sennacherib.{93}

It is therefore my conclusion that the records of the Hittite kings discovered at Boghazkoi relate the history of the Assyrian New Kingdom, from *c*980 through the fall of Ninevah and Harran in 608-604, and continue with the history of Neo-Babylonia, from the accession of Shamash-Shum-Ukin in 666 through the Persian conquest in 519. The history of these two empires would overlap extensively from the accession of Tiglath-Pileser III in 744 through the Neo-Babylonian conquest of Assyria in 604. This 140-year period is the cause of most of the perplexity amongst scholars who not only are attempting to unravel Hittite history *per se* but are also endeavoring to reconstruct it in terms of Velikovskian principles.

That being the case, we must now devise a tentative match-up between the remaining 19 Hittite kings and the Assyrian line of succession. To start with, we should clear away the reigns of Sinsharishkun and Ashur-Uballit II. Two unimportant Hittite kings around the time of Hantilis I and Huzziyas I were Zidantas I and Ammunas. Let Sinsharishkun equal Zidantas I and Ashur-Uballit II equal Ammunas.

Sennacherib was assassinated by his sons. Murshilis I was assassinated by a brother-in-law.{94} Sennacherib conquered Babylonia. Murshilis I conquered Babylonia.{95} Gyges ascended to the throne of Lydia in 695 by murdering his predecessor, Candaules, whom the Greeks called "Myrsilis".{96} Sennacherib ascended to the throne in 695 and was a contemporary of Gyges. The death of Murshilis I took place shortly after his conquest of Babylonia.{97} Sennacherib reconquered Babylonia in 685 and died only six years later. Under Murshilis I, the Hittite kingdom became the leading great power of the Near East, and Sennacherib was one of Assyria's most powerful emperors. These coincidences provoke the conclusion that Sennacherib equals Murshilis I.

Murshilis I was followed by a period of dynastic wrangling involving such kings as Hantilis I, Zidantas I, Ammunas and Huzziyas I.{98} These names have already been correlated with successors of Sennacherib. These kings were supposedly followed by an obscure "dark age", after which a great lawgiver-king named Telipinus emerged to reform the government, restore the grandeur of the empire and inaugurate a new era.{99} The existence of this "dark age" allows us to extract that period of Hittite history from Murshilis I through Huzziyas I and shift it around chronologically. For this reason, the Hittite kings known as Hantilis II, Zidantas II and Huzziyas II (who followed Telipinus) came *before* those of the same names designated as "I".

Murshilis I was immediately preceded by Hattushilis I, a powerful Hittite king who practiced imperialistic policies.{100} Sargon II, who preceded Sennacherib, also practiced imperialistic policies. Therefore, let Sargon II equal Hattushilis I.

Suppiluliumas I wrote in his annals that his father's name was Todkhalijas, his grandfather's name was Hattushilis and his great-grandfather's name was also Todkhalijas. The father of Asshurbanipal was Esarhaddon. Therefore, let Esarhaddon equal Todkhalijas III. This would mean that the grandfather of Ashurbanipal / Suppiluliumas I was Sennacherib / Murshilis I, not a man named Hattushilis. However, Sargon II equals Hattushilis I, and his father, Tiglath-Pileser III, can easily be correlated with Todkhalijas II, giving a sequence of Todkhalijas, Hattushilis, Murshilis, Todkhalijas and Suppiluliumas, instead of Todkhalijas, Hattushilis, Todkhalijas and Suppiluliumas.

By this point, we have synchronized most of the Hittites with all of the Assyrians and Neo-Babylonians from 744 through 519, with two exceptions: Shalmaneser V (the brother and predecessor of Sargon II) and Shamas-Shum-Ukin (the brother of Ashurbanipal). Let Shalmaneser V equal Arnuvantas II and Shamash-Shum-Ukin equal Hattushilis II.

This still leaves 11 Hittite kings who have not been matched up. In addition to these 11, there was the Suppiluliumas of the Amarna Period. Also, Telipinus wrote that he was preceded by three kings named Labarnas, Hattushilis and Murshilis.{101} This gives us an extra Hattushilis and Murshilis, since all of the other known kings by those names have already been synchronized. Adding these three names to the remaining 11 allows us to push the beginning of the Hittite line back three more kings past Tiglath-Pileser II, or back through Ashureshishi II (*c*1000-980) and Ashurabi II (*c*1025-1000) to Ashur-Ninari IV (*c*1050-1025). Exactly how these remaining Hittite-Assyrian correlations should be made is anybody's guess; but I would like to suggest, for reasons too complicated to be discussed here,{102} that Telipinus, the lawgiver-king be matched up with Adad-Nirari III (824-800).

The City of Tyana, Cappadocia, was founded by a local King Nimrod with assistance from the Assyrian Queen Semiramis. Since there was an early Hittite Queen Tuvannanas, and since the City of Tyana was founded in about 825, or just prior to the rule of Adad-Ninari III/Telipinus, then most likely Queen Semiramis equals Queen Tuvannanas; and there is an obvious linguistic similarity between the two names Tyana and Tuvannanas. For additional information the reader is referred to my book *Apollonius Of Tyana & The Shroud Of Turin*, Chapter 8, "The White Syrians Of Aramaean Cappadocia".

It should be emphasized again that a strictly Chaldean-Hittite synchronization, even including the three extra kings noted above, would carry the Chaldean line back only to *c*840, which is still too late for precise mathematical coordination with the height of the Amarna Period in Egypt.

Finally, I would like to pose a bizarre question for general contemplation: If the Hittites were actually the Assyrians, rather than the Chaldeans (that is to say, if the Boghazkoi records written in the Hittite-Chaldean language merely retell the history of the Assyrian New Kingdom from another perspective), then *could it possibly be that Hattushash/Boghazkoi is the true site of the lost city of Ninevah?* An interesting case can be made for such an idea, as radical as it may seem. I urge that this be seriously considered.

*

ASSYRIA & BABYLONIA

1. Although the actual histories of Assyria and Babylonia, up through the Era of Nabonassar, are about as obscure and formidable as the histories of other nations, their precise genealogies are not. Assyrian and Babylonian genealogies, with their detailed, integrated cross-references, are superior to those of neighboring countries. Therefore, when these genealogies are tampered with, incredibly complex problems emerge. However, because of the restructuring of the Egyptian New Kingdom, these genealogies *must* be altered. This, in turn, upsets practically every currently accepted tenet about the evolution of Assyrian and Babylonian culture, language, politics and warfare. Traditional Assyriologists will inevitably prove to be some of the most intransigent, argumentative detractors from the Velikovskian School.

Before I begin this cursory process of Assyro-Babylonian synchronization, I must say that I do not accept Dr. Velikovsky's contention that Shalmaneser III of Assyria was the same king as Burnaburiash II of Babylonia.{103} Burnaburiash II married a daughter of King Ashur-Uballit I of Assyria.

Around the time of Shalmaneser III (unless one first reconstructs the history of Assyria), there was no Assyrian king named Ashur-Uballit I. Moreover, if Burnaburiash II was indeed a king of Assyria, then why would it be said that he married a daughter of the contemporary king of Assyria? Did he marry his own daughter or sister? Obviously not. Such an unnatural act, from the prevailing ancient Asian point of view, could not possibly have gone unnoticed or been covered up by every single historian. Besides, Dr. Velikovsky's reasons for this idea are sketchy at best. Nevertheless, this is a grave problem for Assyriologists because the successful, correct synchronization of Burnaburiash II is crucial to Dr. Velikovsky's entire thesis. If the Burnaburiash problem cannot be resolved satisfactorily, historians, as a whole, will always remain somewhat skeptical of Dr. Velikovsky's theories. Let us, therefore, attempt to resolve it and get it behind us.

2. A predecessor of Burnaburiash II, named Kara-Indash, signed a treaty of friendship with Assyria. When Burnaburiash II ascended to the throne, he and the contemporaneous Assyrian King Ashur-Uballit I reaffirmed their support for this treaty.{104} A few years later, as Assyria grew more and more powerful, Burnaburiash II, for his own protection, decided to cement this alliance by marrying Ashur-Uballit I's daughter, Mubalitat-Sherua. She and Burnaburiash II had two sons named Karakhardash II and Kurrigalzu II. When Burnaburiash II died, his older son, Karakhardash II, ascended to the throne. He was very pro-Assyrian. Two years later, he was ousted and killed in a *coup d'état* by Nazi-Bugash, a militant Chaldean nationalist. But two or three years after that, Ashur-Uballit I invaded Babylonia, killed Nazi-Bugash and installed his younger grandson, Kurrigalzu II, on the throne. Ashur-Uballit I was succeeded in Assyria by his son, Enlil-Nirari; and Kurrigalzu II was eventually followed by his son, Nazi-Maruttash, who was contemporaneous with Assyrian King Adad-Ninari I, a grandson of Enlil-Ninari.{105}

Burnaburiash II was the contemporary of Amenhotep III and Akhnaton in Egypt.{106} Therefore, his reign must be brought down to start before and end after 870, the date of the death of Amenhotep III and the accession of Akhnaton. But, in 870, Babylonia was already being ruled by a king named Marduk-Zakir-Shumi I, the contemporary of Shalmaneser III in Assyria, who was also contemporaneous with the Amarna Period in Egypt.

When Shalmaneser III ascended to the throne of Assyria, toward the end of the reign of Amenhotep III, he and the Babylonian King Nabu-Apal-Iddin renewed an already existing peace treaty between Assyria and Babylonia. About seven years later, Nabu-Apal-Iddin died and his older son, Marduk-Zakir-Shumi I, ascended to the throne. Two years later, Marduk-Zakir-Shumi I's younger brother, Marduk-Bel-Usate, seized power in a *coup d'état.* Marduk-Zakir-Shumi I got military aid from Shalmaneser III and, together, they retook Babylonia. Marduk-Bel-Usate was killed and Marduk-Zakir-Shumi I was restored.{107}

Marduk-Zakir-Shumi I remained in power through the end of the reign of Shalmaneser III. At that time, Assyria began to experience internal revolts and anarchy. The old and dying Shalmaneser III had one of his sons, Shamshi-Adad V, put an end to the trouble and restore the peace.{108} In connection with all of this trouble, Shalmaneser III's other son, Ashur-Danin-Apal, seized power in Babylonia in an attempt to take control of the entire Assyro-Babylonian Empire. Shamshi-Adad V met him in battle and was victorious. Ashur-Danin-Apal was ousted and killed.{109} Marduk-Zakir-Shumi I became King of Babylonia. Shamshi-Adad V returned home and became the Emperor of Assyria. Seven years later, there was more trouble in Babylonia. Shamshi-Adad V invaded again, and Marduk-Balassu-Iqbi fled for his life. He was succeeded by Baba-Aha-Iddina, who ruled only a year or so before Shamshi-Adad V, as reconqueror of Babylonia, temporarily abolished the Babylonian throne. For the next few years, several Babylonian kings, about whom virtually nothing is known, ruled somewhat in exile. They included Nabomukinzeri, Marduk-Belzeri and Marduk-Baladan.

As can be seen, these two historical scenarios, now thought to be separated by several centuries, contain similarities. Because we are *forced* by Egyptian historical reconstruction to correlate these two Assyro-Babylonian scenarios, we must now decide which of the later Assyrian kings equals Ashur-Uballit I and which contemporaneous Babylonian king equals Burnaburiash II.

Let Shalmaneser III (878-843) equal Ashur-Uballit I and let Marduk-Zakir-Shumi I (872-847) equal Burnaburiash II. This immediately satisfies *all* Egyptian cross-references with Amenhotep III (903-870) and Akhnaton (870-853).

That being the case, then the Assyrian Princess Mubalitat-Sherua was the daughter of Shalmaneser III/Ashur-Uballit I. If her marriage to Marduk-Zakir-Shumi I/Burnaburiash II took place *c*865, then their two sons,

Karakhardash II and Kurrigalzu II, could easily have been about 18-20 years old during 847-844. Thus, they would have to be correlated with two of Marduk-Zakir-Shumi I's successors.

Let Karakhardash II equal Baba-Aha-Iddina and Kurrigalzu II equal Marduk-Balassu-Iqbi. This reverses the accepted, though vague, order of succession of Marduk-Balassu-Iqbi and Baba-Aha-Iddina, but this reversal appears to be necessary.

We can now develop the following combined synchronization: Ashur-Dan II / Ashur-Belnisheshu of Assyria and Shamas-Mudammik / Kara-Indash of Babylonia signed a friendship treaty *c*940. This treaty was renewed a second time *c*878 by Shalmaneser III / Ashur-Uballit I and Nabu-Apal-Iddin / Kadashmanenlil II, and a third time *c*872 by Shalmaneser III and Marduk-Zakir-Shumi I / Burnaburiash II. Marduk-Zakir-Shumi's rebellious younger brother, Marduk-Bel-Usate, who staged the first *coup d'état* in 870, would perhaps equal one of the extra, little-known, earlier Babylonians, such as Kudurenlil, Shagaraktishuriash, Enlil-Nadin-Shumi or Kadashmankharbi II.{111}

Following his restoration after the *coup d'état*, Marduk-Zakir-Shumi I / Burnaburiash II married Assyrian Princess Mubalitat-Sherua. Upon his death, his son, Baba-Aha-Iddina / Karkhardash II, took over. He was soon ousted in a second familial *coup d'état* by his maternal uncle, Ashur-Danin-Apal / Nazi-Bugash, the brother of Shamshi-Adad V / Enlil-Nirari. After this revolt was quashed, Shamshi-Adad V's younger nephew, Marduk-Belassu-Iqbi / Kurrigalzu II, became king of Babylonia for about seven years. He was disloyal to his imperial Assyrian uncle, who invaded Babylonia and temporarily ended the monarchy. During this interregnal period, three obscure kings ruled Babylonia from exile. They were Nabomukinzeri I, Marduk-Belzeri and Marduk-Baladan I. These men can easily be correlated with Shagaraktishuriash, Enlil-Nadin-Shumi and Kadashmankharbi II.

Shamshi-Adad V / Enlil-Nirari was succeeded by his wife, Semiramis / Tuvannanas, for five years (during which time she helped to found the Hittite-Cappadocian City of Tyana in 825) and then by his son, Adad-Nirari III, who would equal Arik-Denilu, the son of Enlil-Nirari. Adad-Nirari III / Arik-Denilu placed Eriba-Marduk on the Babylonian throne; and, in the parallel scenario, Nazi-Maruttash, the son of Kurrigalzu II, followed the obscure exile period. Therefore, let Eriba-Marduk equal Nazi-Maruttash.

Adad-Nirari III / Arik-Denilu was followed by his son, Shalmaneser IV / Adad-Nirari "I". This Adad-Nirari "I" was a contemporary of Nazi-Maruttash.

In the Assyro-Babylonian chronological table which follows, several other correlations have also been made; however, at this point in time and research, no more will be attempted. All of the Hittite names, from Pitkhanas of Kussara through Suppiluliumas II, have also been included, although some of them will undoubtedly have to be recorrelated as more precision of cross-synchronization is achieved.

On occasion, I have spent literally days on end trying to unravel and restructure these "Intra-Assyrian Ghost Correlations", as I call them. There are scores of bizarre-sounding names to consider. The most difficult part of the process is trying to decide exactly where to break the highly synchronized Assyro-Babylonian king lists and put them back together again without disrupting the sequences of names in the established cross-genealogies. Entire dynasties get shifted around. Extremely unexpected coincidences pop up to pose peculiar, anachronistic questions, like "Was Sennacherib another name for Tukulti-Ninurta I?" Often it is necessary to triplicate the dynastic lines — or quadruplicate them when the Hittites are also counted. But this is a task which must be completed someday, despite the consequences to art and military history; for if this puzzle cannot be solved, then the entire Velikovskian historical reconstruction will continue to be vulnerable to attack by academic traditionalists.

3. The summary chronology presented in this treatise, based upon the foregoing necessity for Intra-Assyrian Ghost Correlations, *does* allow a lot of harmony with existing historical realities. Moreover, the reigns of the Assyrian kings from Adad-Nirari II through the Babylonian conquest generally correspond to accepted data.{112} Only two reigns have been shortened. The reign of Shalmaneser V (718-714) was shortened from six to four years because Tiglath-Pileser III's final year must overlap Ahaz in Judah and Sargon II's first year in Babylonia must equal 701. The reign of Sennacherib (695-679) was cut from 24 to 16 years, but Sennacherib's own annals go no higher than his 16th year.{113} Sennacherib's third and fourth years must overlap 691, the 14th year of Hezekiah; and Sennacherib died eight years after 687.

Only three reigns have been lengthened, those of Shalmaneser IV, Tiglath-Pileser III and Sargon II. Shalmaneser IV has to be lengthened from nine or ten years to 30 years because of the fixed placements of his father and his son. Eight more years must be added to Tiglath-Pileser III because of the cross-synchronization with Ahaz; and Sargon II's 19 years, as opposed to the usual 16 or 17 years, are mathematically dictated by their relationship to the dates of 714, 702, 701, 699 and 695.

The following are other cross-synchronizations which might be mentioned.

a. Kurrigalzu I was contemporaneous with Amenhotep III because Amenhotep III married his daughter.{114} These two reigns overlap between 903 and 895, so this correlation is validated.

b. According to the monuments, certain Assyrian and Babylonian kings were contemporaneous;{115} and in the chronology presented here, the following pairs of reigns each overlap as they are supposed to do:

Tiglath-Pileser II & Nabomukinpal
Tiglath-Pileser II & Ninurta-Kudurusur
Tiglath-Pileser II & Marbita-Ahheidin
Ashur-Dan II & Shamas-Mudammik
Adad-Nirari II & Shamas-Mudammik
Tukulti-Ninurta II & Nabu-Shum-Ukin
Ashurnasirpal II & Nabu-Apal-Iddin
Shalmaneser III & Marduk-Zakir-Shumi I

c. The reigns from Sargon II and Marduk-Baladan II on down have already been discussed in related historical schemes and will be further partially analyzed in the section on Egypt, so there is no need to reconsider them here.

4. Finally, we are left with only one additional matter to analyze. The so-called "Great Eclipse" occurred in the eighth year of the Assyrian King Ashur-Dan III.{116} Ashur-Dan III ruled for 18 years. His eighth year must be synchronized with the 8th-century cataclysm.

747 + 8 = 755 for the accession of Ashur-Dan III
755 - 18 = 737 for the accession of Ashur-Nirari V
737 - 8 = 729 for the accession of Tiglath-Pileser III
729 - the minimum 18 = 711 for the accession of Shalmaneser V
711 - the minimum 4 = 707 for the accession of Sargon II

As can be seen, a date of 747 for the Great Eclipse cannot be accommodated by the integrated chronology presented here. In fact, an immutable cataclysm date of 747 mathematically cancels out all that has hitherto been discussed.

Dr. Velikovsky dated the Assyrian conquest of Israel in 722, which he was probably considering as Sargon II's (very) first year.{117} Ignore for the moment the contradiction between this date and the 707 computed above for the accession of Sargon II. Now add the minimum four for Shalmaneser V: 722 + 4 = 726. Add the minimum 18 for Tiglath-Pileser III: 726 + 18 = 744. This date coincides with my chronology; but it contradicts the computed date of 729 above, for the accession of Tiglath-Pileser III, where, keep in mind, 747 (not 762) was made to equal the eighth year of Ashur-Dan III. Compare it also with the chronology of Israel. The "Great Earthquake" occurred during the reign of Jeroboam II.{118} Samaria fell in 722. Add to 722 Hoshea's 8 years, Pekah's 20, Pekahiah's 2, Menahem's 11 and no interregnum. The result is 763 to 752 for the reign of Menahem. This is too early for proper correlation with Tiglath-Pileser III; and 747 now falls into the reign of Pekah, not Jeroboam II, whose reign, according to this sequence, would equal *c*817-775 with an interregnum, or *c*806-764 without an interregnum. Also, if Sargon II sacked Samaria as early as 722, then Hezekiah's sixth year was 722: 722 + 5 = 727. Hezekiah ruled for 29 years:{119} 727-29 = 698. By this scenario, Hezekiah would have died 11 years *before* the destruction of Sennacherib's army in 687, which is in clear violation of the established biblical facts.

However, by adding an extra 15-year "Martian Cycle"{120} to 747, we arrive at the year 762, the traditional date for the Great Eclipse.

762 + 8 = 770 for the accession of Ashur-Dan III
770 - (18 + 8) = 744 for the accession of Tiglath-Pileser III
744 - (26 + 4) = 714 for the accession of Sargon II in Assyria

So, if the cataclysm is moved back to 762, everything else falls into place like clockwork, and Dr. Velikovsky's mathematical thesis — that the period between the cataclysm of 7XX and 687 must equal a given number of 15-year "Martian Cycles" — is preserved. Here, we have five such cycles, whereas in *Worlds In Collision* there are only four.

It is my belief that much of the confusion and speculation regarding Dr. Velikovsky's never-published, mysterious century-and-a-half of *Assyrian*

Conquest stems directly from unquestioned adherence to a cataclysm date of 747. It is not a paradox of there being too many or too few years to accommodate the history of that century and a half. Rather, it is a simple problem of precisely coordinated mathematics based around a fixed date of 15 June 762 BCE.

TABLE 8
CHRONOLOGY OF ASSYRIA & BABYLONIA

	Assyria	**Babylonia**
1050	*Ashur-Nirari IV/Buzur-Ashur IV/Pitkhanas of Kussara	*Ninurta-Kudurusur II/Agum II
1045?	Founding of Hattushash	
1025	*Ashurabi II/Enlil-Nirari II/ Anittas of Kussara	*Shiriktu-Shukamuna/ Burnaburiash I
1000	*Ashureshishi II/Ashurabi I/Todkhalijas I	*Marbita-Pileser/Kashtiliash III
980	*Tiglath-Pileser II/Ashur-Nirari I/Pusarrumas	*Nabomukinpal/Ulamburiash
965		*Ninurta-Kudurusur III/ Agum III
960		*Marbia-Ahhedin/ Kadashmankharbi I
957	*Ashur-Dan II/Ashur-Belnishesu/Murshilis	

950		*Shamas-Mudammik/Kara-Indash
940?	Friendship Treaty Between Assyria & Babylonia	
938	*Adad-Nirari "II"/Ashur-Rimnisheshu/Hattushilis	
920		*Naboshumukin/Kurrigalzu I
918	*Tukulti-Ninurta II/ Ashur-Nadinahi/Labarnas	
903	*Ashurnasirpal II/ Eriba-Adad I/Zidantas "II"	
895		*Nabopaladan/ Kadashmanenlil I
878	*Shalmaneser III/ Ashur-Uballit I/ Suppiluliumas	
878	Friendship Treaty Renewal Between Assyria & Babylonia	
872	6 Shalmaneser III	*Marduk-Zakir-Shumi I/ Burnaburiash II

872	Friendship Treaty Renewal Between Assyria & Babylonia	
870	8 Shalmaneser III	**Coup D'Etat* by Marduk-Bel-Usate/Kudurenlil
869	Invasion of Babylonia, 9 Shalmaneser III	*End of Usurpation, Restoration of Marduk-Zakir-Shumi I
865	Marriage of Ashur-Uballit's Daughter & Burnaburiash II	
847	31 Shalmaneser III	*Baba-Aha-Iddina/ Karakhardash II
845	33 Shalmaneser III	**Coup D'Etat* by Ashurdanipal/Nazi-Bugash
844	Invasion of Babylonia, 34 Shalmaneser III	*End of Usurpation
844		*Marduk-Balassu-Iqbi/ Kurrigalzu II
843	*Shamshi-Adad V/Enlil-Nirari III/Huzziyas "II"	
837	Invasion of Babylonia	*Assyrian Interregnum

833		*Nabomukinzeri I/ Shagaraktishuriash (in exile)
832		*Marduk-Belziri/Enlil-Nadin-Shumi (in exile)
830		*Marduk-Baladan I/ Kadashmankharbi II (in exile)
829	*Empress Semiramis/ Tuvannanas	
825	Founding of the Hittite-Cappadocian City of Tyana by Empress Semiramis & Tyanaean Hittite King Nimrod	
824	*Adad-Nirari III/Arik-Denilu/Telepinus	
810		*Babylonian Monarchy Restored
800	*Shalmaneser IV/Adad-Nirari "I"/Alluvamnas	
780		*Noboshumishkun II/ Kadashmanturgu
770	*Ashur-Dan III/ Hantilis "II"	
762	"The Great Eclipse" on June 15, 8 Ashur-Dan III	

752	*Ashur-Nirari V/Arnuvantas I	
747		*Nabonassar/Kadashmanenlil II/Nebuchadnezzar I & Celebrated "Era of Nabonassar"
744	*Tiglath-Pileser III/ Pul/Todkhalijas II	
740	War with Israel, 4 Tiglath-Pileser III, 9 Menahem	
733		*Nabonadinzeri
730	Conquest of Palestine & Arabia	
729	Battle of the Two Sipparas	*Temporary Conquest by Assyria
724	Restoration of Babylonian Independence	*Nabu-Shum-Ukin
719	Conflict with Judah, 1 Ahaz	*Nabomukinzeri II
718	*Shalmaneser V/ Ululai/Arnuvantas II	
717	Temporary Reconquest of Babylonia by Assyria	

714	*Sargon II The Great/ Hattushilis I	
713	Assyro-Babylonian War	
712	Battle of Raphia with Egypt, 14 Piankhi II/Sheshonk/So	
703	More War with Egypt	
702	Reconquest of Babylonia	*Assyrian Rule Restored
701	Siege of Samaria 7 Hoshea, 4 Hezekiah	Sargon II's First Year in Babylonia
699	Conquest of Israel, 9 Hoshea, 6 Hezekiah	
695	*Sennacherib/ Murshilis I/ Candaules/Myrsilus	Independence Rebellion
694		*Marduk-Zakir-Shumi II
694		*Marduk-Baladan II
693		*Belipni
691	Attack on Jerusalem 4 Sennacherib, 14 Hezekiah	*Ashur-Nadin-Shumi

690	Battle of Altaku with Egypt, 14 Shabaka/Takeleth/Sethos	
688	7 Sennacherib/ Murshilis I	*Marduk-Baladan II (restored)
687	Battle of Pelusium, Sinai, with Egypt & Destruction of the Assyrian Army 18 Hezekiah, 3 Osorkon II	
686		Babylonian Diplomats' Visit to Jerusalem, 19 Hezekiah
686		*Nergilishezib
685		*Mushezib-Marduk
685	Conquest of Babylonia 10 Sennacherib	*Assyrian Conquest & Interregnum
679	Murder of Sennacherib/ Murshilis I/Candaules	
679	*Esarhaddon/ Todkhalijas III	
671	Conquest of Egypt, 8 Esarhaddon, 8 Tirhakah	
669	Restoration of Egyptian Independence, 10 Tirhakah	

667	Renewed Attack on Egypt, 12 Esarhaddon, 12 Tirhakah	
666	*Ashurbanipal/ Suppiluliumas I	*Shamash-Shum-Ukin/ Hattushilis II
662	Attack on Egypt, 4 Ashurbanipal, 18 Tirhakah	
656	Attack by Egypt at Kadesh-Carchemish, 4 Seti I	
652	Attack on Egypt, 8 Seti I, 2 Tanutamon	
651	War with Lydia	
651	Defeat by Egypt, 9 Seti I	
646		*Kandalanu/Hantilis I
630	War with Media, 6 Cyaxares	
624	*Ashur-Etel-Ilani/Arnuvantas III	
622		*Nabopolassar/Murshilis II
618	*Sinsharishkun/ Zidantas "I"	
611		Alliance with Media

608	Destruction of Ninevah by Babylonia & Media	Conquest of Ninevah, 14 Nabopolassar
608	*Ashur-Uballit II/ Ammunas (in Harran)	Start of 23-Year War with Egypt
604	Alliance with Egypt, 2 Ramses II	
604	*Fall of the Assyrian Empire	Defeat of Assyria by Babylonia & Egypt
602		*Nergilissar I/Muwattalis
601		First Battle of Kadesh/ Carchemish, 5 Ramses II
590		*Labash-Marduk/Urkhi-Teshub
585		Second Battle of Kadesh/ Carchemish & Peace Treaty with Egypt, 21 Ramses II
584		*Nebuchadnezzar II The Great/Hattushilis III
572		State Visit to Egypt by Hattushilis III, 12 Nebuchadnessar II, 34 Ramses II

566	Conquest of Judah, 11 Zedekiah, 19 Nebuchadnezzar II
560	Renewed Warfare with Egypt, Start of Siege of Tyre
547	Conquest of Egypt, End of Siege of Tyre, Nebuchadnezzar II's Visit to Aswan, 37 Nebuchadnezzar II
541	*Evil-Merodach/Todkhalijas IV
540	Release of Jehoiachin after 36 Years' Imprisonment in Babylon
538	*Nergilissar II/Huzziyas I
535	*Nabonidus/Arnuvantas IV/ & Belshazzar/Suppiluliumas II
519	*Conquest by Persia, 19 Cyrus II

*

EGYPT

1. The state visit by Queen Ma'atkhare Hatshepsut Makeda Sheba of Egypt and Ethiopia to Jerusalem/Punt occurred in her ninth year.{121} This corresponded to the beginning of the last half of Solomon's 40-year reign.{122} Herein, this state visit to Jerusalem is dated for convenience in Queen Hatshepsut's 11th-12th year, 975 BCE. The attack on Jerusalem by Thutmose III/Shishak occurred in the fifth year of Rehoboam.{123} The times of Solomon and Rehoboam are mathematically fixed by the rest of the Hebrew chronology, with its links to the chronologies of the other countries already discussed and their combined chronological synchronizations with the cataclysm dates of 762 and 687. The fifth year of Rehoboam must equal 950 BCE. This attack on Jerusalem occurred during the 22nd year of Thutmose III.{124} Thus, Thutmose III ascended to the throne in 971 BCE. These are dates which cannot be altered. Therefore, the 35-year reign of Amenhotep III began *c*905. Going forwards through Akhnaton, Smenkhare, Tutankhamen and Ay puts us at 842. These dates are already synchronized with the chronologies of Israel, Judah, Assyria and Babylonia, and adhere to the principles of Dr. Velikovsky. The death of Ay is the point at which Dr. Velikovsky's unpublished historical gap begins, following the book *Oedipus & Akhnaton*. See Table 11, at the end of this section, for my fully reconstructed Egyptian chronology.

2. As was determined from the earlier analyses, Ramses II/Necho II ascended to the throne as sole ruler in 605 and died in 570. Part of his overall 66-year reign was as co-regent with his father Seti I/Psammetichus I: 570 + 66 = 636.

Seti I/Psammetichus I ruled for 54 years: 605 + 54 = 659. Seti I ascended to the throne in 659. This is the point at which Dr. Velikovsky's historical gap ends before the book *Ramses II & His Time* commences. Thus, this gap of the unpublished *Assyrian Conquest* can be exactly dated during 842-659 BCE, slightly fewer than 200 years. So when scholars mention a 150-year-long gap, they are referring to this period; and a *lot* of Egyptian history must go into this period. How should it be arranged?

3. The Ethio-Egyptian King Tirhakah, who was never fully analyzed by Dr. Velikovsky, is nevertheless mathematically fixed to an accession date of 679. See the discussion of Hatti above, item 8. Tirhakah recorded that he was the son of Piankhi The Great, who was also an Ethio-Egyptian king.{126} Tirhakah was preceded, however, not by Piankhi but by his

nephew, Shabataka, whom he murdered after about a 13-year reign.{127} Thus, Shabataka ascended to the throne *c*692. Shabataka was preceded by his father, Shabaka, who also claimed to be a son of Piankhi.{128} This was the second of two Shabakas.{129} This Shabaka II ruled for 12 to 15 years:{130} 692 + 15 = 707.

So, if Ay died in 842 and Shabaka II began to rule *c*707, we have automatically narrowed the missing gap from 183 years to 135 years. Into this 135 years, four more Ethiopian kings and about a dozen Libyans must be squeezed; but one of these Libyans, Sheshonk III, ruled for 51 years:{131} 135 - 51 = 84. Only 84 years remain for about 15 kings; but one of these, Sheshonk IV, ruled for 37 years: 84 - 37 = 47. Now only 47 years are left for the remaining four Ethiopians and ten Libyans: 47 ÷ 14 = 3.36 years per king. This is highly unrealistic, assuming that each name represents only one monarch and that there were no long periods of a divided kingdom in Egypt.

Moreover, the 15th year of Sheshonk III must be synchronized with the cataclysm of 7XX, when "the heavens devoured the moon".{132} This is a peculiar statement from Egyptian history, and clearly it refers to what the Israelites called "The Great Earthquake" and the Assyrians remembered as "The Great Eclipse". If that date signifies the commencement of the departure sequence of Planet X Nibiru, then the Egyptian "moon" may refer to Planet X and this "devouring" by "the heavens" could refer to an almost instantaneous disappearance of the electromagnetic tethering beam from Planet X to our North Pole, "The Cosmic Treetrunk". When this "rainbow bridge" suddenly existed no more, it might have appeared to those who were watching in Egypt that it had been "devoured" by the background sky. But as much as I would like to believe that, it is pure speculation at this point.

If 747 *does* equal the cataclysm, then Sheshonk III took over in (747 + 14 =) 761 BCE: 761 - 51 = 710. He would have to be followed at once in 710 by Sheshonk I, who reigned for at least 21 years, because he conquered Palestine in his 21st year{133} and was the Pharaoh So, the contemporary of Hoshea:{134} 710 - 21 = 689. We immediately have to follow this with Osorkon II because his third year *must* synchronize with 687.{135} As he took over in 689, this scheme superficially appears to work out perfectly.

But wait a minute. We know that Shabaka II and Shabataka, as late as they were in the Ethiopian line, were *not* contemporaneous with Sheshonk

I and Osorkon II, as early as they were in the Libyan line. This dynastic overlapping is absolutely prohibited by all of the evidence of history and architecture. We hit a snag, a true blind-alley-type paradox, which forces us to adopt a different method of analysis.

4. It has already been shown that Thutmose III was called Shishak in the Scriptures.{136} It has also been proven that Akhnaton, who went blind, is identical with the blind King Bocchoris or King Anysis.{137} Anysis was a predecessor of an Ethiopian king named Shabaka.{138} A war between Egypt and Ethiopia occurred during the reign of Tutanhkamen.{139} Clearly this Amarna Period "Shabaka" could not have been the father of Shabataka because their two eras are separated by a full century. Thus, this Shabaka I followed soon after the Amarna Period, since he *did* follow Anysis, that is, Akhnaton.

The father of the blind King Bocchoris was named Tefnakhte.{140} Therefore, Amenhotep III, Akhnaton's father, becomes the equivalent of Tefnakhte.

As can be seen, three instances of duplications already exist in Dynasty 18: Thutmose III = one of the Sheshonks, but not Sheshonk I or Sheshonk III; Amenhotep III = Tefnakhte; and Akhnaton = Bocchoris-Anysis. Much later on, as was shown by Dr. Velikovsky in *Ramses II & His Time*, all of Dynasty 19 equals all of Dynasty 26; and Dynasty 20 is also a duplicated dynasty. Since duplications exist at both ends of the New Kingdom, is it not logical to further assume that duplications could exist throughout the entire period, including the 150-year gap? If so, then the number of kings would be cut in half, and the lengths of the reigns could be stretched beyond an average of 3.36 years to more realistic proportions.

Two lists of kings are correlated in Table 9. The left-hand column contains kings from Dynasties 18, 24-25 and 19. The right-hand column contains the kings of Dynasties 22, 23 and 26. Also included, separately, are six miscellaneous names. These lists were compiled from four reliable sources{141}, and these kings are listed in their generally accepted order of succession. In parentheses next to each king's name, except for those of Dynasties 19 and 26, is the length of the reign as determined from historical or monumental sources. The letters A, B, C and D indicate that these four pairs of names already refer to identical monarchs. For Ramsside Dynasty 20 see Table 11.

The total number of years in the left-hand column is 274 + X, not counting Horemhab. In the right-hand column the total is 210 + Y. The letters X and Y represent the combined lengths of the unknown reigns in each column.

TABLE 9
EGYPTIAN GHOST CORRELATIONS

Dynasties 18, 24-25 & 19			Dynasties 22, 23 & 26
(54) Thutmose III	A	L	Sheshonk I (21)
(26) Amenhotep II	B	B?	Osorkon I (15)
(9) Thutmose IV	?	M	Takeleth I (25-29?)
(35) Amenhotep III	C	N	Osorkon II (23)
(17) Akhnaton	D	P	Sheshonk II (7+)
(1) Smenkhare	E	R	Takeleth II (14)
(7) Tutankhamen	F	F	Takelothis (13)
(2-4) Ay	G	K	Sheshonk III (51)
(50) Shabaka I	H	G	Pamai (2+)
(?) Kashta	J	?	Sheshonk IV (37)
(?) Piankhi I	K	H	Petasebast (?)
(21+) Piankhi II	L	J	Osorkon III (?)
(12-15) Shabaka II	M	E	Peshee (2)
(12-14) Shabataka	N	C	Tefnakhte (?)
(26) Tirhakah	P	D	Bocchoris-Anysis (?)
(2+) Tanutamon	R		
(?) Sethos	M	A	Sethos (?)
(?) Harmais	P	B	Zerah (?)
(8-59?) Horemhab	P	L	So (?)

Ramses I	=	Necho I
Seti I The Great	=	Psammetichus I The Great
Ramses II The Great	=	Necho II The Great
Seti II & Tuosri	=	Psammetichus II
Merneptah I	=	Hophra-Apries
Amenmeses	=	Amasis
Merneptah II/Siptah	=	Psammetichus III

Now we must look for additional "ghost correlations". "Ay" could be a syllable of "Pamai", which could also be spelled "Pam-ay"; both kings reigned about 2+ years. Therefore, let Ay equal Pamai, designated by the letter G.

Osorkon II must synchronize with 687. But this date falls into the approximate period for Shabataka, from 692 to 679, as computed above. Therefore, let Osorkon II equal Shabataka. Mathematically, this reign must be fixed at (687 + 2 =) 689 through 679, so that the third year coincides with 687. Its duration can only be ten years; and the references to the 12-14 years of Shabataka, or the 23 years of Osorkon II, can only refer to a previous point in time, such as the start of a revolution or a co-regency. This correlation is designated N.

Petasebast was followed by Osorkon III. Both reigns preceded Piankhi I, but Osorkon III was somehow connected with Piankhi I.{142} Therefore, let the two predecessors of Piankhi I, in the left-hand column, become the equivalents of Petasebast and Osorkon III in the right-hand column. Shabaka I = Petasebast, and Kashta = Osorkon III; these pairs are indicated by the letters H and J. Kashta, incidentally, was the father of a Piankhi.{143}

As was noted previously, the crucial 51-year reign of Sheshonk III must be linked to the first cataclysm; using a date of 747 leads us up that mathematical blind-alley paradox. So let us try 15 June 762 BCE. Sheshonk III's 15th year was 762.

762 + 14 = 776 for his accession
776 - 51 = 725 for his death

This reign ended in 725, about 20 years before we are required to start the reign of Shabaka II. Preceding Shabaka II was Piankhi II The Great, whose reign lasted for 21+ years. These two periods harmonize quite well. Therefore, Piankhi I, the predecessor of Piankhi II, would become the equivalent of Sheshonk III. Yet another connection can be made (letter K), and it is reinforced again by the fact that a flood *did* occur during the reign of Piankhi I.{144}

Shabaka II was a contemporary of Sargon II.{145} Sargon ruled from 714 to 695. Shabaka II died in 689, after a reign of 12-15 years: 689 + 14 = 703. An accession date of 703 for Shabaka II makes him the contemporary of Sargon II during the years 703-695; and since the reign of Hoshea was 706-699, Hoshea was also contemporaneous with Shabaka II during the years 703-699.

The father of Shabaka II was Piankhi II The Great. Piankhi II conquered all of Egypt in his 21st year, reuniting the foreign-held north with the south once again.{146} Sheshonk I (the contemporary of Hoshea and the So of the Scriptures) marched on Palestine and took scores of cities by his 21st year. Sheshonk I was a great king, as was Piankhi II. Both conquered territory in their 21st years. Therefore, let Piankhi II The Great = Sheshonk I = So. This duplication is designated by the letter L. Piankhi II followed Piankhi I, who must equal Sheshonk III. Thus, Piankhi I/Sheshonk "III" reigned 776-725, and Piankhi II/Sheshonk "I"/So ruled 725-703, after which year his son, Shabaka II, came to power, as computed above.

Sargon II took over in Assyria in 714. In his second year, he fought a "Prince Shabaka" of Egypt at Raphia, but this Shabaka was not the actual ruler of Egypt:{147} 714 - 2 = 712. In 712, Piankhi II/Sheshonk I/So was the king of the Nile, so this "Prince Shabaka" was most likely his heir and co-regent, Shabaka II. Then, in his seventh year, Sargon II battled Egypt and its ally, Queen Tsamsi of Arabia:{148} 714 - 7 = 707, which still falls within the first 20 years of Piankhi II/Sheshonk I/So.

In 706, Hoshea ascended to the throne in Samaria. In the following year, 705 (the 21st year of Piankhi II/Sheshonk I/So), all of Egypt was conquered by Piankhi. Hoshea hurriedly switched his allegiance from Assyria to So, the Egyptian king, causing an invasion of Israel by Assyria.{149} Hoshea undoubtedly switched his allegiance because of Piankhi's successes in Egypt. Then Sheshonk marched across Palestine between 705 and 704, conquering over 100 cities. A third battle between Sargon and Egypt occurred in Sargon's 11th year:{150} 714 - 11 = 703. This date coincides with Sheshonk's march through Asia. After this battle with Sargon in 703,

Piankhi II/Sheshonk I/So returned home a hero but died a few months later. Or, perhaps, he died just prior to this third battle with Assyria because in 703 Shabaka II *must* ascend to the throne of Egypt.

Sargon II conquered Babylonia in 702 and marched on Samaria in 701. Samaria fell in 699, and Sargon died four years later. He was followed by Sennacherib in 695. Sennacherib attacked Jerusalem in 691, on his way to do battle with the Egyptians at Altaku (El-Tekeh).{151} Egypt was soundly defeated, and Sennacherib turned his attention back to the north. A few years later, toward the end of the siege of Jerusalem, Sennacherib moved on Egypt again. He camped at Pelusium to do battle with an Egyptian king whose troops were about to be reinforced by an Ethiopian army commanded by Tirhakah; but before this battle could get underway, the Assyrian army was annihilated by a mysterious "plague", the origin of which was attributed to "cosmic forces". Sennacherib retreated to Ninevah in dismay, ending his military adventures against Egypt and Jerusalem.{152} Eight years later in 679, Sennacherib was murdered, and Esarhaddon ascended to the Assyrian throne. That same year, Tirhakah murdered Shabataka and took control of all of Egypt.

Now we are faced with a peculiar dilemma. In an extract from his unpublished *Assyrian Conquest*,{153} Dr. Velikovsky wrote that Sennacherib was responsible for the installation of Horemhab as King of Egypt because, at that same time, Horemhab married Sennacherib's daughter, Mutnodjme. Horemhab was the brother of an Egyptian King Sethos, who was an adversary of Sennacherib in the years preceding the disaster at Pelusium.{154} Thus, in order to determine the "ghost correlation" of Horemhab, we need to determine exactly who his brother Sethos was.

Sennacherib encountered the Egyptians twice: in 690 at Altaku and in 687 at Pelusium. In 687, Shabataka/Osorkon II was King of Egypt. In 690, Shabaka II was king. Thus, one of these two men must equal Sethos. Which of them had a brother named Horemhab-Harmais, who would have married an Assyrian princess? Shabataka was the son of Shabaka II, who was a son of Piankhi II The Great. Tirhaka was also a son of Piankhi II. Thus, Shabaka II and Tirhakah were brothers — just as were Sethos and Horemhab-Harmais. If Shabaka II equals Sethos, then the encounter between Sethos and Sennacherib was at Altaku in 690, just prior to the mathematically mandatory accession in 689 of Shabataka/Osorkon II.

Since Egypt was defeated at Altaku, the death of Shabaka II could possibly have occurred as a result of this battle.

But according to Dr. Velikovsky, at about this same time, Sennacherib installed Horemhab and Mutnodjme on the Egyptian throne, and Horemhab was a contemporary of Tirhakah. Shabataka was also Tirhakah's contemporary and became King of Egypt during Sennacherib's reign. If Shabataka equals Horemhab, then he certainly could not also equal Horemhab's brother, Sethos; but if Sethos equals Shabaka II, then Shabataka could not equal Horemhab because Shabataka was the son, not the brother, of Shabaka II. Also, if Shabataka equals Sethos, then Shabaka II could not equal Horemhab because Horemhab followed Sethos, whereas Shabaka II preceded Shabataka; and if Shabataka equals Sethos, a subsequent king would have to equal Horemhab — and in this case it could not be Tirhakah because Horemhab did not assassinate Sethos, nor could it be Tanutamon because Tanutamon followed Sennacherib by over 25 years. Therefore, Shabataka cannot equal either Sethos or Horemhab.

There is only one logical avenue of escape before we reach yet a second blind alley: Shabaka II, the brother of Tirhakah, equals Sethos, the brother of Horemhab-Harmais. Thus, Tirhakah and Horemhab were *one and the same person*, not just "contemporaries" — notwithstanding the arcane fact that the two of them are supposedly pictured together on a wall relief at Karnak.{155} In Table 9 these two correlations are designated by the letters M and P.

Horemhab married Mutnodjme, an Assyrian princess. After the death of Tirhakah, his widow, Dakhatamon, wrote a letter to Hittite King Suppiluliumas I (that is, Ashurbanipal of Assyria), requesting that he send her a new husband. If Dakhatamon equals Mutnodjme, an Assyrian princess, then why should she not have appealed to her royal Assyrian nephew, Ashurbanipal, for a new Assyrian husband to replace her dead one, the Ethio-Egyptian general and king Tirhakah/Horemhab? Moreover, pursuing Dr. Velikovsky's logic, if Shabataka *cannot* equal Horemhab and if Tirhakah *does not* equal Horemhab, then *where do we place a separate Horemhab*, because we have completely run out of years, as well as kings, to duplicate? But if Horemhab and Tirhakah were not the same man, then who was Horemhab's brother, Sethos? If we make that brother equal to Seti I or Seti II, as Dr. Velikovsky has done,{156} then where do we get an extra 19th Dynasty king to match up with Psammetichus II of Dynasty 26? Regarding this final point, Dr. Velikovsky has provided no solution. But if this is true, then where would we squeeze an extra Seti into the very tight Ethiopian period that has no room for Horemhab in the first place? There is no other conclusion than that Horemhab *must* equal Tirhakah!

Now we must return to the list of Libyan "ghosts". Who were Shabaka I/Sethos and Tirhakah/Horemhab amongst the Libyans? Coincidentally, and quite conveniently, there is one pair of brothers in the Libyan line. Takeleth II and Sheshonk II were brothers and, supposedly, followed Osorkon II.{157} Since Osorkon II, also coincidentally, must correlate with Shabataka in 689, then Shabaka II and Tirhakah may also be called Takeleth and Sheshonk. And since the order of the Libyans seems so confused and distorted already, I believe that Shabaka II/Sethos was Takeleth I (not Takeleth II) and that Tirhakah/Horemhab was Sheshonk II. This allows the father of Osorkon II to be named Takeleth I, as he is generally known, and it does not change the paired brothers' names. If we now let Tanutamon equal Takeleth II (the letter R), then the only one of the first six Libyans out of synch with the generally accepted order of succession is Osorkon I; and even these Libyans no longer seem so confusing. Tanutamon was not the son of Tirhakah. His father was probably Shabataka; no one really knows for sure. But if Takeleth II was the son of Osorkon II, then, at least here, Takeleth II, as Tanutamon, follows Osorkon II, just as Shabataka follows Shabaka II. Yet another traditional correlation is preserved and integrated.

At this point, in the left-hand column, only Thutmose IV, Smenkhare and Tutankhamen have not been synchronized with other names. Smenkhare reigned for one year. Peshee reigned for two years. Let Smenkhare = Peshee (letter E). Let Tutankhamen = Takelothis (letter F).

Only Thutmose IV remains on the left, whilst only Osorkon I and Sheshonk IV remain on the right. Three Apis bulls were born in the reign of Sheshonk IV, the last one in his 37th or final year. Another Apis bull, born in the 28th year of Sheshonk III, died in the second year of Pamai.{158} Since Ay must equal Pamai, then a Sheshonk connected with Apis bulls must precede him, as does Thutmose IV. I believe that the Apis bull story of Sheshonk "III" actually refers to the second of the three Apis bulls of Sheshonk "IV". The reign of Thutmose IV was not 37 years long, but that of Amenhotep III was 35 years long. More than likely, Amenhotep III equals Sheshonk IV, not Tefnakhte (who was probably Thutmose IV, the grandfather, not the father, of the blind Bocchoris-Anysis). So if it turns out that Amenhotep III equals Sheshonk IV and that the Apis bull of Pamai's reign was misinterpreted, then, from the 28th year of Amenhotep III to the second year of Ay, there are 34 years for the life of this Apis bull. All of this leaves only Osorkon I on the right side, but he was probably Amenhotep II, who has also been correlated with the Biblical name Zerah. These problematic "ghost correlations" are designated by a question mark in Table 9.

And that wraps up the present hypothesis of one-on-one Egyptian Ghost Correlations, which are summarized in Table 10.

TABLE 10

REVISED EGYPTIAN GHOST CORRELATIONS

Thutmose III	=	Shishak
Amenhotep II	=	Osorkon I = Zerah = Menelik I{159}
Thutmose IV	=	Sheshonk IV (?)
Amenhotep III	=	Tefnakhte
Akhnaton	=	Bocchoris-Anysis
Smenkhare	=	Peshee
Tutankhamen	=	Takelothis
Ay	=	Pamai
Shabaka I	=	Petasebast
Kashta	=	Osorkon III
Piankhi I	=	Sheshonk III
Piankhi II	=	Sheshonk I = So
Shabaka II	=	Takeleth = Sethos
Shabataka	=	Osorkon II
Tirhakah	=	Sheshonk II = Horemhab = Harmais
Tanutamon	=	Takeleth II
Ramses I	=	Necho I
Seti I	=	Psammetichus I
Ramses II	=	Necho II
Seti II & Tuosri	=	Psammetichus II
Merneptah I	=	Hophra-Apries
Amenmeses	=	Amasis
Merneptah II	=	Psammetichus III = Siptah

The 15 Libyans are exactly matched with all the Thebans and Ethiopians, except for Thutmose III — but even he was called Shishak in the Scriptures, a name that is linguistically similar to Sheshonk. It is, perhaps, significant that this process of duplication begins with the reign of Queen Hatshepsut's son, Thutmose III. There was a tremendous backlash against Queen Hatshepsut's memory after her death, with her own son being one of her chief defamers. This era of political and cultural upheaval, coming as it did at the beginning of the Assyrian New Kingdom and just prior to the divisive Egyptian Amarna Period, was probably one of the primary causes for the subsequent linguistic confusion in the Egyptian New Kingdom — confusion that led, in turn, to the eventual creation by perplexed historians of two separate but equal dynastic lines.

For the record, the succession of kings named Ramses who followed Merneptah II can be matched up one-on-one with all of the so-called "Priest Kings" who followed Psammetichus III. See Table 11 for these correlations, which will not be discussed here, since they are adequately covered by Dr. Velikovsky in his *Peoples Of The Sea.*

5. Several minor problems of synchronization still remain. Tirhakah/Horemhab did not ascend to the throne of Egypt until 679, which also corresponded to the accession of Esarhaddon in Assyria. If, as Dr. Velikovsky proposes, Sennacherib installed Horemhab and Mutnodjme on the Egyptian throne, then Sennacherib must have done this just after Tirhakah murdered Shabataka, but before he, himself, was murdered in Ninevah. Perhaps the reason for Sennacherib's murder was his installation of Tirhakah/Horemhab and Mutnodjme/Dakhatamon in Egypt.

On the other hand, Hittite history tells us that a sister of Suppiluliumas I married the King of Hayasa, a nation which practiced incest.{160} Egypt practiced incest. Suppiluliumas I was Ashurbanipal. His sister, like himself, was a child of Esarhaddon. An Assyrian king gave his daughter to Horemhab, who equals Tirhakah, the contemporary of Esarhaddon. Was it Esarhaddon, not Sennacherib, who installed Horemhab / Tirhakah and Mutnodjme / Dakhatamon on the throne, immediately following the murders of Sennacherib and Shabataka, in order to change the course of Egyptian history?

Regardless of the mathematically precise course of events in 679, by 671 Assyria and Egypt were fighting again. Esarhaddon invaded and conquered Egypt in the latter part of his first ten years.{161} Tirhakah fled to Ethiopia, and Esarhaddon placed Necho I (that is, Ramses I) on the throne.{162} After Esarhaddon returned to Ninevah, Tirhakah regrouped his forces and counterattacked. With the help of certain Assyrian-appointed governors, Tirhakah retook Egypt. Necho I escaped to Ninevah for safe haven,{163} taking his son, Psammetichus I, with him.{164} Two years later, Assyria marched on Egypt again. The Delta was captured only temporarily, because Esarhaddon's death at the height of the campaign forced a return to Ninevah.{165} Ashurbanipal, who had assisted in this invasion, became King of Assyria in 666.

What happened to Necho I/Ramses I? Dr. Velikovsky believes that Ramses I/Necho I returned to govern Egypt but was murdered by the Ethiopians who had been fighting Assyria for control of Egypt for over 50 years,{166} a period which, in this chronology, would be *c*716-666. But who murdered Ramses I/Necho I? We have already seen that Tirhakah murdered Shabataka. Was Shabataka the same as Ramses I/Necho I? No. It is futile to attempt that correlation because it disrupts literally scores of related synchronisms. And, Herodotus wrote that a "Shabaka" murdered Necho I, but he equated this "Shabaka" with an earlier successor of Anysis.{167} Since Shabaka II died in 689, he could not have murdered someone 20 or so years later, even if he were the one intended by Herodotus. So — whodunit?! Consider this scenario.

Following Esarhaddon's and Ashurbanipal's joint invasion of Egypt in 667, Necho I was brought back from exile and re-enthroned in Sais.{168} Esarhaddon's sudden, premature death caused an unexpected Assyrian military retreat which left Necho I defenseless. Tirhakah attacked Sais from Thebes and Ethiopia in 666 and killed Necho I. Thus, for the third time in 13 years, Tirhakah battled and won control over all of Egypt. This would mean that *Horemhab both preceded and followed Ramses I!*

After establishing himself on the Assyrian throne, in his fourth year Ashurbanipal resumed the Assyrian war with Egypt. Ashurbanipal was completely victorious and sailed down the Nile to Aswan.{169} This was in 662, only three years before the required accession date of 659 for Seti I/Psammetichus I. Tirhakah's documented reign of 26 years ended in 653. Either Ashurbanipal allowed him to continue at Thebes until his death, or Tirhakah ruled from exile in Ethiopia.

Seti I/Psammetichus I ascended to the throne of Sais in 659. Although he and his father had collaborated with the Assyrians during 671-659, once on the throne, Seti I/Psammetichus I turned against Assyria. In his first year, Seti I campaigned in Palestine and conquered Asia soon thereafter.{170} *Circa* his fourth year, Seti I attacked Kadesh.{171} After the death of Tirhakah, Tanutamon revolted against the continued Assyrian influence at Thebes.{172} This revolt by Tanutamon prompted a second attack on Egypt by Ashurbanipal in his 14th year, during which Tanutamon was ousted and chased into Ethiopia.{173} But Psammetichus I seized this opportunity, attacked and defeated the Assyrian army, and conquered all of Egypt in his ninth year.{174}

By this reckoning, Seti I/Psammetichus I warred in Asia from 659 to 656. Tanutamon revolted in 653. Ashurbanipal's second attack on Egypt was in 652, and Seti I/Psammetichus I defeated Assyria and conquered all of Egypt in 651. Gyges of Lydia, as was noted earlier, sent troops to assist Seti I/Psammetichus I. Gyges died between 657 and 649. Thus, Gyges' and Psammetichus I's reigns overlapped. Ashurbanipal attacked Gyges after his second assault on Egypt, no later than 652. This suggests that Gyges probably died between 651 and 650.

This brings the Egyptian chronology down to 651, the ninth year of Seti I / Psammetichus I and the 15th year of Ashurbanipal / Suppiluliumas I. Three more points for correlation still remain. During the reigns of Psammetichus I and Ashurbanipal, a 29-year siege of Ashdod finally ended, with Egypt victorious.{175} If this siege ended in 652, then it began in (652 + 29 =) 681. This date seems too early. On the other hand, if the siege began around 669, when Tirhakah counterattacked against Necho I and Assyria, then it would have ended in 640, the 18th year of Psammetichus I and the 26th year of Ashurbanipal. It is also known that Psammetichus I attacked Palestine *circa* his 23rd year but gave up when the Scythians invaded the area:{176} 659 - 22 = 637, and 637 + 29 = 666, the year of Ashurbanipal's accession. Did the siege of Ashdod begin in 666? In 630, the Scythians conquered Media. Should 29, therefore, be added to 631 for a siege date of 660? After the Scythians conquered Media, they moved against Psammetichus I but were repulsed in Syria.{177} Was *this* coincidental with the end of the siege of Ashdod?

The elusive "59th year of Horemhab"{178} may or may not have been his last, but it puzzles almost everybody. Horemhab did not rule for 59 years, and Tirhakah ruled for only 26 years. Adding 58 years to Tirhakah's death in 653 yields the date of 711, which falls into the reign of Tirhakah's father, Piankhi II The Great. Since this 58-year period cannot be computed from any date *later than* 711, it must refer to a period that began at an earlier date. Piankhi II ascended to the throne in 725: 725 - 58 = 667, the year of the joint Assyrian invasion of Egypt by Esarhaddon and Ashurbanipal. Was this the mysterious 59th year?

In the latter part of his reign, Seti I fought Hittite King Murshilis II.{179} Subsequently, he made war and peace with Hittite King Muwattalis.{180} Seti I/Psammetichus I overlapped Nabopolassar / Murshilis II from 622 to 605. He did not overlap Nergilissar I/Muwattalis, who came to power in 602. So if Seti I made war and peace with Muwattalis, then Muwattalis was not actually a king at the time, but only a crown-prince and co-regent perhaps.

6. At this point we can end the discussion of Egyptian chronology because the subsequent period, from the accession of Ramses II/Necho II in 605 through the Persian conquest in 504, has been analyzed and synchronized in the earlier sections of this treatise. The following chronological Table 11, encompassing years 983-322, includes my suggested outline of Egyptian history for years 504 through 331. The sequence of events for this period takes into account the necessity for shortening the histories of Persia and Greece by 21 years, overlapping the histories contained within five adjacent Greek Olympiads around the time of the Battle of Cyprus and the Peace of Callias in 449; but after the restoration of Egyptian independence by Amyrtaeus II/Ramses "XI" in 404, the chronology easily follows the scenario presented by Dr. Velikovsky in *Peoples Of The Sea.*

Why, one might ask, is this final 21-year period of Greco-Persian duplication really necessary? Why is it that the Persian conquest of Egypt simply cannot be pushed back to its traditional date of 525 and everything else before it be adjusted by 21 years? The reason is that the reigns of Osorkon II, Sennacherib and Hezekiah are immutably linked to 687. Pushing everything back 21 years destroys this *fixed* synchronism. The only alternative, therefore, would be to shorten by 21 years the period between 687 and 504, as described in this synchronology, without upsetting the balance of events surrounding the Eclipse of Thales on 28 May 584. Such a alternative seems impossible to me, because no mathematical flexibility exists within this synchronology in order to permit it.

In conclusion, those who, after reading this treatise, are still reluctant or unwilling to change the cataclysm date of 747, or to duplicate the Libyan and Ethiopian Egyptian dynasties, should keep in mind that the version of history presented here places the Persian conquest of Egypt in 504, not 525. Here, there are 21 extra years between 687 and 322 to work with than in Dr. Velikovsky's rendition, which adheres to the traditional date of 525. Even with these additional 21 years, there is just barely enough time for all the known Egyptian kings, and even these were reduced in half by the process of ghost correlation. To shorten this period by 21 years and, on top of that, to stick with the date of 747 and make no one-on-one duplications amongst the Egyptians, will pose formidable, probably insurmountable, obstacles to other would-be researchers, because any adjustments to this chronology of Egypt automatically force auxiliary adjustments in all the related chronologies.

If anyone else can present an alternative scenario to my synchronology, I'll be the first to read it and carefully compare notes. But as far as I know, to date, even after all this time, I am the only one who has ever attempted to refine this period of *Assyrian Conquest* on such a comprehensively synchronized basis as this. And let me tell you, it wasn't easy.

TABLE 11
CHRONOLOGY OF EGYPT

As usual, * denotes throne data.

983 *Queen Ma'atkhare Hatshepsut Makeda Sheba (as sole ruler)

975 Queen Hatshepsut's State Visit to Jerusalem/Punt

972 Completion of Deir El-Bahri

971 *Thutmose III/Shishak

950 Attack on Jerusalem, 22 Thutmose III, 5 Rehoboam

932 *Amenhotep II/Osorkon I/Zerah/Menelik I

924 Attack on Jerusalem, 12 Asa

916 *Thutmose IV/Sheshonk IV

905 *Amenhotep III The Magnificent/Tefnakhte

870 *Amenhotep IV/Akhnaton/Bocchoris-Anysis

865 Foundation of Akhetaton, City of the Sun

858 Banishment of Queen Nefertiti

853 Banishment & Exile of Pharaoh Akhnaton & Princess Beketaten

853 *Smenkhare/Peshee

852 *Tutankhamen/Takelothis

848 War with Ethiopia

845 *Ay/Pamai

842 *Shabaka I/Petasebast

792 *Kashta/Osorkon III

776 *Piankhi I/Sheshonk III

762 June 15, "The day the heavens devoured the Moon",
15 Sheshonk III

725 *Piankhi II The Great/Sheshonk I/So

712 Battle of Raphia with Assyria, 2 Sargon II

705 21st Year of Conquest by Piankhi II/Sheshonk I

703 More War with Assyria

703 *Shabaka II/TakelethI/Sethos

690 Battle of Altaku with Assyria, 5 Sennacherib

689 *Shabataka/Osorkon II

687 Battle of Pelusium with Assyria, 3 Osorkon II

679 Murder of Shabataka by Tirhakah

679 Marriage & Installation of Tirhakah/Horemhab
& Princess Dakhatamon/Mutnodjme
by the King of Assyria

679 *Tirhakah/Sheshonk II/Horemhab-Harmais

671 Conquest by Assyria & Flight of Tirhakah to Ethiopia

671 *Ramses I/Necho I

670 Civil War

669 *Tirhakah (restored)

669 Flight of Necho I to Assyria

667 Attack by Assyria, 12 Esarhaddon

667 *Ramses I/Necho I (restored)

666 Renewed Civil War

666 Murder of Ramses I/Necho I by Horemhab/Tirhakah?

666 *Tirhakah (restored)

662 Attack by Assyria, 4 Ashurbanipal

659 *Conquest of Sais by Seti I/Psammetichus I The Great

656 Attack on Assyria at Kadesh/Carchemish, 4 Seti I

653 Death of Tirhakah at Thebes

653 *Tanutamen/Takeleth II (only at Thebes)

652 Attack by Assyria, 14 Ashurbanipal

651 *Conquest of Thebes, 8 Seti I/Psammetichus I

648 Birth of Ramses II/Necho II The Great

636 *Co-Regency of Seti I/Psammetichus I & Ramses II/Necho II

608 Start of a 23-Year War with Babylonia/Hatti

605 *Ramses II/Necho II The Great (as sole ruler)

604 Alliance with Assyria against Babylonia, 2 Ramses II
Babylonian Conquest of Egypt and Assyria &
End of the Assyrian Empire

601 First Battle of Kadesh/Carchemish, 5 Ramses II

598 Capture of Ashkelon, 8 Ramses II/Necho II

588 Invasion of Palestine & Death of Josiah,
18 Ramses II/Necho II

585 Second Battle of Kadesh/Carchemish &
Peace Treaty with Babylonia/Hatti, 21 Ramses II

572 State Visit to Egypt by Nebuchadnezzar II/Hattushilis III,
34 Ramses II

570 *Seti II/Psammetichus II & Queen Tuosri

568 *Merneptah I/Hophra-Apries

567 Military Assistance to Judah, 9-10 Zedekiah

564 First War with Libya, 5 Merneptah I

561 Peak of Hebrew Refugee Flights into Egypt

560 Renewed Warfare with Babylonia &
Start of Siege of Tyre by Nebuchadnezzar II

548 Second War with Libya, 21 Merneptah I

547 End of Siege of Tyre &
Conquest by Babylonia, 37 Nebuchadnezzar II

547 Mob-Killing of Merneptah I/Hophra-Apries

547 *Amasis/Amenmeses

505 *Merneptah II/Psammetichus III/Siptah

504 *Conquest by Persia

455 Revolt of Inaros/Ramses IX

449 End of Six-Year Egyptian War & Execution of Inaros

449 Battle of Cyprus & Peace of Callias

444? Revolt of Amyrtaeus I/Ramses "VIII"

404 *Restoration of Egyptian Independence

404 *Amyrtaeus II/Ramses "XI"

398 *Nepherites I/Setnakhte

392 *Psammuthis/Ramses "V"

391 *Achoris/Ramses "X"

380 *Nepherites II/Ramses "VIII"

379 *Nectanebo I/Ramses III

372 First Invasion of the Sea People, 8 Ramses III

368 Second Invasion of the Sea People, 12 Ramses III

360 *Tachos/Ramses IV

355 *Nectanebo II/Ramses VI

340 *Reconquest by Persia, 19 Artaxerxes III

332 *Conquest by Alexander III The Great of Macedonia

331 Founding of Alexandria

322 Death of Alexander III The Great

*

EPILOGUE

Ultimately, one might ask why all of this is so important in the first place. Well, as Dr. Velikovsky has stated in *Ages In Chaos*, one cannot definitively determine the chronology for the last major cosmic cataclysms without first making sure that the history of the region can be "reconstructed" to accommodate such cataclysmic events. When the end of the Middle Kingdom is brought down by 600 years to coincide with the Exodus in the spring of 1587 BCE, there are not enough years remaining between then and Alexander The Great to accommodate all of the traditional Egyptian history in its customary order. Dr. Velikovsky understood this problem immediately; and in trying to sort it out, he uncovered the duplications or "ghost correlations" within Egyptian history. The reign of King Solomon and the visit to Jerusalem by Ethio-Egyptian Queen Ma'atkhare Hatshepsut Makeda Sheba are the keys to this puzzle, because when Queen Hatshepsut is brought down by 600 years, as the Queen of Sheba, she automatically overlaps the reign of King Solomon and the mysterious land of "Punt" becomes the land of Israel. Moreover, Josephus wrote{181) that the Exodus took place 592 years before the Founding of the Temple by King Solomon in 991. 991 + 592 = 1583 BCE. But Solomon ascended to the throne in 995. Add 592 years to that and you arrive at the date of 1587. Close enough!

This revised Epilogue is mostly a repeat of the Epilogue to *Slow-Motion Doomsday*. But it is worth re-appending here in light of the foregoing analyses. Most of the chronology past 687 is fairly straightforward, and there is no dispute about anything following Alexander The Great.

This Epilogue comprises the entire previous "Millennium Of The Gods", when Planet X Nibiru arrived, docked, tethered and finally departed. All of the people who lived during this period would actually have seen The Cosmic Tree with their own eyes. Not until late in the 600s would all those people have died out, leaving no more eyewitnesses, only strange legends and myths of a "Golden Age Of The Gods" in a "World Tree" or a "Sacred Mountain" in a "Land Beyond The North". Thus, if Egyptian King Thutmose III depicted a "Winged Disk" atop a "Pyramid", then he was depicting an image of what he saw himself, or what his artists and sculptors saw.

No attempt is made here to mathematically justify these dates prior to about 1000 BCE except to state that all of the Hebrew dates precisely follow the Old Testament information, except possibly for Samuel, about whom there is controversy.{182} Here, Samuel's sole leadership, between the death of Eli and the coronation of Saul, lasts for 21 years, 1096-1075 BCE. This chronology is based entirely upon the principles of the Historical Reconstruction of the Velikovskian School (see the *Ages In Chaos* series). Dr. Velikovsky's "duplicated kings" or "ghost correlations" appear in parentheses after the more acceptable names of the monarchs. For example, Egyptian King Shabataka of Thebes was known as Osorkon II in Memphis/ Sais; and Assyrian King Sennacherib, also referred to as Candaules, was the Hittite King Murshilis II, called Myrsilus by the Greeks. Also, a notation such as "14 Piankhi II" or "3 Hoshea" means that this event took place in the 14th or 3rd year of the respective monarch.

For some subsequent chronology after 687 BCE, you are referred to Chapters 23 and 24 of *Apollonius Of Tyana & The Shroud Of Turin.*

E = Egypt	C = Chaldea
I = Israel	A = Assyria
J = Judah	B = Babylonia
P = Persia	M = Media
G = Greece	L = Lydia
R = Rome	

* denotes throne data, such as accession dates.

1588 Commencement of the Arrival Sequence of Planet X Nibiru
Eruption & Explosion of Volcanic Greek Isle Thera/Santorini

1587 E/I: Israelite Exodus from Egypt in March (First Passover)
E: *Death of the Pharaoh in the Whirlpool at Pi-Khiroth
C: *Ismi-Dagon

1586 I/A: Battle between the Israelites and "Amalekites" (Hyksos) in Northern Sinai
E/A: *Conquest of Egypt by Hyksos (Hittite-Assyrians) & End of the Egyptian Middle Kingdom

1585 E: *Salatis (approximate)

1570 E: *Beon (approximate)

1547 I: *Death of Moses & Leadership of Joshua
C: *Gurguna (approximate)

1536 Completion of the Arrival Sequence of Planet X Nibiru
I: "The Day The Sun Stood Still"
(Joshua 10:12-14, approximate)

1525 E: *Apachnas (approximate)
C: *Chushan-Rishathaim (approximate)

1522 I: *Death of Joshua &
Commencement of 8-Year Interregnum when Israel was ruled by Mesopotamian-Chaldean King Chushan-Rishathaim

1514 I: *Othniel & Commencement of the Period of the Judges

1500 I: *Eglon, King of Moab
C: *Sagaraktiyas

1482 E: *Apopi I (approximate)
I: *Ehud & Shamgar
C: *Naram-Sin

1460 C: *Queen Bilatat (approximate)

1440 E: * Jonias (approximate)
C: *Sin-Shaba (approximate)

1420 C: *Zur-Sin (approximate)

1402 E: *Assis (approximate)
I: *Jabin, King of Canaan
C: *Nur-Vul (approximate)

1380 I: *Barak & Deborah

1370 E: *Apopi II (approximate)
C: *Rim-Sin (approximate)

1340 I: *Midianite Rule
C: *Arabian Conquest of Chaldea & Era of Khammurabi

1333 I: *Gideon

1330 E: *Apopi III (approximate)

1300 E: *Banan (approximate)
C: *Samsuiluna (approximate)

1293 I: *Abimelech

1290 I: *Tola

1267 E: *Abehenkhepes (approximate)
I: *Jair
C: *Ammidikaga (approximate)

1245 I: *Period of Philistine and Ammonite Oppression

1240 E: *Apopi IV (approximate)

1235 E: *Seqenenra I's Theban Revolt against the Hyksos & Restoration of Egyptian Rule at Thebes (approximate)

1227 I: *Jephthah
C: *Sinbarsikhu (approximate)

1221 E: *Seqenenra II (Thebes, approximate)
I: *Ibzan

1214 I: *Elon

1204 E: *Seqenenra III (Thebes, approximate)
I: *Abdon

1200 E: *Khenddel (Memphis, approximate)
C: *Kharbisikhu (approximate)

1196 I: *Period of Philistine Oppression

1180 E: *Kames (Thebes, approximate)

1170 E: *Nubti (Memphis, approximate)
C: *Ulam-Puriyas (approximate)

1156 I: *Samson

1146 I: Samson's Betrayal by Delilah (approximate)

1140 E: *Queen Ahotep (Thebes, approximate)

1136 I: *Eli

1130 E: *Khian (Memphis, approximate)
C: *Nazi-Urdas (approximate)

1110 E: *Ahmessepari (Thebes, approximate)

1100 E: *Apopi V/Agag (Memphis, approximate)
A: *Commencement of the Assyrian Empire,
Independent of Chaldea,
Under Sargon I of Akkad (= Tlabarnash)
C: *Nilisikhu (approximate)

1096 I: *Samuel

1080 E: *Amasis/Ahmose I The Great (Thebes, approximate)

1075 I: *Saul

1065 E: *End of Hyksos Rule in Avaris, Gaza &
Conquest of All Egypt by Amasis I The Great
I: Saul's Defeat of the "Amalekites" in Gaza,
in Military Alliance with Egypt
C: *Karakharbi (approximate)

1050 A: *Belsumilikapi (approximate)

1043 E: *Amenhotep I

1035 I: *David
A: *Irba-Vul (approximate)
C: *Tsibir-Deboras (approximate)

1025 A: *Ashurabi I/Ashurabumar/Enlil-Nirari II
(= Anittas of Kussara), Conquest of Chaldea &
Commencement of the Old Babylonian Dynasty
B: *Shiriktu-Shukamuna/Burnaburiash I

1013 E: *Thutmose I (as sole ruler)

1000 E: *Thutmose I & Queen Hatshepsut Sheba (co-regency)
A: *Ashureshishi II/Ashurabi II (= Todkhalijas I)
B: *Marbita-Pileser/Kashtiliash III
M: *Parattarna (approximate)

995 I: *Solomon

983 E: *Queen Hatshepsut, Queen of Sheba,
Ma'atkhare Makeda Saba of Ethiopia (as sole ruler)

980 A: *Tiglath-Pileser II/Ashurnirari II (= Pusarrumas)
B: *Nabomukinpal/Ulamburiash
M: *Shutarna I (approximate)

975 E/I: Queen Hatshepsut's Visit to Jerusalem/Punt, 20 Solomon

972 E: Completion of Deir El-Bahari

971 E: *Thutmose III (= Shishak of Old Testament)

965 B: *Ninurta-Kuduruzur III/Agum III

960 B: *Marbita-Ahheidin/Kadashmankharbi I
M: *Saushattar I (approximate)

957 A: *Ashurdan II/Ashur-Belnishesu (= "Murshilis"?)

955 I: *Death of Solomon &
Division of the Kingdom of Israel into Israel & Judah
I: *Jeroboam I
J: *Rehoboam

950 E/I: Attack by Egypt on Israel, 5 Rehoboam
B: *Shamas-Mudammik/Kara-Indash

940? A/B: Treaty of Friendship between Assyria and Babylonia

938 J: *Abijah, 18 Jeroboam I
A: *Adad-Nirari II/Ashur-Rimnisheshu (= "Hattushilis"?)

935 J: *Asa, 20 Jeroboam I

934 I: *Nadab, 2 Asa
M: *Shattuara I (approximate)

932 E: *Amenhotep II (= Osorkon I, Old Testament Zerah,
Menelik I of Ethiopia [son of Nighisti Saba & Solomon])
I: *Baasha, 3 Asa

924 E/I: Attack by Egypt on Judah, 12 Asa

920 B: *Naboshumukin/Kurrigalzu I

918 A: *Tukulti-Ninurta II/Ashur-Nadinahi (= Labarnas)

916 E: *Thutmose IV (= Sheshonk IV)
M: *Artatama I (approximate)

909 I: *Elah, 26 Asa

908 I: *Zimri & Omri, 27 Asa

905 E: *Amenhotep III The Magnificent (= Tefnakhte)

903 A: *Ashurnasirpal II/Eriba-Adad I (= Zidantas II)

897 I: *Ahab, 38 Asa
M: *Shutarna II (approximate)

895 B: *Nabopaladan/Kadashmanenlil I

894 J: *Jehoshaphat, 4 Ahab

878 A: *Shalmaneser III/Ashuruballit I (= Suppiluliumas)
A/B: Renewal of Friendship Treaty
Between Assyria and Babylonia
M: *Tushratta (approximate, contemporary with Akhnaton)

872 B: *Marduk-Zakirshumi I/Burnaburiash II, 6 Shalmaneser III
A/B: Renewal of Friendship Treaty
Between Assyria and Babylonia

870 E: *Amenhotep IV (= Akhnaton, Bocchoris-Anysis)
B: **Coup d'Etat* by Marduk-Belusate/Kudurenlil,
8 Shalmaneser III

869 J: *Jehoram, 28 Ahab
A/B: Assyrian Invasion of Babylonia & End of Usurpation,
9 Shalmaneser III
B: *Marduk-Zakirshumi I/Burnaburiash II (restored)

868 I: *Ahaziah, 2 Jehoram

866 I: *Jehoram, 4 Jehoram

865 E: Foundation of Akhetaton, City of the Sun
A/B: Marriage of Ashuruballit I's Daughter to Burnaburiash II

862 J: *Ahaziah, 5 Jehoram

861 I: * Jehu
J: *Queen Athaliah

858 E: Banishment of Queen Nefertiti

855 J: *Joash, 7 Jehu
A/M: *Assyrian Conquest of Media (approximate)

853 E: *Smenkhare (= Peshee) &
Banishment and Exile of Akhnaton and Princess Beketaten

852 E: *Tutankhamen (= Takelothis)

848 E: War with Ethiopia

847 B: *Baba-Ahaiddina/Karakhardash II, 31 Shalmaneser III

845 E: *Ay (= Pamai)
B: **Coup d'Etat* by Ashurdanipal/Nazi-Bugash,
33 Shalmaneser III
M: *End of Assyrian Occupation under Mattiwaza
(approximate)

844 A/B: Assyrian Invasion of Babylonia & End of Usurpation
B: *Marduk-Balassuiqbi/Kurrigalzu II

843 A: *Shamshi-Adad V/Enlil-Nirari III (= Huzziyas II)

842 E: *Shabaka I (= Petasebast)

837 A/B: Assyrian Invasion & Conquest of Babylonia

833 I: *Jehoahaz, 23 Joash
B: *Nabomukinzeri I/Shagaraktishuriash (in exile)

832 B: *Marduk-Belzeri/Enlil-Nadinshumi (in exile)

830 B: *Marduk-Baladan/Kadashmankharbi II (in exile)

829 A/B: *Empress Semiramis/Tuvannanas

825 A: Approximate Founding of the City of Tyana,
Named for Empress Semiramis/Tuvannanas,
in Cooperation with Cappadocian Hittite King Nimrod

824 A: *Adad-Nirari III/Arik-Denilu (= Telipinus)
M: *Shattuara II (approximate)

818 I: *Jehoash, 37 Joash

816 J:Amaziah, 2 Jehoash

811 G: Birth of Atreus

810 B: *Restoration of the Babylonian Monarchy
by Eriba-Marduk/Nazi-Maruttash

802 I: *Jeroboam II, 15 Amaziah
M: *Wasashata (approximate)

800 A: *Shalmaneser IV/Adad-Nirari I (= Alluvamnas) &
Approximate Founding of the Kingdom of
Phrygia/Cappadocia by King Minos of Macedonia

792 E: *Kashta (= Osorkon III)

791 G: Birth of Agamemnon

790 G: Period of Orpheus and The Muses &
Origin of the Orphic Music Mystery Cult
by Apollo on Mount Olympus

786 J: *Uzziah, 17 Jeroboam II

782 G: Period of the Expedition of Jason and the Argonauts
for the "Golden Fleece" (approximate)

780 B: *Naboshumishkun II/Kadashmanturgu
M: *Saushattar II (approximate)

776 E: *Piankhi I (= Sheshonk III)

772 G: Marriage of Menelaus & Helen (approximate)

771 G: *Death of Atreus, Accession of Agamemnon &
Abduction of Helen by Paris of Troy/Ilium

770 G: Commencement of the Trojan War
A: *Asshurdan III (= Hantilis II)

764 J: Vision of Amos

15 June 762
Beginning of the Departure Sequence of Planet X Nibiru
E: The Day The Sun Devoured The Moon, 15 Sheshonk III
I/J: The Great Earthquake, 25 Uzziah
A/G: The Great Eclipse, 8 Asshurdan III

761 G: Period of Homer's *Iliad*

760 I: *11-Year Interregum, 27 Uzziah
G: End of the Trojan War

759 G: *Murder of Agamemnon

758 G: Period of Homer's *Odyssey*

756 G: Commencement of the Greek Olympiads

752 R: *Founding of Rome by Romulus & Remus
A: *Ashurnirari V (= Arnuvantas I)

749 I: *Shallum & Menahem, 38 Uzziah
M: *Artatama II (approximate)

747 B: *Era of Nabonassar/Nebuchadnezzar I/Kadashmanenlil II
G: *Accession of Orestes I

744 A: *Tiglath-Pileser III/Pul (= Todkhalijas II)

740 I/A: War Between Israel & Assyria,
9 Menahem, 4 Tiglath-Pileser III

737 I: *Pekahiah, 50 Uzziah

735 I: *Pekah, 52 Uzziah

734 J: *Jotham, 2 Pekah

733 B: *Nabonadinzeri

730 A: Assyrian Conquest of Media, Palestine & Arabia

729 A: Battle of the Two Sipparas with Babylonia
B: Temporary Conquest by Assyria

725 E: *Piankhi II The Great (= Sheshonk I/So)

724 B: *Restoration of Independence by Nabushumukin

719 J/A: *Ahaz, 17 Pekah & Conflict with Assyria,
25 Tiglath-Pileser III

718 A: *Shalmaneser V/Ululai (= Arnuvantas II)

717 A/B: Temporary Reconquest of Babylonia by Assyria

714 I: *8-9 Year Interregum, 4 Ahaz
A: *Sargon II (= Hattushilis I)

713 A/B: Assyro-Babylonian War

712 E/A: Battle of Raphia between Egypt and Assyria,
2 Sargon II, 14 Piankhi II

710 G/R: Period of Pythagoras of Samos in Crotona, Italy &
His Theory of "The Counter Earth"

708 A/M: First Revolt of the Medes against Assyrian Occupation

707 I: *Hoshea, 12 Ahaz

706 R: Visit by Pythagoras to Ancient Rivarta in India?
(approximate?)

705 E: Famous 21st Year of Conquest by Piankhi II/Sheshonk I

704 J: *Hezekiah, 3 Hoshea

703 E: *Shabaka II (= Tekeleth I, Sethos)
A: More War with Egypt

702 A/B: Assyrian Conquest of Babylonia

701 I/A: Assyrian Siege of Samaria, 4 Hezekiah,
1 Sargon II (as King of Babylonia)

699 I/A: Conquest of Israel by Assyria, 9 Hoshea & 6 Hezekiah

695 A: *Sennacherib/Candaules (= Murshilis I/Myrsilus)
B: Independence Rebellion against Assyria
L: *Founding of the Kingdom of Lydia by Gyges

694 B: *Marduk-Zakirshumi II & Marduk-Baladan II (restored)

693 B: *Belipni

691 J/A: Attack by Assyria on Jerusalem,
14 Hezekiah, 4 Sennacherib
B: *Ashur-Nadinshumi

690 E/A: Battle of Altaku, 5 Sennacherib, 14 Shabaka

689 E: *Shabataka (= Osorkon II)

688 J: Vision of Isaiah
B: *Marduk-Baladan II (restored again), 7 Sennacherib

687 Completion of the Departure Sequence of Planet X Nibiru
E/I/A: Destruction by Cosmic Forces of the Assyrian Army
at Pelusium, Sinai, Egypt
18 Hezekiah, 8 Sennacherib, 2 Osorkon II

679 Historical Commencement of the Mayan Calendar, with
End-Time Date of 21 December 2012 CE

1588 BCE + 2012 CE = 3,600 Earth Years =
1 "Shar" Planet X Nibiru

*

FOOTNOTES

1. Immanuel Velikovsky (A), *Worlds In Collision* (New York, 1950), pp. 210-211
2. *Ibid*, p. 365.
3. George Rawlinson (A), *The Five Great Monarchies of the Ancient Eastern World*, 3 Vols. (New York, 1871), Vol. 2, p. 52.
4. Theodor Ritter von Oppolzer, *Canon of Eclipses*, transl. by Owen Gingerich (New York, 1962), solar eclipse chart #21.
5. Velikovsky (A), *Op. Cit.*, p. 215. The footnote to this sentence states that "Oppolzer and Ginsel arranged canons of the solar eclipses in antiquity on the premise that there was no change in the movement of the Earth or the Moon."
6. Rawlinson (A), *Op. Cit.*, p. 156.
7. Immanuel Velikovsky (B), *Ages In Chaos* (New York, 1952), pp. 255-262.
8. 2 Kings 15:1-2.
9. Amos 1:1.
10. Rawlinson (A), *Op. Cit.*, p. 124.
11. 2 Kings 16; Rawlinson (A), *Op. Cit.*, p. 133.
12. 2 Kings 17:1.
13. 2 Kings 18:1-2.
14. 2 Kings 18:9-10.
15. Rawlinson (A), *Op. Cit.*, p. 138.
16. *Ibid.*, pp. 147-148.
17. 2 Kings 18-19; Isaiah 36-37.
18. William Smith (A), *A Dictionary of the Bible*, 3 Vols. (Boston, Massachusetts, 1863), Vol. 3, pp. 1195-1196.
19. Rawlinson (A), *Op. Cit.*, p. 50.
20. 2 Kings 20; 2 Chronicles 32.
21. Rawlinson (A), *Loc. Cit.*
22. Immanuel Velikovsky (C), *Ramses II and His Time* (New York, 1978), pp. 1-3, 254.
23. *Ibid.*, pp. 37-38.
24. *Ibid.*, p. 43.
25. *Ibid.*, pp. 49-53, 254.
26. Jeremiah 39:1-3.
27. 2 Kings 25:8-11.
28. Velikovsky (C), *Op. Cit.*, pp. 189-202.

29. Jeremiah 37:1-8 and Ezekiel 17:11-18, as interpreted by George Rawlinson (B), *The Five Great Monarchies of the Ancient Eastern World*, 3 Vols. (New York, 1871), Vol. 3, pp. 52-53.
30. Velikovsky (C), *Op. Cit.*, p. 189
31. E.A. Wallis Budge, *The Mummy* (London, England, 1972), pp. 58-59.
32. Velikovsky (C), *Op. Cit.*, pp. 180-184.
33. *Ibid.*, p. 181.
34. *Ibid.*, pp. 180-184.
35. J.E. Manchip White, *Ancient Egypt: Its Culture and History* (New York, 1970), p. 200.
36. Budge, *Loc. Cit.*
37. 2 Chronicles 36:22-23; Flavius Josephus, *Antiquities of the Jews*, transl. by William Whiston (Philadelphia, Pennsylvania, 1957), XI.3.1.
38. Ezra 6:15.
39. Josephus, *Op. Cit.*, XI.1.1.
40. Rawlinson (B), *Op. Cit.*, pp. 400, 406.
41. *Ibid.*, pp. 358, 400.
42. *Ibid.*, p. 393.
43. *Ibid.*, p. 388.
44. *Ibid.*, p. 65.
45. *Ibid.*, pp. 368-70.
46. William Smith (B), *A Dictionary of Greek ad Roman Biography and Mythology*, 3 Vols. (London, England, 1890), Vol. 3, p. 1338.
47. Rawlinson (B), *Op. Cit.*, pp. 380,384.
48. Josephus, *Loc. Cit.*
49. William Smith (C), *A Dictionary of Greek and Roman Biography and Mythology*, 3 Vols (London, England, 1890), Vol. 1, p. 122.
50. Rawlinson (B), *Op. Cit.*, pp. 364-369.
51. Herodotus (A), *Histories*, transl. by Aubrey de Selincourt (New York, 1954), Vol. I, p. 111.
52. *Ibid.*, p. 130.
53. *Ibid.*, pp. 103-108.
54. *Ibid.*, pp. 103, 108, 130.
55. Rawlinson (A), *Op. Cit.*, pp. 384-385.
56. *Ibid.*, pp. 391-392, 414.
57. Herodotus (A), *Op. Cit.*, p. 103.
58. Velikovsky (C), *Op. Cit.*, p. 142.
59. Rawlinson (A), *Op. Cit.*, pp. 384-385.
60. Daniel D. Luckenbill, *Ancient Records of Assyria and Babylonia*, 2 Vols. (Chicago, Illinois, 1927), Vol. 2, secs. 1172-1174.
61. Herodotus (A), *Op. Cit.*, pp. 103-108.

62. Rawlinson (A), *Op. Cit.*, pp. 147-151, 379.
63. Josephus, *Op. Cit.*, X.2.
64. Smith (C), *Op. Cit.*, p. 953.
65. Rawlinson (A), *Op. Cit.*, p. 410.
66. *Ibid.*, p. 412.
67. Herodotus (A), *Op. Cit.*, p. 410.
68. *Ibid.*, p. 412.
69. *Ibid.*, pp. 7-19, 86.
70. *Ibid.*, pp. 7-19.
71. White, *Op. Cit.*, pp. 194-195; Budge, *Op. Cit.*, pp. 55-56.
72. J.H. Breasted, *A History of Egypt* (New York, 1905), pp. 410-411.
73. Rawlinson (A), *Op. Cit.*, pp. 201-204.
74. *Encyclopedia Britannica Micropaedia*, 10 Vols. (London, England, 1952), Vol. 8, p. 75; Budge, *Loc. Cit.*, pp. 55-56.
75. Robert H. Hewsen, "Eastern Anatolia and Velikovsky's Chronological Revisions 1", *KRONOS Journal*, I:3 (1975), p. 23.
76. Velikovsky (C), *Op. Cit.*, pp. 94, 102.
77. White, *Op. Cit.*, p. 192.
78. Velikovsky (A), *Op. Cit.*, p. 350.
79. Smith (A), *Op. Cit.*, p. 1514.
80. C.W. Ceram, *The Secret of the Hittites*, transl. by Richard and Clara Winston (New York, 1956), pp. 163-165.
81. Velikovsky (C), *Loc. Cit.*; Luckenbill, *Op. Cit.*, secs. 1185-1186.
82. *Ibid.*, p. 97; *Ibid.*, secs. 1177-1178.
83. James G. MacQueen, *Babylon* (New York, 1964), pp. 133-134.
84. Luckenbill, *Op. Cit.*, pp. 408-410.
85. Ceram, *Op. Cit.*, p. 260.
86. Velikovsky (C), *Op. Cit.*, pp. 104-105.
87. Ceram, *Op. Cit.*, pp. 257-260.
88. MacQueen, *Op. Cit.*, Appendix.
89. Velikovsky (B), *Op. Cit.*, Chapter 8.
90. Velikovsky (C), *Op. Cit.*, p. 91.
91. *Ibid.*, p 92.
92. *Ibid.*
93. *Ibid.*, p. 143.
94. Ceram, *Op. Cit.*, p. 127.
95. *Ibid.*
96. Herodotus (A), *Op. Cit.*, pp. 7-13.
97. Ceram, *Op. Cit.*, p. 150.
98. *Ibid.*, p. 127.
99. *Ibid.*, pp. 128, 258.

100, *Ibid.*, p. 258.
101. *Ibid.*, p. 124.
102. A discussion of this matter inevitably leads one into an analysis of the origins of Sparta in Greece. Before venturing into this treacherous theoretical territory, I plan to wait until after I have read Dr. Velikovsky's *Assyrian Conquest.* [Unfortunately, I wrote those words before Dr. Velikovsky died, and his *Assyrian Conquest* was never published. I stand by whatever else that I have written about Greece here. Its history *must* be shortened by 20 years! Simple.]
103. Velikovsky (B), *Op. Cit.*, p. 321.
104. George Rawlinson (C), *The Five Great Monarchies of the Ancient Eastern World*, 3 Vols. (New York, 1871), Vol. 1, pp. 169-171.
105. MacQueen, *Op. Cit.*, pp. 108-110.
106. Velikovsky (B), *Op. Cit.*, Chapter 8.
107. MacQueen, *Op. Cit.*, p. 121.
108. *Ibid.*
109. Rawlinson (A), *Op. Cit.*, pp.109-110.
110. MacQueen, *Op. Cit.*, p. 122.
111. *Ibid.*, pp. 111-112.
112. See Luckenbill, *Op. Cit.*, pp. 441-442 and Rawlinson (A), *Op. Cit.*, p. 49.
113. Rawlinson (A), *Ibid.*, pp. 155, 177.
114. MacQueen, *Op. Cit.*, p. 105.
115. Luckenbill, *Op. Cit.*, p. 424.
116. Rawlinson (A), *Op. Cit.*, p. 52.
117. Velikovsky (B), *Op. Cit.*, p. 4.
118. Amos 1:1.
119. 2 Kings 18:2.
120. Velikovsky (A), *Op. Cit.*, pp. 362-363.
121. Budge, *Op. Cit.*, p. 119.
122. Compare 1 Kings 6:1, 37-38 & 7:1, 9-10 with 2 Chronicles 3:1-2, 8:1 & 9:1.
123. 1 Kings 14:25.
124. Velikovsky (B), *Op. Cit.*, pp. 163-164; *Encyclopedia Britannica Micropaedia*, *Op. Cit.*, p. 72.
125. Budge, *Op. Cit.*, p. 39.
126. Breasted, *Loc. Cit.*; White, *Op. Cit.*, p. 192.
127. Breasted, *Ibid.*
128. White, *Op. Cit.*, p. 191.
129. Rawlinson (A), *Op. Cit.*, p. 159.
130. White, *Loc. Cit.*; Smith (A), *Op. Cit.*, p. 1337.

131. White, *Ibid.*, p.187; Budge, *Op. Cit.*, p. 51.
132. Smith (A), *Op. Cit.*, pp. 1287-1288.
133. Velikovsky (A), *Op. Cit.*, p. 355.
134. Velikovsky (B), *Op. Cit.*, pp. 174-176.
135. Velikovsky (A), *Op. Cit.*, pp. 209, 335.
136. Velikovsky (B), *Op. Cit.*, Chapter 4.
137. Immanuel Velikovsky (D), *Oedipus & Akhnaton* (New York, 1960), pp. 119-125.
138. Herodotus (B), *Histories*, transl. by Aubrey de Selincourt (New York, 1954), Vol. II, pp. 137-141.
139. Velikovsky (D), *Op. Cit.*, pp. 119-120.
140. White, *Op. Cit.*, p. 191.
141. See J.H. Breasted's *A History of Egypt*; E.A. Wallis Budge's *The Mummy*; William Smith's *A Dictionary of the Bible* and *A Dictionary of Greek and Roman Biography and Mythology*; and J.E. Manchip White's *Ancient Egypt: Its Culture and History* — all cited throughout this treatise.
142. Breasted, *Op. Cit.*, pp. 536-543.
143. *Ibid.*, p. 539.
144. Donovan A. Courville, "Limitations of Astronomical Dating Methods", *KRONOS Journal*, I.2 (June 1975), p. 69.
145. Rawlinson (A), *Op. Cit.*, pp. 138-147.
146. White, *Op. Cit.*, pp. 189-190; Budge, *Op. Cit.*, p. 52.
147. Rawlinson (A), *Loc. Cit.*
148. *Ibid.*
149. 2 Kings 17:1-6.
150. Rawlinson (A), *Loc. Cit.*
151. *Ibid.*, pp. 158-160.
152. *Ibid.*, pp. 167-168.
153. Immanuel Velikovsky (E), "The Correct Placement of Horemhab in Egyptian History", *KRONOS Journal*, IV.3 (February 1978), pp. 3-22.
154. See also Herodotus (B), *Op. Cit.*, p. 141; Rawlinson, *Loc. Cit.*; and Velikovsky (C), *Op. Cit.*, p. X.
155. Velikovsky (E), *Op. Cit.*, p. 5.
156. Velikovsky (C), *Loc. Cit.*
157. Breasted, *Op. Cit.*, pp. 531-533.
158. Budge, *Op. Cit.*, p. 51.
159. According to Ethiopian tradition, Menelik I was the son of Solomon and Sheba. If Sheba equals Hatshepsut, then one of her sons or successors was Menelik I, a King of Ethiopia. I am tentatively making this additional correlation with Amenhotep II because of what Dr. Velikovsky wrote

in *Ages In Chaos*, page 210: "Was Amenhotep II an Ethiopian on the Egyptian throne? In the veins of the Theban dynasty there was Ethiopian blood. Was the royal wife of Thutmose III a full-blooded Ethiopian, and did she bear him a dark-skinned son? Or was Amenhotep II not the son of Thutmose III at all? He called himself son of Thutmose III, but this claim need not have been literally true. He called his mother Hatshepsut. Is it possible that before ascending [to] the throne of Egypt, he was a viceroy in Ethiopia? Conventional chronology identifying Zerah with Osorkon of the Libyan Dynasty encounters difficulty in the Biblical reference to Zerah as an Ethiopian."

160. Ceram, *Op. Cit.*, p. 159.
161. Rawlinson (A), *Op. Cit.*, pp. 192-194.
162. White, *Op. Ctt.*, p. 193; Budge, *Op. Cit.*, pp .54-55.
163. White, *Ibid.*, pp. 193-194; Rawlinson (A), *Op. Cit.*, pp. 195-196.
164. White, *Ibid.*, p. 194.
165. *Ibid.*, p. 193; Rawlinson (A), *Op. Cit.*, p. 201.
166. Velikovsky (C), *Loc. Cit.*
167. Herodotus, *Op. Cit.*, p. 154.
168. Luckenbill, *Op. Cit.*, sec. 771.
169. Rawlinson (A), *Op. Cit.*, pp. 202-203.
170. Breasted, *Op. Cit.*, pp. 410-411.
171. *Ibid.*, p. 412.
172. White, *Op, Cit,*, p. 194; Budge, *Op. Cit.*, p. 55.
173. White, *Ibid.*; Budge, *Ibid.*, pp. 55-56; Luckenbill, *Op. Cit.*, secs. 776-778.
174. White, *Ibid.*; Budge, *Ibid.*; *Enclyopaedia Britannica Micropaedia, Op. Cit.*, p. 75.
175. Herodotus (B), *Op. Cit.*, p. 160.
176. Breasted, *Op. Cit.*, pp.580-582.
177. Herodotus (A), *Op. Cit.*, pp. 103-108.
178. Breasted, *Op. Cit.*, pp. 407-408.
179. *Ibid.*, p. 412.
180, *Ibid*, p. 43; Ceram, *Op. Cit.*, p. 168.
181. Josephus, *Op. Cit.*, VIII.50.
182. See, for example, Smith (A), *Op. Cit.*, Vol. 1, p. 323, and elsewhere. Also:
http://www.bible.ca/ef/expository-1-samuel-7-2.htm

Appendix A

Dark Star
The Planet X Evidence
By Andy Lloyd
Copyright 2005, All Rights Reserved

Commentary By Rob Solàrion

*

During the preparation of *Osiris, Isis & Planet X,* I had the good fortune and opportunity to read a new book titled *Dark Star, The Planet X Evidence* by Andy Lloyd (Santa Barbara, California, 2005). This is a 328-page book with illustrations and index. It has a cover price of US$21.95 and is available through Amazon.com and other booksellers.

http://www.darkstar1.co.uk/ (Andy's website)
http://www.timelessvoyager.com/ (publisher's website)

Andy and I have been acquainted over the Internet for several years, and both of us were interviewed by Hollywood film-producer Robert Sepehr for the second of his 2003 *Planet X Videos.*

http://www.planetxvideo.com/

Andy and I have agreed to disagree on certain matters pertaining to Planet X for reasons which will become apparent in this review of his book. Andy presents his arguments in a logical and efficient manner, starting with the simpler anomalies of our Solar System and then gradually working into more complex discussions of Dwarf Stars in general and Planet X in particular. By Chapter 12, Andy has actually *overly complicated* his theory, in my opinion; but he has certainly covered all the bases, to use an American sports metaphor. For a couple of hundred pages, Andy speculates and theorizes about one aspect of Planet X or another; and I think that he would agree with me that we are basically at a dead-end in terms of purely "theoretical" analyses. What we need now is actual *physical* proof of Planet X, either its telescopic discovery or its sudden passage through the "mainstream" Solar System.

In reviewing Andy's book, I shall do it in the order that Andy presented his material.

Most researchers of Planet X Nibiru, myself included, tend to follow the postulation by Zecharia Sitchin in *The Earth Chronicles*, notably *The Twelfth Planet*, that this "tenth" or "unknown" planet is approximately the size of Uranus and Neptune, or about 4-5 times larger than the Planet Earth, and therefore that it is merely an as yet "undiscovered" planet within our Solar System. Andy, by contrast, equates Planet X with a Brown Dwarf Star, a distant, unseen binary companion of our Sun several times larger than the Planet Jupiter, with a planetary system of its own. Whereas I suggest that Planet X is accompanied by an "entourage" or "host" of planetoids and moonlets, in Andy's scenario these bodies, seven in all, orbit the Dark Star. The innermost planet of the Dark Star, the warmest and most hospitable for life, is the Home Planet of the Anunnaki. The planet farthest from the Dark Star is what becomes visible to peoples on Earth during the perihelial passage of the Dark Star's system, leading in turn to all of our ancient "myths" about this "perturber" or "interloper" planet.

Andy's Dark Star itself does not actually enter the boundaries of the other planets. However, its "Seventh Moon" (Sitchin's Nibiru, or "Planet of the Crossing") does "cross over" into that part of the Solar System between Neptune and Pluto, close enough and bright enough to be visible to people on Earth, at least for such a sufficiently lengthy time that cosmic legends could be born and later develop around it. In my own Cosmic Tree Theory, of course, I have Planet X Nibiru coming as close to the Earth as about 60,000 miles (about 100,000 kilometers) and then anchoring itself to our

North Pole by an electromagnetic "tether" beam. As it approaches close enough to Earth, its South Magnetic Pole is attracted to our North Magnetic Pole, like the opposite poles of all magnets, locking it into place above our North Pole for 900 years, approximately a "Millennium of the Gods", after which time it "detethers" and returns in its orbit to an aphelion somewhere between here and the Oort Cloud, or even beyond as discussed earlier in this book.

On page 48, Andy writes the following about Sitchin's Nibiru: "The passages [of Planet X] also present us with evidence that Nibiru/Marduk appeared to the Mesopotamians as a red star during historical times, and that its heavenly passage was unusual. It was faint, red, stood still in the sky and then wandered like a planet. This is highly unusual, to say the least. It is no wonder that the nature of Nibiru remains controversial."

When I refer to Planet X Nibiru's "standing still" over our North Pole, I use the expression literally: It stopped and "stood still", tethered to Earth as a Winged Disk atop a Cosmic Tree or World Tree or Sacred Tree. Andy's and Sitchin's ideas that Nibiru "stood still" refer to that optical illusion we get when any planet seems to "stop" and "go backwards" in its orbit, which we refer to as its "retrograde movement". Neither Andy nor Sitchin would agree with me on the meaning of the concept that it "stood still" in the sky.

As an aside here, let me add that amongst the Velikovskian School there is a group of researchers, most prominently amongst them David Talbott and colleagues, who believe, as Dr. Immanuel Velikovsky suggested in "On Saturn And The Flood" (see Chapter 14) published by *KRONOS Journal* (Volume V, Number 1) in the fall of 1979, that the object which "stood still" over our North Pole was the Planet Saturn. In 1996 David Talbott produced a video regarding this idea, titled *Remembering The End Of The World*. Some of Talbott's animated computer graphics at the end of the video are quite well-done, but his absolutely preposterous "Saturn Theory" does not depict the Planet Saturn. It depicts the Planet X Nibiru standing above our North Pole as "The Cosmic Tree"! But Planet X notwithstanding, the flaw in Talbott's video is obvious. Dr. Velikovsky explicitly stated in his *KRONOS* article that this hypothetical "Saturn Theory" *preceded* the so-called "Birth Of Venus" that he described in *Worlds In Collision*. Yet for Talbott's scenario to work, it *includes* a Planet Venus revolving around Saturn and thus intrinsically contradicts Dr. Velikovsky's original idea. David Talbott's video can be obtained from the Kronia Group.

http://www.kronia.com/

On page 53 Andy writes the following, and I certainly agree with him. His thoughts are worth repeating here.

"The idea that there is a massive undiscovered planetary body orbiting the sun is almost 100 years old now. It is certainly not a new idea, but is one whose popularity has fluctuated down the years. At the moment, it is a possibility that is regaining a certain amount of scientific credibility. An idea, perhaps, whose time has arrived.

"Our science and technology seem to progress at an accelerating rate, and this tends to make us all a little complacent about what remains to be discovered. It seems common sense that any scientific endeavour lasting 100 years would have certainly reached a conclusion by now, as the means to discover the answer has improved. Yet, many of the most important scientific questions remain unanswered: a cure for cancer; a renewable energy source; a unified field theory in physics, to name but a few. These problems remind us that our knowledge of the cosmos, the Earth and ourselves is far from complete, and that science has much to learn.

"And so it is with our knowledge of the solar system. Because we are looking further and further into space with larger and more technologically refined telescopes, we have a tendency to assume that everything in-between has been discovered, catalogued and understood. This is far from the truth in reality.

"Astronomy is only as good as its ability to pick up light sources, or sources of other types of radiation, and distinguish them from other similar sources. Our eyes, searching the heavens at night, perform the most simple form of astronomy, detecting the light from distant stars. Yet we cannot see closer objects, including the outer planets of the solar system beyond Saturn, nor the asteroids and distant comets."

Andy then makes an analogy to a garden in front of a house. If one were standing at the gate to the garden at night, one could see the lighted house beyond the garden (stars) but not see all the dark details of the garden itself (outer solar system). And finally, regarding this hunt for Planet X and other undiscovered objects, on page 80 Andy concludes, rather lamentably, "This is a hunt for a needle in a haystack, with the lights turned off."

Throughout his book, Andy cites references to Dr. Carl Sagan, who was one of the bitterest philosophical enemies of Dr. Immanuel Velikovsky, I might add. Andy wrote me in an email that he was trying to present his material as much as possible in a way to incorporate mainstream astronomical thought, and he does that well. On page 57 Andy remarks, "The late Carl Sagan, a popular and brilliant scientist from Cornell University [in New York, RS], described the potential for a dark sister companion orbiting the sun back in 1985. Sagan acknowledged the speculation surrounding a proposed Nemesis 'star' orbiting the sun at a great distance. He even proposed a fictional scenario where ancient peoples mythologized this 'Death Star' as the sun's Dark Sister. The 'Death Star' — presumably taking its name from the equally fictitious moon-like battle station of George Lucas's *Star Wars* trilogy — could periodically bombard the solar system with comets, when its elliptical orbit caused it to brush through the comet clouds. This, in turn, could create a periodic extinction cycle."

For the record, here is an original quote (source unknown) from Carl Sagan regarding this "Demon Sun": "There is another Sun in the sky, a Demon Sun that we cannot see. Long ago, even before great-grandmother's time, the Demon Sun attacked our Sun. Comets fell, and a terrible winter overtook the Earth. Almost all life was destroyed. The Demon Sun has attacked many times before. It will attack again."

Chapter 4 ("Binary Companion") was particularly appealing to me. On pages 77-78, Andy writes the following:

"In 1986, a rather diligent researcher named William Corliss published his book 'The Sun and Solar System Debris — A Catalog of Astronomical Anomalies'. Several observed anomalies are cited which may allude to Planet X, or even a dark companion to the sun. These anomalies remain unconfirmed, of course, but make for interesting reading nonetheless. One of them describes an object captured by the IRAS [infrared telescope, RS] survey which sounds very much like the 'Orion' sighting, only this time it is located in the zodiacal constellation of Sagittarius, in the opposite half of the sky ...

"The article was published in 'New Scientist' on 10th November 1983, and discusses the discovery of an object in space whose temperature is 230K, which is too cool for a star, but too warm for a dust cloud. It was spotted by the infrared space telescope in the constellation of Sagittarius, and fit the bill for an object 'several times heavier than Jupiter'. Remarkably,

British scientists at this time accused their American colleagues of keeping the information of this find to themselves. The British scientists publicly questioned why the Americans had 'been keeping quiet about it in recent weeks'. Speculation was rife, that the discovery was nothing but an intriguing ploy to bolster the chances of further funding from NASA for a new infrared space observatory.

"Those few weeks of silence which followed the report of a new Jupiter-sized planet in the solar system have now extended to 22 years! ... Without this article in New Scientist, no one would have known any different. There is usually some fire behind the smoke, after all. But, why would anyone want to shelve such an incredibly important discovery?"

Yes, indeed — unless to release details of this potentially catastrophic "interloper" would fuel worldwide panic and turmoil!

The Orion sighting mentioned above refers to the detection of a "mini-galaxy" or group of "rogue planets" in the direction of Orion in 1983. An additional group of "rogue planets" was discovered (or rediscovered?) in Orion in the fall of 1997. See Chapter 11 ("Rogue Planet Crossings") of my book *Planet X Nibiru: Slow-Motion Doomsday*. In the Mayan legends, their "Demon Sun" is first sighted in the Constellation of Sagittarius, after which it travels along "The Black Road" to its ultimate stationary position atop "The Sacred Tree".

Then Andy continues, "In his analysis of ancient texts, Zecharia Sitchin offered a number of constellations as probable points along the path trodden by Nibiru. These include, in order, the Great Bear (Ursa Major); Orion (along with the star Sirius); then, Taurus and Aries; before heading towards Sagittarius. The last of these is not listed as a constellation that Nibiru visits, but rather one that it usually disappears from, in its course away from our solar system."

Ursa Major is a North Polar Constellation. Planet X Nibiru is stationary over our North Pole as The Winged Disk atop The Cosmic Tree for a "Golden Age" of 900 Earth Years. It seems to come and go from the direction of Sagittarius, the Center of the Galaxy, exactly opposite from the direction of Sirius, its original point of origin. Although Sitchin did not mention The Cosmic Tree, nevertheless it is noteworthy that he included Ursa Major as a station on the path of Planet X Nibiru.

In June 2001 NASA discovered a strange "microlensing object" in the direction of Sagittarius. This object is located between here and globular cluster M22. It can't be seen because it is "microlensing" the light of M22 behind it. NASA never followed up with additional information about this apparently important discovery. Was it yet another sighting of Planet X? For additional details, see:

http://science.nasa.gov/headlines/y2001/ast29jun_1htm?list553051

What Andy writes on page 81 sounds a bit like this "microlensing object" at M22: "It turns out that [engineer and amateur astronomer John] Bagby was interested in the work of one E.R. Harrison who, in 1977, postulated the existence of a massive nearby body, lying in Sagittarius, required to explain observational anomalies regarding a 'pulsar period time derivative'. This sounds like a bit of a mouthful, doesn't it? Simply put, pulsars are highly regular emitters of strong radiation. If a gravitational field comes between a pulsar and us, as observers on Earth, then the highly specific data from the pulsar will be altered slightly. This will allow us to imply the existence of a dark gravity field, which is what Harrison proposed in Sagittarius. His finding may thus imply the location of the Dark Star."

The 1997 discovery of "rogue planets" in Orion seemed to demonstrate once and for all that a planet doesn't necessarily have to orbit a star. Some of them wander in clusters about the galaxy, and these starless planets have been termed "rogue planets". See *Astronomy* Magazine, December 1997, "On The Trail Of Rogue Planets" by Peter Catalano. Andy touches on this matter on page 83: "There is no known mechanism to help us understand how a planet could form so far away from a star in the latter case, and many think it unlikely that a 'free-floating' planet might be captured into such an extended orbit. However, if firm data pointed to the existence of such an orbit, the scientific community would quickly figure out a plausible mechanism to explain its presence, I'm sure."

Andy pursues the idea further on page 101: "We now know that many of the newly discovered 'extra-solar planets' have eccentric orbits, indicating that noncircular orbital arrangements in star systems might be fairly normal. In at least one case, a brown dwarf has been found embedded within a 'normal' extra-solar planetary system, without its presence seeming to create chaos among the other planets. The birth of planetary systems appears to be anything but simple.

"In relation to the Dark Star Theory, the modern understanding of these failed stars appears to offer an ideal platform to explore the concept of an inhabitable world in our comet-cloud, as described by the Sumerians. A world orbiting a dark star that is essentially invisible to us, but that emits massive amounts of heat and enough low-frequency light to support life, whilst not subjecting the denizens of that world to the sort of harmful radiation we are subject to from our sun.

"Could this also explain the almost immortal life-spans that Sitchin claims for the Anunnaki? One might speculate that our woefully short life-spans are due to our constant exposure to high energy particles radiated from the sun. Astronaut 'Gods' coming to our world might find their life-spans significantly shortened, as well as the subsequent life expectancies of their children. Life on Earth is necessarily mortal. Perhaps the less hostile environment of a habitable moon orbiting a brown dwarf would help to extend the human life cycle."

That is a point that is very well taken, and I must admit that until I read these words by Andy, I'd never considered such a scenario for "immortality", but it makes perfect sense! *Bravo, Andy!*

In Chapter 5 ("Brown Dwarfs") on page 106 Andy writes, "But it is also possible that the Dark Star lies on the edge of the brown dwarf spectrum. It is too large to be simply a massive gas giant, but its stellar properties may be too minimal to allow it to be classed as a brown dwarf. It would fit into a class of objects that have yet to be properly defined or studied. However, astronomers are contemplating what these sub-brown dwarfs might be like, with accompanying speculation that there might be at least one more stellar class beyond the T-dwarfs.

"If the Dark Star was to be discovered here in our solar system, this would clearly be the opportunity that astronomers have been waiting for. At the present time, the knowledge of these small sub-brown dwarfs is limited, even at a theoretical level. We do not know the extent of their stellar characteristics; how warm they are, how active their atmospheres are, and how much light they emit, if any.

"Their extensive magnetic fields are a mystery, and they may or may not form like regular stars. With so many unknowns, we cannot predict what scientists will discover next about these objects, and what this will tell us about a possible Dark Star orbiting our own sun. But what we can comfortably predict is that new discoveries will be forthcoming in the near future, and that, based on the history of the brown dwarf studies so far, those findings will contain the unexpected."

You can say that again, Andy! When Planet X Nibiru returns, if it does not destroy our science and technology in the process, we shall certainly have our hands full with new astronomical data. I can't wait! Because I am one of those who are placing my bets on 21 December 2012 or even earlier.

According to Sitchin in *The Twelfth Planet*, the Earth's moon named "Kingu" was previously a satellite of the proto-Earth "Tiamat", both of which were catapulted into Earth's present orbit following the "Marduk-Tiamat Celestial Battle" when one of Marduk-Nibiru's moonlets, referred to in the *Enuma Elish* as a "North Wind", crashed into Tiamat-Earth, creating the asteroid belt and propelling Earth to its newer and warmer orbit closer to the Sun. It is in this context that Andy mentions Dr. Velikovsky on page 123:

"The similarity between the Earth and the Moon's rocky constituents answers those who have hypothesized that the Moon is a relatively recent companion to the Earth. The writer Immanuel Velikovsky tried to explain various ancient myths that hinted at a previous absence of the Moon and infamously promoted the idea that the Moon had been recently captured by the Earth following a catastrophe, and that the time-scale for this event was relatively recent. If we can take the evidence presented by NASA scientists at face value, then it seems that Velikovsky was wrong. Yet this evidence is in accordance with Sitchin's version of events in that the Moon was formed by the cosmic collision very early on in the history of the Earth."

By way of reference, Andy cites the treatise *In The Beginning* by Dr. Velikovsky. This work can be found online at the following URL:

http://www.varchive.org/itb/index.htm

This Prague-based website is maintained by archivist Mr. Jan Sammer, who was Dr. Velikovsky's personal secretary at the time of his death in November 1979. One of the chapters of that treatise is titled "The Earth Without The Moon". I quote from that chapter:

"The period when the Earth was Moonless is probably the most remote recollection of mankind. Democritus and Anaxagoras taught that there was a time when the Earth was without the Moon. Aristotle wrote that Arcadia in Greece, before being inhabited by the Hellenes, had a population of Pelasgians, and that these aborigines occupied the land already before there was a moon in the sky above the Earth; for this reason they were called Proselenes.

"Apollonius of Rhodes mentioned the time 'when not all the orbs were yet in the heavens, before the Danai and Deukalion races came into existence, and only the Arcadians lived, of whom it is said that they dwelt on mountains and fed on acorns, before there was a moon.'

"Plutarch wrote in *The Roman Questions*: 'There were Arcadians of Evander's following, the so-called pre-Lunar people.' Similarly wrote Ovid: 'The Arcadians are said to have possessed their land before the birth of Jove, and the folk is older than the Moon.' Hippolytus refers to a legend that 'Arcadia brought forth Pelasgus, of greater antiquity than the moon.' Lucian in his *Astrology* says that 'the Arcadians affirm in their folly that they are older than the moon.'

"Censorinus also alludes to the time in the past when there was no moon in the sky."

Since we cannot merely sit back and ignore these "myths" from Greece and Rome, and presuming that all of these chroniclers weren't making up the same story, which is most unlikely, we must attempt to explain this to ourselves. Supposedly the "celestial battle" occurred in extremely remote antiquity, about a half-million years ago, even before Cro-Magnon *Sapiens* emerged. Clearly such an event could not have been remembered by anyone. If, however, by "moon" these ancient writers were referring to *Nibiru*, or The Winged Disk that appeared at the beginning of the last *shar* in 1588 BCE (Exodus and Santorini Cataclysm), then they were referring to the absence of what might be more precisely defined as a "Night Sun" rather than a "moon". Tethered to our North Pole at a distance of about 60,000 miles (100,000 kilometers), Planet X Nibiru may more resemble a "moon" than a "sun" when viewed from Earth. Beyond that, we simply have no other explanation. See Chapter 6 ("The Night Sun") of my book *Planet X Nibiru: Slow-Motion Doomsday.*

In a subchapter section titled "The 3-Body Solution" on pages 177-179, Andy writes the following:

"The solution I am proposing neatly answers a number of other problems. In fact, everything seems to fall in to place quite neatly.

"Nibiru is seen to enter the planetary solar system moving backwards through the sky (the so-called 'retrograde motion' of Nibiru). This is one of the puzzling aspects of Sitchin's account. The backwards motion of this body has always implied that it could not have been an original member of the solar system, making its initial capture nothing short of miraculous. Is there a way that a body can appear to move backwards, even though it is actually moving in the 'normal' direction through the sky?

"Any student of the stars will recognize this pattern. The outer planets are sometimes seen to undergo retrograde motion, particularly Mars. This was a major puzzle for early astronomers, who charted the movements of the wandering planets across the heavens.

"Why did some of the planets seem to stop, and then, for a short while, move backwards? This motion was due to a phenomenon called 'parallax'. As the Earth spun relatively quickly around the sun, an observer looking out into the solar system would see planets overtaken in a relative sense. Their motion was seemingly negated, and from an observational point of view, temporarily reversed by the actual movement of the Earth around the sun.

"Before Copernicus released that the sun was the centre of the solar system, this effect was quite inexplicable. It resulted in models of the solar system that allowed for additional movements of the outer planets around their own 'spheres'.

"I think that something similar is going on with Nibiru. Let us say that Nibiru is a rocky planet at the edge of the Dark Star system, rather like Pluto is in the sun's. Let us say that Nibiru's orbit is quite extended. It seems quite possible then, that as the two halves of the binary star system move towards each other at perihelion, that the outer rims of each system would overlap. The outermost planet of the Dark Star system might enter the planetary zone of the solar system, becoming a visible comet.

"One might also conclude that Pluto, and perhaps other outer Solar planets temporarily enter the Dark Star system, moving within the orbit of Nibiru. Perhaps that is why tiny Pluto's orbit is eccentric and inclined: such a 'crossing' alters its orbit over time. The other planets would be too large to significantly perturb, being significant gas and ice giants bound more heavily to the sun.

"Such a scenario affects the way the outer planet of the Dark Star system would be perceived by an observer on Earth. In the same way that the outer planets appeared to pre-Copernican star-gazers to be moving backwards when they weren't, Nibiru also seems to be moving backwards. But this, too, is an illusion. ...

"This removes the difficulty posed by a 'capture' scenario, which is statistically unlikely, although not impossible. The pro-grade orbit is also in keeping with the discovery of Sedna, which also has a pro-grade orbit. I strongly suspect that there is a relationship between the orbits of Sedna and the Dark Star; probably taking the form of a resonant orbit. Indeed, the movement of a brown dwarf through the Edgeworth-Kuiper [Comet] Belt at perihelion would explain many of the apparent anomalies of the bodies found in its scattered disc. It makes sense of the science."

Personally I have no quarrel with the idea of this Brown Dwarf binary's being the "sun" of Planet X Nibiru, if indeed this proves to be the case. After all, in my own estimation Planet X is the center of a "mini-system" of planetoids and moonlets, its accompanying "host" or "entourage". It is Andy's opinion that our complete solar system evolved from the very beginning as a binary system, and seven "moons" or planets subsequently formed around the Dark Star just as they did around the Sun, removing the "difficulty" of the "capture scenario" proposed by Sitchin. However, since the purpose of *Osiris, Isis & Planet X* is to propose and describe the events of this "capture scenario", I must disagree with Andy on this point. See introductory Chapter 1 ("Osiris & Isis") earlier in this book.

On pages 184-185 Andy briefly discusses the idea that the word "nibiru" has the meaning of "ferry" or "ferryman", as I have already mentioned earlier in Chapter 11. He writes: "The 'Planet of the Crossing' is thus a ferry of sorts. This has made little sense up until now, because the implication is that Nibiru takes travellers onto another place. That place was never defined by Sitchin, who insisted that Nibiru was itself the homeworld of the Anunnaki, the gods of ancient Mesopotamia. Yet with our new insight, the meaning behind the name 'ferry' becomes crystal clear. The transit of the Dark Star around the sun at perihelion is still a very remote event. At its closest, the Dark Star is still twice as far away as Pluto. To rendezvous with the Dark Star would take many years of space travel, with the risk of missing an object too remote to observe. Yet, Nibiru acts as an intermediary. It swoops into the planetary solar system, and then returns to the comet clouds. It would provide space travellers with the ideal stepping stone to the Dark Star. It literally acts as a ferry."

However, that idea is plausible only within the framework of this Dark Star Theory; and any space travellers that Nibiru might "ferry" to other planets within the Dark Star System would have been the Anunnaki themselves, certainly not any ancient earthly space travellers. However, as I have postulated in Chapter 11, if Planet X Nibiru has an aphelion within the adjacent Sirius System, rather than at the Oort Cloud, then such a Dark Star might provide enough "slingshot momentum" to "ferry" Nibiru from the gravity of our system into the gravity of the Sirius System, whence it originated. Yet, neither of these ideas is provable at present.

Andy devotes Chapter 11 to a discussion of Sedna, which was discovered in 2004. It is a small planet like Pluto which appears to be orbiting the Sun. To date, there is not much definitive information available regarding Sedna. Certainly if one googled for Sedna, one could find all that is known about it.

On page 201 we read, "When confronted by the twin problems of an astronomer burying her real conclusion within her paper, and the scientific news media subsequently reporting only half the story, one could be forgiven for wondering whether the possibility of a rogue brown dwarf companion to the sun is just a little too much for everyone's reputations to withstand. One must wonder whether such a notion is tantamount to a modern scientific heresy."

Exactly! They are scientists with vested interests who are afraid of the truth.

Then on page 228: "Whether this is the case or not, I suggest that Sedna's discovery draws us ever closer to that of the Dark Star's, and that this parent body will be found somewhere in the sky north of Sagittarius, probably within some of the dense star fields ignored by IRAS. It is quite possible that it has already been catalogued, but incorrectly defined as a more distant stellar object. (It is interesting to note that a faint 'red dwarf" star was recently identified as the third closest star to the sun, at a mere 7.8 light years.)"

Andy cites the following URL as the source of information about this "red dwarf":
http://www.spacedaily.com/news/stellar-o3a.html
Space Daily, 26 May 2003

In Chapter 12 ("The Dark Star System"), Andy quite surprised me with something. He suggested that we *triple* the length of Nibiru's orbit from 3,600 years to 10,800 years! That is in direct contradiction to what we know about the length of its *shar*, and I cannot accept it. However, oddly enough and not mentioned by Andy, we find on page 29 of *Worlds In Collision* by Dr. Velikovsky in a subchapter titled "The World Ages" this idea: "Anaximenes and Anaximander in the sixth pre-Christian century, and Diogenes of Apollonia in the fifth century, assumed the destruction of the world with subsequent recreation. Heraclitus (-540 to -475) taught that the world is destroyed in conflagration after every period of 10,800 years." It is certainly no "coincidence" that Heraclitus' cycle is three times one *shar* in length! Perhaps every third *shar* is a particularly cataclysmic Crossover of Planet X Nibiru.

And here, as mentioned earlier, is where I think that Andy begins to overcomplicate his general theory. In connection with this on page 244 he mentions the Mayan Calendar and the Mayan End-Time Date of 21 December 2012. Half of 10,800 is 5,400, which in terms of years, going back in time, equals the approximate beginning of the current Mayan Calendar. Andy writes, "The period between then and now roughly fits in with the current Mayan Age, which will come to an end on 21st December 2012. This date may be associated with changes in the sun's activity, or possibly even a reversal of the solar system's neutral sheet. Does that Age coincide with half an orbit of the Dark Star?"

Since Andy feels that the Dark Star is now near its aphelion in Sagittarius, it will not return to our vicinity until 1,800 years or even 5,400 years from now. He postulates earlier in this chapter that the previous perihelial passage of the Dark Star probably coincided with the so-called "Star of Bethlehem" in about 6-3 BCE. Thus, he would date the next perihelial passage of Planet X Nibiru, at a minimum, in about the year 3600 CE, long after we are dead and gone and all our writings long forgotten. If so, then what was the "cosmic object" associated with the legends of the Exodus and such like, as documented by Dr. Velikovsky in *Worlds In Collision*? Thus, I simply cannot accept the possible validity of these of Andy's ideas.

Andy continues his remarks about the time-scale on page 263 of Chapter 13 ("The Dark Star & Mass Extinctions"): "To explore this idea, we must immediately get to grips with a problem of time-scale. I am often confronted with e-mails that state that Planet X could not have appeared in our skies on such-and-such a date, because there was no massive catastrophe associated with its arrival. The implication is that every time the Dark Star system was to brush past the planetary zone, the Earth (and presumably some other planets too) would be subject to fundamental change. So, if the Dark Star exhibits an orbit analogous with Sitchin's 3,600 years, the implication is that Nibiru causes devastation on a highly regular basis — extremely often, when viewed on a geological scale. However, I don't accept this argument: it does not fit with the evidence at our disposal."

Again, I beg to differ. As Robert Sepehr pointed out in the first *Planet X Video*, we have evidence of catastrophes occurring regularly at intervals of 3,600 years. In my opinion, Polar Axial Displacement accompanies every single "Crossover" of the "Planet of the Crossing". I have presented some of this evidence in *Planet X Nibiru: Slow-Motion Doomsday*, particularly in Chapter 2 ("The Polar Pivotal Axis").

Then on pages 266-267 we find: "If the cycle of these extinction events is to be believed (and it remains controversial among scientists), then any direct extraterrestrial cause must be coincident with that enormous time-scale. So it would not be satisfactory, then, to associate a 26 million-year extinction cycle with a planet whose orbit is measured in thousands of years only. The Dark Star's relatively short orbit (Zecharia Sitchin's 'Sar' of 3,600 years, or even a multiple-Sar orbital period of, say, 10,800 years) could only produce a random pattern of extinction events distributed thinly over this time-scale.

"Putting this another way, if the Dark Star is directly accountable for extinction level events on Earth, then it must either pass very close to Earth during a transit actually into the inner solar system, or else it must have brought with it a comet, or swarm of comets, that happened to collide with Earth. Since both these possibilities are statistically unlikely given the sheer size of the solar system, then they could not occur during each perihelion passage. Instead, they might occur very, very occasionally throughout geological history, and the pattern of these events would be effectively random over that time-scale, even if it was closely associated with a cyclical event that was more frequent, like the perihelion passage of the Dark Star."

Once again, I completely disagree. We cannot refer to these ancient cataclysms and mass extinctions in terms of *millions* of years if we have any hope at all of reconciling written historical records with these events. Take the dinosaur extinction, for example. Did it really occur 65 million years ago? Don't people actually comprehend what a ***long*** period of time that is?! All of these establishment geological time-scales are terribly overestimated in length.

On page 280: "There is a common adage in science that the more you study a phenomenon, the more confusing it becomes. I think it is self-evident that the material I have presented here is complex and by no means clear-cut. Each of the three examples I have offered provide their own mystery, but taken together they lead to even greater obfuscation."

Andy obviously understands our collective lack of hard evidence in this research. Until we can actually catalogue scientifically, from our "modern" perspective, all of the events associated with Crossover, we shall never be able to write about it with complete clarity and certainty.

Andy estimates that the Dark Star's mass is several times that of Jupiter. I disagree. I think that Planet X is only about 5 times the size of Earth. Only time will tell us for sure. We are in a "waiting game" at this point in our history.

Finally, Andy ends his book on a rather philosophical note. "Our modern thinking has long since rubbished the warnings of the ancients about catastrophe. By ignoring the ever-present dangers — our modern society — through its misplaced scepticism, has foolishly turned its back on the wisdom handed down to us from the past. We should learn from this. There is great wisdom to be found in the writings of the ancients, and the orally transmitted tribal teachings. These teachings cannot replace our science, but they can, and should, complement our modern framework of knowledge. ...

"These are high stakes indeed."

Hear, hear! High stakes indeed!

Andy's *Dark Star* is an invaluable and indispensable addition to the library of anyone who is serious about *The Planet X Evidence.* I wish Andy Lloyd much success with his book!

Appendix B

Jesus Goes To Hollywood
By William Bramley

Commentary By Rob Solàrion

*

William Bramley was writing *Jesus Goes To Hollywood* at the same time that I was writing *Apollonius Of Tyana & The Shroud Of Turin*. It is regrettable that I was not able to include Bramley's excellent book in my Apollonius Chronological Historical Bibliography. Bramley has thoroughly researched all aspects of controversy surrounding the legitimacy of "The Jesus Christ". This book is highly recommended both for its literary quality and for its value as a general reference source. William Bramley is a terrifically talented writer in my opinion. I also recommend his spellbinding book *The Gods Of Eden* regarding Planet X Nibiru.

The indented chapter titles refer to an inserted novelette regarding a man from New York City who claimed that he was the reincarnation of Jesus and was subsequently flown to Hollywood for interviews and talent evaluation. The historical material is woven around this novelette, which is enjoyable but ultimately not important to Bramley's overall book, a fact which he mentions in his Introduction.

*

TABLE OF CONTENTS

Pages 54-57

What if a fictional "Jesus Christ" was based on a person who had actually lived during that era? What if that person was a man known as Apollonius of Tyana?

A few skeptics believe that Apollonius was a fictional character because of the obvious myths that eventually surrounded him, but most historians acknowledge him as a real individual. He is mentioned by several reliable independent sources, including early Christian leaders who tried to diminish his fame.

Apollonius was born into a wealthy Greek family in the city of Tyana about the time that Jesus was supposedly born. As an adult, Apollonius became a widely-traveled philosopher and spiritual teacher. He journeyed to Palestine, Babylon, Rome, Spain, and as far east as India and Nepal where he studied Hinduism and Buddhism. Apollonius was probably one of the most footloose prophets of his day, and he gained admission into a variety of religious sects and spiritual brotherhoods where he both learned and taught. ...

Of his personal qualities, we are told that Apollonius was extremely beautiful as a youth, gentle, and had a powerful memory and studious disposition. He enjoyed good-natured (and sometimes ironic) humor. Apollonius was a vegan — a person who abstains from eating or using anything of animal origin. He lived a simple life style by letting his hair grow, wearing plain white linen cloth, and walking barefoot. When he received an inheritance from his father, Apollonius kept only a small portion for himself and distributed the remainder to poor relatives. He never married or had children. His primary companion was a devoted male friend named Damis. Damis deeply loved Apollonius and kept a written record of the prophet's activities. ...

Proponents of the Jesus Was Apollonius Theory point out a curious fact. Damis never mentioned Christ in his writings despite Apollonius' widespread travels within the Roman Empire. Two hundred years later, Philostratus was also silent about the existence of Christ even though he pointed out the errors of other religions. It seems that Damis and Philostratus had never heard of Jesus or were unconcerned about him, and this may be further evidence that Christ never existed. Apostle Paul probably knew about Apollonius and may have used him as the model for a Christ myth. Apollonius was in Paul's home city of Tarsus when they were children, and the last city in which Apollonius established a religious school was Ephesus — a key center of earliest Christianity. ...

The theory in this chapter turns the spotlight to a fascinating spiritual teacher who came along during the tumultuous years of the first century A.D. Even if the Apollonius theory does not provide us with conclusive answers about Christ's identity, it offers another glimpse into the many spiritual byways that humanity has traveled. It also reminds us that itinerant preachers have been a part of humanity's spiritual mosaic for millennia. Some of them today fly first class on luxury airliners, stay at deluxe hotels, and attract thousands of devotees to events that are staged in enormous meeting places with sophisticated video and audio technology. The barefoot preachers of yore have come a long way.

[COMMENT: The reader is referred to my book *Apollonius Of Tyana & The Shroud Of Turin* for my complete commentary on the life of Apollonius of Tyana. RS]

Pages 97-99

"Nazarite", "Nazarene", and "Nasorean" look very much alike, so they are often confused. They have all been used at some time to label Christ, but not always correctly. Clarifying this word may be a step towards understanding Jesus.

[COMMENT: This matter is covered in depth in *Apollonius Of Tyana & The Shroud Of Turin*, Chapter 11 ("Nazareth, Nazarenes, Nazarites & Essenes"). RS]

As discussed in chapter 3, "Nazarite" comes from the Hebrew *nazîr* ("to consecrate" or "separate oneself"). Any Jew, including a woman or slave, could become a Nazirite by following the steps spelled out in *Numbers* 6:1-

21: (1) The candidate must not consume wine or anything else that comes from grapevines, (2) must not shave, (3) must no go near dead bodies, and (4) must make certain animal offerings. At the end of a trial period, the candidate's hair was shaved from the head and placed in a ritual fire. The Nazirite would then be allowed to drink wine again.

Was Jesus a Nazirite? Probably not. He drank wine (*Matthew* 26:26-9, *Luke* 7:34) and touched dead bodies (*Matthew* 9:18-25). There is no report in any gospel that he made animal sacrifices or placed spiritual significance on hair. The word "Nazirite" is not used anywhere in the New Testament.

[**COMMENT:** "Nazirite" is a variant spelling of "Nazarite". Even though this word may not be used in the New Testament, nevertheless John The Baptist was considered to be a Nazarite. Only two other men in Hebrew history were also given the lofty title of "perpetual" Nazarite: Samson and Samuel. Why was not "Jesus Christ" considered to be a Nazarite? Well, if the Hebrew Jesus was simply Jesus Bar Abbas (see below), then this bandit, revolutionary and murderer was not "spiritual" enough or ascetic enough to be placed into such a meritorious category. RS]

"Nazarene" and "Nasorean" ("Nazorean") were often used interchangeably to mean the same thing. Paul was said to be "a ringleader of the sect of the Nazoreans". (*Acts* 24:5, NAB [New American Bible]). (Other English translations use the term "Nazarene" in this passage.)

What was a Nasorean, and why should we care?

As noted in chapter 3, "Nasorean" might come from *ne-ser*, meaning bud or branch. Some scholars, however, trace "Nasorean" to the people called *Nâsôrayê* (or *Nasurai*) who are known today as the sect of the Mandaeans. ("Mandaean" comes from *manda* meaning either "knowledge" or "dwelling".)

[**COMMENT:** The Greek word *gnosis*, from which the Gnostic sect derived its name, also means "knowing" or "knowledge". Clearly all these sects were philosophically related to one another as well as to Apollonius of Tyana. See also the last 10-11 pages of Chapter 6. RS]

Today's Mandaean high priests are still called Nasoreans (or "Nasurai", meaning "observers" in the Mandaean language) which was the original name of their sect. They are often classified as a race due to certain physical characteristics and their own ancient language.

About twenty thousand Mandaeans still exist, mostly in southern Iraq and Iran, although political problems in the 1990s caused many to emigrate, and they can now be found as far away as Australia and San Diego, California. Modernization is taking its toll as many younger members find little appeal in Mandaean ways.

There are strong similarities between Mandaeans and early Christians. Mandaeans practice baptism in flowing water, and so they tried to live by moving rivers. They have savior figures with the names Manda-d-Hiia ("Knowledge of Life"), Hibil Ziwa ("a light bearer") and Yawar-Ziwa ("Awakening Light"). Good and evil spirits exist in their cosmology, including those that can be driven out of people's bodies.

A remarkable Mandaean claim is that John the Baptist was an early member of their sect, *i.e.*, he was a Nasurai (Nasorean). As a result, Mandaeans are sometimes called "Christians of St. John". Some of them proclaim John as their "prophet", although this may have started in the seventh century to avoid being forced to convert to Islam after the Moslem conquest. (Sects that declared an acceptable holy book and prophet were exempt from conversion, including some Christian sects.) However, most Mandaeans do not believe that John the Baptist started the Nasorean sect: he was simply a famous member. According to Epiphanius, a sect of "Nasaraeans" existed before Christ as an offshoot of Judaism.

Can Mandaean claims be authenticated? Not entirely. It is impossible to find artifacts of Mandaean existence earlier than the second century A.D. During the centuries that followed, Mandaeans adopted new ideas from surrounding religions, and so it is difficult to pinpoint Mandaean origins by following their belief system. On the other hand, Mandaeans have their own written language which appears traceable to the first century A.D., and that was the century in which John the Baptist and Jesus lived. When their geographic and linguistic history is investigated further, we find that the Mandaeans were probably in the Jordan River area when John the Baptist performed his baptisms there.

The New Testament confirms that John the Baptist was already some sort of sect leader before Jesus went to him, which bolsters the Mandaean claim that he belonged to their group. (See *Matthew* 3:5, *Mark* 1:5, *Luke* 3:7, *John*

1:6-28.) The Mandaeans say that Christ was also a member, and his baptism by John was his initiation into the sect. This provides a good solution to the riddle of the term "Nazarene" and why early Christians were called "Nasoreans" (a.k.a. "Nazoreans", "Nazarenes") by contemporaries.

Does this mean that the Mandaeans revere Christ? Not at all. They denounce him as a false Messiah who betrayed their sect, led people astray, and wrongly revealed secrets that were entrusted to him.

FOOTNOTE: Mandaeans also put a negative spin on Moses and Muhammad.

A Mandaean version of Christ's baptism alleges that John the Baptist was unwilling to perform the ceremony but relented when he received a command from heaven to "baptize the liar in the Jordan". *Matthew* 3:13-15 confirms that John was hesitant to baptize Jesus, but for different reasons. John the Baptist's uncertainty about Jesus persisted: when he was in prison he sent his disciples to ask Christ, "Are you the one who was to come, or should we expect someone else?" (*Matthew* 11:1-3, NIV [New International Version]). John had already performed the baptism long before, and it is unlikely that he would ask such a question unless he still doubted Jesus' Messiahship.

[COMMENT: If Jesus Bar Abbas (see below) were truly the cousin of John, then John would not have hesitated to baptize him, unless his "sins" were so heinous as to forbid it. Furthermore, it is stated in the New Testament that John did not recognize the "Messiah" that he baptized. Surely John would have recognized his cousin Jesus, but John would *not* have recognized Apollonius of Tyana. RS]

These Mandaean stories, along with the gospel accounts that several disciples left John to join Jesus, suggest that Christ led a splinter group of Nasoreans. Christ's followers continued to be called Nasoreans for a while, especially by the Jews, but eventually everyone would call them Christians.

Pages 102-103

As it turns out, the Dionysian ritual was modeled after the Osiris myth taught in Egyptian Mystery Schools and guilds.

[COMMENT: Earlier Bramley cites the book *The World's Sixteen Crucified Saviors (or Christianity Before Christ)* by Kersey Graves, who wrote about the similarities of birth, life and death of various historical figures: Krishna, Buddha, Prometheus, Dionysus, Osiris, Mithra and Jesus, to name some of them. This matter will not be discussed further here, and I can only refer you to Bramley's book for additional fascinating details that I have not transcribed. RS]

Osiris was a wise king of Egypt who traveled for three years to teach other nations the arts of civilization. He was slain and physically resurrected, whereupon he became ruler of the dead and passes judgment on the departed: the righteous receive eternal life and wicked souls are destroyed. Osiris was part of a holy Trinity (Osiris, Isis, and Horus), and his wife Isis was a virgin when she conceived Horus (who is symbolized by the Great Eye).

Isis was sometimes portrayed in a veil similar to the one used in the earliest images of Virgin Mary. *The Book of the Dead* says that Osiris will be with the creator god Amun when Amun destroys the world and begins a new creation. These beliefs can be documented to approximately 3,000 B.C., *i.e.*, long before the creation of the Dionysian Mysteries and Christianity. In light of this, it should come as no surprise that Christianity's most important symbol — the cross — was also significant in Egypt where it was called the *ankh* and symbolized "life". The *ankh* looks like the Christian cross except that it has a loop at the top and is often depicted in the hands of a god or goddess.

[COMMENT: Both the cross and the ankh are ultimately symbols referring to The Cosmic Tree. RS]

The Egyptian Mysteries initially worshipped multiple gods, but that was changed by Pharaoh Amenhotep IV, a.k.a. Akhenaten (*r.* 1375-1358 B.C.) who demanded worship of a single supreme God symbolized as a disc or sun-disc.

[COMMENT: Bramley either is unfamiliar with the Historical Reconstruction of Dr. Immanuel Velikovsky in the *Ages In Chaos* series or chose to ignore it. Akhnaton actually ruled during 870-853 CE. The Amon-Ra that Akhnaton worshipped was none other than the treacherous, rebellious Baron Marduk of Planet X Nibiru. And keep in mind that The

Cosmic Tree was still in place and the "Gods" were still here during the lifetime of Akhnaton. RS]

Although Akhenaten may have acknowledged the existence of several other gods, he is generally viewed as a monotheist, perhaps the first in the world. Akhenaten declared himself the only son and human intermediary of the supreme God. A famous Egyptian artwork from that era depicts Akhenaten under a solar disc that emits rays with *ankhs* (crosses). Was there a connection between Akhenaten's teachings and Jesus?

[COMMENT: The above remark is valid *only* if one looks at Egyptian history in the traditional way, wherein Akhnaton would have *preceded not followed* the time of Moses. RS]

By the second century, Jew were accusing Christ of having learned sorcery in Egypt. They said that he had been born into poverty and went to Egypt to work as a servant. There he acquired magical powers taught by Egyptian mystics. Elated with his new abilities, Jesus returned home and held himself out to be a god.

[COMMENT: This can be referring *only* to Jesus Bar Abbas (see below). Bramley provides no citation for this statement, but clearly it comes from Celsus who wrote in *Logos Alethes*, around 180 CE: "It was Jesus himself who fabricated the story that he had been born of a virgin. In fact, however, his mother was a poor country woman who earned her living by spinning [thread]. She had been driven out by her carpenter-husband when she was convicted of adultery with a [Roman] soldier named Panthera. She then wandered about and secretly gave birth to Jesus. Later, because he was poor, he hired himself out in Egypt where he became adept in magical powers. Puffed up by these, he claimed for himself the title of God." RS]

Page 143

The "Jesus Studied in India" theory is not necessarily as controversial as it sounds. It simply means that Christ did what so many other people have been doing for centuries: he studied the religions, histories, and cultures of other people. Today, large bookstores and libraries have religion sections where all of us can go to learn about other people's beliefs. In Christ's time, people had the same interests, but they needed to work harder and travel farther to get the information. That Jesus may have done so hardly seems remarkable.

Pages 168-169

An apocryphal work titled *Letter of Pubulus Lentulus* contains a description of Christ that is similar to Epiphanius and even to the image in the Shroud of Turin:

"At this time, there hath appeared, and still lives, a man endued with great powers, whose name is Jesus Christ. Men say that he is a mighty prophet; his disciples call him the Son of God. He restores the dead to life, and heals the sick from all sorts of ailments and diseases. He is a man of stature, proportionately tall, and his cast of countenance has a certain severity in it, so full of effect, as to induce beholders to love, and yet still to fear him. His hair is of the colour of wine [*i.e.*, probably red], as far as to the bottom of his ears, without radiation, and straight; and from the lower part of his ears, it is curled, down to his shoulders, and bright, and hangs downwards from his shoulders; at the top of his head it is parted after the fashion of the Nazarines.

"His forehead is smooth and clean, and his face without a pimple, adorned by a certain temperate redness; his countenance gentlemanlike and agreeable, his nose and mouth nothing amiss; his beard thick, and divided into two bunches, of the same colour as his hair; his eyes blue, and uncommonly bright. In reproving and rebuking he is formidable; in teaching and exhorting, of a bland and agreeable tongue. He has a wonderful grace of person united with seriousness. No one hath ever seen him smile, but weeping indeed they have. He hath a lengthened stature of body; his hands are straight and turned up, his arms are delectable; in speaking, deliberate and slow, and sparing of his conversations; — the most beautiful of countenance among the sons of men."{1}

Pages 308-309

{1} As quoted in Taylor, Robert, *The Diegesis* (1834, Boston), Abner Kneeland, pp. 379-380. A slightly different translation of this letter is reproduced by author John Iannone on page 149 of his book *The Mystery of the Shroud of Turin* (1998, New York, Alba House). Iannone cites Ernst von Dobshutz's *Christusbilder* (pub. 1899) as his source:

"There has appeared in our times, and there still lives, a man of great power (virtue), called Jesus Christ. The people call him prophet of truth; his disciples, son of God. He raises the dead, and heals infirmities. He is a man of medium size; he has a venerable aspect, and his beholders can both fear and love him. His hair is the color of the ripe hazel-nut, straight down to the ears, but below the ears wavy and curled, with a bluish and bright reflection, flowing over his shoulders. It is parted in two on the top of the head, after the pattern of the Nazarenes.

"His brow is smooth and very cheerful, with a face without wrinkle or spot, embellished by a slightly reddish complexion. His nose and mouth are faultless. His beard is abundant, of the color of his hair, not long, but divided at the chin. His aspect is simple and mature, his eyes are changeable and bright. He is terrible in his reprimands, sweet and amiable in his admonitions, cheerful without loss of gravity. He was never known to laugh, but often to weep. His stature is straight, his hands and arms beautiful to behold. His conversation is grave, infrequent, and modest. He is the most beautiful among the children of men."

On page 140 of his book *Jesus Now* (1973, New York. E.P. Dutton, & Co., Inc.) Malachi Martin offers this excerpt of the letter:

" ... nut-brown hair that is smooth down to the ears and from the ears downward forms soft curls and flows on his shoulders in luxuriant locks, with a parting in the center of his head after the fashion of the Nazarenes, a smooth clear brow, and a reddish face without spots or wrinkles. Nose and mouth are flawless, he wears a full luxuriant beard with is the same color as his hair and is parted in the middle; he has blue-gray eyes with an unusually varied capacity for expression."

The Archko Volume contains two apocryphal writings that purport to give eyewitness descriptions of Jesus. *Gamaliel's Interview With Joseph and Mary and Others Concerning Jesus* begins by telling us about Jesus' father Joseph: "Joseph is a wood-workman. He is very tall and ugly. His hair looks as though it might have been dark auburn when young. His eyes are gray and vicious. He is anything but predisposing in his appearance, and he is as gross and glum as he looks. He is but a poor talker, and it seems that yes and no are the depth of his mind. I am satisfied he is very disagreeable to his family. His children look very much like him, and upon the whole I should call them a third-rate family. ... I discovered that all of Joseph' ideas were of a selfish kind. All he thought of was himself."

(McIntosh, Dr. and Dr. Twyman, *The Archko Volume* (1887), republished by The Book Tree (Escondido), pp. 79-80, 82).

Gamaliel then writes about mother Mary: "Mary is an altogether different character, and she is too noble to be the wife of such a man. She seems to be about forty or forty-five years of age, abounds with a cheerful and happy spirit and is full of happy fancies. She is fair to see, rather fleshy, has soft and innocent-looking eyes, and seems to be naturally a good woman. ... I asked her if Jesus was the son of Joseph. She said he was not." (*Ibid.*, p. 82).

Gamaliel finally provides us with a description of Jesus, as told to him by a Jewish teacher who was personally acquainted with Christ: "He is the picture of his mother, only he has not her smooth, round face. His hair is a little more golden than hers, though it is as much from sunburn as anything else. He is tall, and his shoulders are a little drooped; his visage is thin and of a swarthy complexion, though this is from exposure. His eyes are large and a soft blue, and rather dull and heavy. The lashes are long, and his eyebrows very large. His nose is that of a Jew. In fact, he reminds me of an old-fashioned Jew in every sense of the word. He is not a great talker, unless there is something brought up about heaven and divine things, when his tongue moves glibly and his eyes light up with a peculiar brilliancy; though there is this peculiarity about Jesus, he never argues a question; he never disputes." (*Ibid.*, p. 92).

[**COMMENT:** It could well be that the Jesus described by Gamaliel was in fact Jesus Bar Abbas. Contrast the preceding description with the earlier one, which clearly matches the Naples Bust of Apollonius, as well as the one recently discovered at Old Tyana by Venetian Professor Asim Tanis and colleagues. In the Naples Bust, there is no resemblance to a Jew. Yet in those days, if the hair and beard colors and shapes were fairly similar, it might have been possible for the Nazarenes to have initially confused Apollonius with their rebellious native-son Jesus Bar Abbas, only to come to their senses later and then try to throw this "impostor" Jesus off a cliff, as is recorded in the New Testament. Certainly the Nazarenes would not have tried to throw the *real local Jesus* off that cliff. See Chapter 11 of my book about Apollonius. RS]

In *Report of Pilate to Tiberius* (*Archko* version), Pilate claims to have personally seen Jesus: "One day in passing by the place of Siloe, where there was a great concourse of people, I observed in the midst of the group a young man who was leaning against a tree, calmly addressing the multitude. I was told it was Jesus. This I could easily have suspected, so great was the difference between him and those listening to him. His golden-colored hair and beard gave to his appearance a celestial aspect. He appeared to be about thirty years of age. Never have I seen a sweeter or more serene countenance." (*Ibid.*, p. 131)

[COMMENT: Apollonius of Tyana was a handsome Greco-Aramaen who looked nothing like a Jew. It is not surprising to read that there was a "great difference" between him and the Jews of Jerusalem. RS]

Pages 179-181

If Jesus was romantically active, did he have children?

The official and apocryphal gospels are silent on this question, which is commonly interpreted to mean that he did not. But if Christ's inner circle had hidden the fact of his marriage, they would have also stayed mum about any offspring, perhaps to protect them. As discussed in chapter 7, any claim by Jesus or his followers that he was the Messiah from the House of David made him a political threat to the ruling House of Herod and the Romans. His children, especially sons, would be next in succession to Jesus, and so they had the potential to pose the same political threat as their father. The best way to protect Christ's children during the decades after the crucifixion was to pretend that they never existed.

A related theory suggests that the Roman Church hid the fact of Christ's children because they or their descendants could gain political power and use it against the Church. Jesus therefore became a "perpetual virgin" like his mother to prevent anyone from claiming kinship (or kingship) on the basis of Biblical authority.

Is there any affirmative evidence that Jesus had children?

In the story of Christ's trial and crucifixion, all four gospels mention a Roman prisoner named Barabbas. (*Matthew* 27:16-26, *Mark* 15:6-15, *Luke* 23:13-25, and *John* 18:39-40). Pilate told a crowd that he would release either Jesus or Barabbas in accordance with the custom of freeing a single Jewish prisoner at Passover. The crowd chose Barabbas, and so Jesus went to the cross.

Who was Barabbas? Was he the middle-aged homicidal goofball portrayed in Mel Gibson's movie *The Passion of the Christ*?

Barabbas was one of the "insurrectionists who had committed murder in the uprising". (*Mark* 15:7, NIV [New International Version]). In other words, he was in prison on charges similar to those leveled at Christ, although Jesus had not been accused of murder. What many people do not know is that early gospel manuscripts called Barabbas *bar'Abba-*, which means "son of the father". According to early Christian tradition, Barabbas was known as "Jesus bar'Abba-" which can be interpreted as, "Jesus son of the Father [Jesus]", *i.e.*, "Jesus, Jr."

Barabbas may have been Christ's son.

This theory is strengthened by an apocryphal writing called the *Gospel of the Nazoreans* (a.k.a. *Gospel of the Hebrews*). This gospel no longer exists in its original form but is briefly quoted by three early Christian writers. *Nazoreans* was probably written during the first half of the second century, and Jerome makes the following comment about it:

<< In the so-called Gospel of the Hebrews the name of the man who was to be condemned for sedition and murder [*i.e.*, Barabbas] is interpreted as "son of their teacher." >>

Son of *whose* teacher? We cannot be certain, but the reference might be to Christ's followers, so Barabbas could once again mean "son of Jesus".

A number of commentators have suggested that Jesus bar-Abbas was another man named Jesus who was also claiming to be the son of God, *i.e.*, "son of the Father". In other words, the Romans had two competing Jesus Christs and were intent on crucifying one of them. This interpretation postulates too much of a coincidence for some people, and if it were true, then one would expect more commentary about it in the gospels or apocrypha. The "Jesus Jr." theory seems to fit the gospel evidence better.

If Christ was in his early-to-mid thirties when he went to trial, he was old enough to have an adolescent or young adult child. Marriage in Christ's day often occurred in he teens, *i.e.*, about the time the body reached sexual maturity. If that had been true of Jesus and he had sired a boy, his son would have been in his late teens by the time of the crucifixion. That was old enough to join a group of insurrectionists, and to be arrested and executed for it. That makes it possible for Barabbas to have been both Christ's child and a political prisoner of the Romans.

Barabbas catches our attention for another reason. As the *New Jerome Biblical Commentary* points out, "There is no extrabiblical evidence for the annual custom of releasing a prisoner at Passover." *Jerome* adds, "Also disputed is whether the Barabbas incident and the supposed custom underlying it are historical or the creation of Christian tradition." In other words, some people doubt the very existence of Barabbas because there was no known "custom" of releasing a Jewish prisoner at Passover as the New Testament claims, yet the Barabbas story cuts across all four gospels. Why would all of the gospel authors invent a phony custom like that?

[COMMENT: Indeed! RS]

Perhaps they did so to disguise the true reason for Barabbas' release: Pilate had Christ's son in custody and was using him as a bargaining chip to coerce Jesus into a voluntary surrender. Jesus may have yielded to a horrible Roman ultimatum: "Surrender, and we'll release your kid. If you don't, your kid dies." This might explain the riddle of why Jesus did not flee after the Last Supper when he knew that his arrest was imminent, even though he had fled to Galilee on prior occasions when faced with danger. Christ's weak and ineffectual defense at trial and the offhand attitudes of Herod and Pilate suggest that the trial was a charade — the deal had already been made. Jesus allowed himself to be arrested, and his son was released as promised.

If Barabbas was Christ's son, then it means that Jesus had a child with someone other than Mary Magdalene, or he had met (and possibly wed) Magdalene as a teenager during his Missing Years. The gospels do not tell us when Christ and Magdalene first became acquainted. She is simply introduced as a member of his entourage early in the ministry.

[COMMENT: This "Jesus Jr." theory is a real stretch of the imagination. Surely if Barabbas had been the son of the Hebrew Jesus Christ, then such a relationship would not have gone unmentioned by everybody, then and later. Any children that might have been attributed to the Hebrew Jesus would have been children of Jesus Barabbas. However, when one accepts the irrefutable fact that the image on the Shroud of Turin is the image of Apollonius of Tyana, then all of this can be explained quite easily. Jesus

Bar Abbas was the Hebrew Jesus, the cousin of John The Baptist, the son of Mary and the "Father Superior" or High-Priest Joseph, friend or relative of the High-Priest Zacharias, husband of Elizabeth (cousin of Mary) and father of John The Baptist. It would have been much more in the interests of the Romans to crucify Barabbas than Apollonius, but the Jews forced Pilate to spare the life of their own, local revolutionary and crucify the "false messiah" who was uncircumcised and therefore clearly a Gentile, from north of the Galilee, and not a Jew. Pilate undoubtedly realized that Apollonius was not a Jew, but a citizen of Rome, and was thus reluctant to sentence him to death. Significantly the Roman soldier put in charge of the crucifixion was Longinus, a fellow-Cappadocian of Apollonius, and either this "premeditated plan" or "sheer coincidence" is certainly one of the reasons that Apollonius was able to survive the crucifixion.

[Earlier, Bramley mentioned that Apollonius travelled in Palestine. Bramley certainly did not obtain this information from the current edition of Philostratus' biography of Apollonius, as it does not exist. However, St. Jerome wrote that he'd read the Philostratus biography and that Apollonius had indeed travelled to Palestine. Obviously, Apollonius never returned to Palestine after the crucifixion, because in June 69 CE soon-to-be Roman Emperor Vespasian, stationed with Roman troops outside Jerusalem, heard that Apollonius was in nearby Alexandria and invited Apollonius to come to Jerusalem to consult with him. However, as Philostratus relates, Apollonius "refused to enter a country which its inhabitants polluted both by that they did and by what they suffered." Would *you* return to a country whose citizens had tried to kill you?! So, Vespasian had to travel to Alexandria to meet with Apollonius there. These matters are dealt with at great length in my book about Apollonius and the Shroud. RS]

Pages 214-222

In Turin, Italy, sits a cathedral which houses Catholicism's most famous relic: a large linen cloth curiously imprinted with the life-size image of a bearded man who appears to be dead. Many people believe that this cloth — commonly known as the Shroud of Turin — was the burial linen of Christ. Although the Vatican has never officially declared the Shroud to be genuine, many people believe that the image was created as the result of Christ's oiled skin coming into contact with the cloth or by an unusual radiance from the body at the moment of resurrection.

The Shroud image looks like a photographic negative. The hands and feet bear crucifixion wounds, and the face appears swollen. The nose is disjointed and elongated as though it had been broken. A hundred scourge marks cover the back from an apparent flogging. A side injury corresponds to the rib wound of Jesus as described in the New Testament.

Does the Shroud of Turin prove the existence of Christ and the story of his crucifixion?

The authenticity of the Shroud has been disputed almost since it appeared. In a written report to Pope Clement VII dated 1389 — about 32 years after the Shroud was first known to be on public display — Bishop Pierre D'Arcis said that his predecessor (Bishop Henri of Poitiers) had determined the cloth to be a fake; the offending artist had admitted to the fact.

In 1973, the Vatican allowed the Shroud to be scientifically tested. Bloodstains were analyzed, and the initial results seemed to come up negative for human blood. A later analysis revealed human DNA in a blood sample, but it was pointed out that any person who had ever touched or cried over the Shroud left a potential DNA signal.

A five-day battery of tests was conducted on the Shroud in 1978. Some scientists concluded that the image was caused by contact with a human body and that the blood was real, but other colleagues accused them of "scientific zealotry".

A small piece of the Shroud was cut away in 1988 and put through a carbon-14 test by three different laboratories. The unanimous result was a 99.9% certainty that the cloth originated from the period between A.D. 1000 and 1500, and a 95% certainty that it dated between 1260 and 1390. In other words, the cloth was spun more than a thousand years after Christ. This would place its origin at a time when Bishop D'Arcis said that an artist had admitted to hoaxing the image. It is also consistent with the historical evidence that the Shroud in its double-imprint image (*i.e.*, front and back of the body) did not appear in history until *c.*1355 when it was duplicated on a pilgrim's medallion to commemorate the first-known display in France.

Other problems exist with the Shroud. The head appears to be separate from the body and has a strange "base" line. This could mean that the cloth was draped over a bust to create the facial image, and a real human body was used for the remainder. (On the other hand, the base line image could have been caused by a rope or cloth secured at the neck — a common burial practice in Jesus' day.) The hair hangs as though the man is standing rather than reclining, but hair should not even be showing up if the image was caused by contact with the skin. The blood on the head is unrealistic: blood should spread and clot on the cloth instead of running down the hair in artistic rivulets. If the face image was caused by contact with a real body, then the face should be horizontally oblong (moon-shaped) instead of being in natural proportions. The fact that some "blood" remains bright red indicates that paint was used to create it, and scientist Walter McCrone of the 1978 Shroud of Turin Research Project speculates that the "blood" was a mix of red ocher and vermilion tempera paint.

If the Shroud is a hoax, then who created it, and why?

The face is almost identical to several portraits of Leonardo Da Vinci. (See *Turin Shroud* by Lynn Picknett and Clive Prince.) Did he create the Shroud? Da Vinci was born in 1452 and died in 1519, and this would have placed him within the time parameters established by the carbon-14 results. Da Vinci was extremely inventive, and he may have been familiar with the basics of photography. Art historian Nicholas Allen of Port Elizabeth University conducted an experiment showing that the Shroud may be the world's first known photograph that was created by using materials available during the era established by the carbon-14 test: a large linen cloth soaked in silver sulphate, dual glass lenses, a closed room ("camera obscura"), and a body hanging outside in the sun for at least eight hours. With this technique, Allen was able to create a full-size linen imprinted like the Turin Shroud. (See the PBS documentary *Secrets of the Dead: Shroud of Christ?*) The Turin Shroud has never been tested for the presence of silver sulphate.

[**COMMENT:** Photography was invented in the very early 1800s, and it took off like a rocket. In my family photo albums, I have photos from that time-period onwards. Originally photos were printed on tin plate, not paper, but paper photographs came soon thereafter. The overall technology of the early 1800s was not all that advanced over the technology of Da Vinci's era. If photography had truly been "discovered" in Da Vinci's time, you can bet the bank that it would have taken off like a rocket back then, too. The image on the Shroud is *not* a photographic imprint or negative. Needless to say, this matter is discussed in great detail in my *Apollonius Of Tyana & The Shroud Of Turin.* RS]

If the Shroud image was caused by contact with a human body, it may have been the body of Knights Templar Grand Master Jacques de Molay who was tortured in 1307 by the Catholic Inquisition. ...

[COMMENT: This short discussion of Jacques de Molay will be omitted here, as irrelevant. RS]

The Shroud remained in storage for decades and was eventually provided to a local church for its first undisputed public exhibition. By then, the true identity of the image had been lost with the death of de Charney's nephew [Jacques de Molay]. People jumped to the conclusion that the image was of Jesus.

Believers in the Shroud's authenticity have responded with counter-arguments.

1. The carbon-14 results were contaminated by a process known as "bioplastic coating" caused by bacterial buildup. The Shroud had been publicly displayed many times during the previous six centuries. The small carbon-14 sample was cut from a corner of the Shroud that had been frequently gripped in human hands when displayed to the crowds. (The Shroud is now encased behind glass.) UV fluorescence photography of the Shroud reveals that the carbon-14 cloth sample was dramatically different in some ways than the rest of the Shroud. Bioplastic coating can cause carbon-14 dating errors of a thousand years or more.

2. The carbon-14 results were affected by a fire that had occurred in 1532 where the Shroud was stored. The fire caused a chemical modification of the linen, and this would have "rejuvenated" the cloth. In 1994, Dr. Dmitri Kouznetsov of the Biopolymer Laboratory in Moscow experimented on a linen sample that was indisputably created between 100 B.C. and A.D. 100 (*i.e.*, the time of Christ). The cloth was carbon-14 tested with a result of 385-107 B.C. Kouznetsov then exposed the sample to similar conditions that might have been caused by the 1532 fire, followed by another carbon test. This time the result said A.D. 1044 to 1272 — an error of over 1,000 years. Using the same margin of error, Kouznetsov concluded that the Turin Shroud must have originated in the first or second century A.D.

3. Many details in the Shroud differ from the manner in which fourteenth-century artists were depicting the crucifixion. For example, Jesus was always shown with an open-topped halo of thorns, but the head wounds

on the Shroud indicate that a cap of thorns had covered the entire top of his head. Artists depicted Christ with nails through the palms, but the Shroud wounds are at the wrists just below the palms which are more likely locations: nails in the palms would rip the hands open unless the body was supported by ropes or straps, but nails in the wrists could support the entire body. Driving nails into the wrists at the points indicated on the Shroud would damage nerves and cause the thumbs to go towards the palms, and that is what we see in the Shroud. These are details that few hoaxers would have known.

4. The flagellation wounds are consistent with a common whip used in ancient Rome. If the person in the Shroud was not Jesus, then it was another crucifixion victim or a masochistic model.

5. Pollen found on the Shroud would be consistent with a cloth that originated in first-century Jerusalem, traveled to several other places (including Edessa [modern Sanli-Urfa, Türkiye, RS] where it became known as the Cloth of Edessa), and finally arrived in Europe courtesy of the Templars. (The pollen traces, however, do not rule out the Da Vinci or de Molay theories: Da Vinci or the Templars may have obtained imported linen from the Middle East.) In *The Blood and the Shroud*, Ian Wilson presents a believable path that the Turin Shroud could have traveled from Christ's body to the cathedral at Turin.

[**COMMENT:** According to Kersten and Gruber (see below), the Shroud was taken from Jerusalem to Edessa in about 50 CE by the disciple Thaddeus. It remained in Edessa, sometimes on public display and sometimes hidden away for its own protection, until 944 CE, when it was taken to Constantinople on the orders of Byzantine Emperor Romanus Lecapenus, arriving in Constantinople on August 15. Then in 1356 CE the Shroud was moved from Constantinople to France, where the Holy See was headquartered at Avignon during the 1300s. It was kept first at Lirey and later at Chambery and occasionally was brought out for religious purposes. Finally in 1578, the Shroud was moved to Turin, Italy, where it has remained until the present day. This "holy relic" is currently on public display at St. John The Baptist Cathedral in Torino. Moreover, the "St. Thomas" of the New Testament, the so-called "twin" (*didymus*) of "Jesus", retired to Edessa as an older man, and it was in Edessa that he wrote his *Gospel of Thomas*. But "Thomas" was in fact Damis of Ninevah, the companion and scribe for Apollonius. Did Damis/Thomas retire in Edessa because he knew that the *sindon* of the crucified Apollonius/Christ was being kept there? RS]

6. When the Shroud was repaired in 2002, restoration leader Mechthild Flury-Lemberg observed that the linen was sewn in a style that she had only seen once before in cloth found at Masada that dated from within seventy years of Christ's birth. It was a style not used in medieval Europe. Unless the Shroud linen was already more than a thousand years old before it was used to create a hoax (an unlikely but not impossible scenario), the Shroud probably originated from Jesus' time.

7. At least two centuries before the Shroud was first known to be on public display, texts were already mentioning the existence of a life-size image of Christ on a cloth. Ordericus Vitalis wrote in his *Ecclesiastical History* (*c.*1141) that the Edessa Cloth was miraculously created when Jesus used it to wipe sweat from his face, but an image of his entire body was formed on it. In *Otia Imperialia* (*c.*1211), Gervase of Tilbury said that records from an ancient archive report that Jesus had prostrated himself on a linen and left a full-body image through divine power. A Hungarian manuscript dated to 1190 contains an apparent illustration of the Shroud with its herringbone weave and four small holes that are situated in a pattern identical to the four small "poker holes" in the Shroud.

8. Tests were conducted on a smaller linen known as the Cloth of Oviedo which is said to be the material that covered Christ's head when he was entombed. This smaller type of linen is called a *sudarim* — a fabric placed by Jews over the face of a deceased person at the moment of death. *John* 20:8 reports that Christ's *sudarim* was discovered in the empty tomb. This cloth reportedly survived the centuries to reach the Camara Santa Church in Oviedo, Spain. A blood stain on the cloth is similar in shape to a Shroud stain, indicating that the Cloth and Shroud were once wrapped around the same body at the same time. Tests performed in the 1990s indicate that bloodstains on the Oviedo Cloth are the same rare AB blood type that may have been confirmed on the Turin Shroud, and only three percent of the human population shares this blood type. Body fluids on the Cloth are consistent with discharge that would come horizontally. The Oviedo Cloth can be traced with confidence to at least the sixth century, which is four hundred years before the earliest date indicated by the Shroud carbon-14 test.

[**COMMENT:** Dr. Peter J. D'Adamo in his book *Eat Right For Your Type* (New York, 1996), which is an analysis of blood types and dietary habits, has the following to say about blood type AB: "Type AB is rare. Emerging from the intermingling of Type A Caucasians with Type B Mongolians, it is found in less than 5 percent of the population, and it is the newest of the blood types. Until ten or twelve centuries ago, there was no Type AB blood. Then barbarian hordes sliced through the soft underbelly of many collapsing civilizations, overrunning the length and breadth of the Roman Empire. As a result of the intermingling of these Eastern invaders with the last trembling vestiges of European civilization, Type AB blood came to be. No evidence for the occurrence of this blood type extends beyond nine hundred to a thousand years ago, when a large western migration of eastern peoples took place. Blood Type AB is rarely found in European graves prior to A.D. 900. Studies on exhumations of prehistoric graves in Hungary show a distinct lack of this blood group into the Longobard age (fourth to seventh century A.D.). This would seem to indicate that up until that point in time, European populations of Type A and Type B did not come into common contact, or if so, did not mingle or intermarry."

[And in their book *You Are Your Blood Type* (New York, 1983), Toshitaka Nomi and Alexander Besher write the following: "AB people make up only four percent of the American population. The lowest concentration of AB people in the world has been found among the Basques in Northern Spain and Southern France, where studies have shown entire villages where the frequency of AB type is — yes — an unbelievable zero percent. This had led to the speculation that the aboriginal Celtic population of Europe, who were mostly replaced by the Aryan invasions starting over 5,000 years ago, are largely type O and A people. Lower frequencies of AB types can also be found in remote parts of Ireland, Scotland and Wales, all places with populations tracing their roots to the ancient Celtic civilization responsible for such marvels of neolithic engineering as Stonehenge and the stone temples of France. The AB types seem to have filtered into Europe, as mentioned, with the onslaught of the Aryan invasions. The last of the Aryan migrants are the Gypsies who began appearing in Europe from Asia in the twelfth century. Not surprisingly, European Gypsies show frequencies of AB types in their populations running as high as thirteen percent. The Gypsy connection may also have something to do with the extraordinary spiritual sensitivity of AB people. According to a Japanese survey, AB people are often found in the ranks of fortune tellers, shamans, witches, clairvoyants, faith-healers, and mediums. In fact, these professions are almost twice as likely to be ABs as any other type. You might find a lot of AB people going into preaching as religious leaders — not your normal neighborhood pastor, but more the guru-master type, who demands strict loyalty from his adherents. Is it any surprise then that the most illustrious AB in history was Jesus Christ? Christ was identified as an AB type through chemical analyses of blood stains on the famous Shroud of Turin, which has been tested extensively for thousands of different purposes over the last decade. ... Unfortunately, the blood types of other great religious leaders remain unknown. But odds are that the founders of the world's great religions like the Gautama Buddha, Mohammed, and mystics from St. Francis to Mahatma Gandhi were likely AB types."

[So, draw your own conclusion here. Asian Aramaean Apollonius could hypothetically have had blood type AB. It cannot be automatically ruled out. RS]

A curious conspiracy theory has emerged about the Shroud. There are those who believe that the cloth is completely authentic, but someone in the Catholic Church deliberately sabotaged the carbon-14 test by substituting a piece of medieval cloth for the real one (or knowingly caused a contaminated piece to be used). (See, *e.g.*, *The Jesus Conspiracy* by Holger Kersten and Elmar R. Gruber.) The accusation seems preposterous. Why would anyone in the Catholic Church deliberately cast doubt on an object that proves the existence of Christ and the reality of the crucifixion? The surprising answer might be this: if the Shroud is genuine, then it actually shows that Jesus did not die on the cross. He was still alive when he was wrapped. Proof of his survival would be a blow to Christian belief about his resurrection from death.

[COMMENT: This book is highly recommended! The complete title is *The Jesus Conspiracy: The Turin Shroud & The Truth About The Resurrection.* Part III ("The Secrets Of Golgotha") contains a chronological examination of every single event that occurred on "Good" Friday. This can also be found at my website.

http://www.apollonius.net/golgotha.html

Then in Part IV ("Fraud Of The Century") the authors actually created a linen "shroud image" of their own by duplicating the process described in primarily the Gospel of John. They provide photographs of their result. Although not perfect, their "shroud image" certainly can be clearly seen on their cloth. These photographs speak for themselves.

[Given that Apollonius of Tyana was the one who was crucified, then he did not die on the cross. He *almost* died, but he didn't. Does the Vatican actually realize that the image on the Shroud is that of Apollonius, with that being the reason why they have never acknowledged that the Shroud is "genuine"? But read on. RS]

When a body is dead, the blood stops flowing and obeys the law of gravity by sinking. The Shroud image shows an actively bleeding body on all sides, and this would mean that the heart was still pumping when it was being taken to the tomb. The Cloth of Oviedo also reveals active bleeding. The inescapable conclusion would be that the crucifixion did not kill Christ.

[COMMENT: And without a resurrection from the dead, Christianity collapses like a house of cards. Surely the Vatican understands this simple concept, no matter whose image is reflected on the Shroud. RS]

Further scientific support for this idea was provided in 2003 by Dr. Stephen J. Mattingly, a microbiologist at the University of Texas Health Science Center. He proposed that the Shroud image was caused by the common skin bacterium *Staphylococcus epidermidis*. According to Mattingly's theory, the bleeding body was alive for a number of hours and acted as an incubator for the bacterium. Mattingly performed an experiment by coating his own face with excess skin bacteria and covering it with a damp linen cloth. The result was an image surprisingly similar to the Shroud image, albeit moon-shaped.

The best evidence that Jesus did not die on the cross is provided by the New Testament.

According to the orthodox gospels, Christ's body was placed in a tomb after being removed from the cross. Several days later, the body was no longer there and Jesus started making personal appearances to his supporters. The obvious conclusion is that Jesus' executioners had failed to kill him. This conclusion is supported by the Biblical statement that he was on the cross for less than a day — perhaps as few as three hours and probably not more than six or seven. As *The NIV Study Bible* tells us, "Crucified men often lived two or three days before dying, and the early death of Jesus was therefore extraordinary." When Joseph of Arimathea asked Pilate for Christ's body, "Pilate was amazed that he [*i.e.*, Jesus] was already dead." (*Mark* 15:44, NAB [New American Bible]). Despite his surprise, Pilate granted Arimathea's request after being assured by a Roman centurion that the victim had perished. The act of surrendering the body to Joseph was unconventional in itself, as *The NIV Study Bible* points out: "The release of the body of one condemned for high treason, and especially to one who was not an immediate relative, was quite unusual." (On the other hand, if Arimathea was Jesus' great uncle as discussed in chapter 13, such a release is entirely plausible.) These circumstances suggest that Christ's backers had gotten him off the cross quickly enough to save him.

[**COMMENT:** Pilate himself might well have been a part of some "plot" to make sure that such a great man as Apollonius did indeed not die on the cross. But as noted previously, this is discussed with much greater detail in my book about Apollonius and the Shroud. RS]

How could Pilate and the Roman soldiers be fooled into thinking that Jesus had died so soon? One theory is that Christ suffered from a form of epilepsy. Some animals with this condition can be pushed into a state of hypothermia (low body temperature) and appear dead for about three days if they are given the drug reserpine and restrained for a few hours. This may have been the cause of Jesus appearing dead so quickly: he was an epileptic who was administered reserpine or a similar drug while restrained on the cross. A consequence of this might have been brain damage leading to behavioral changes such as anorexia or being unkempt. The New Testament reports that the resurrected Christ was not instantly recognized by some of the people who had known him.

[**COMMENT:** There is absolutely no evidence to support the idea that Apollonius was either epileptic or anorexic, or that he suffered any sort of brain damage as a result of the crucifixion. RS]

Muslims hold it as an article of faith that Jesus did not die on the cross. The religion of Islam was founded by the prophet Muhammad about six hundred years after Christ. The Muslim holy book, the *Koran*, honors Jesus as one of God's (Allah's) prophets. In fact, mother Mary is a much-respected woman in Islam with a chapter in the *Koran* named after her.

{FOOTNOTE: Then why is there such a big problem between Muslims and Christians? Muslims believe that Moses, Jesus, and Muhammad were three prophets sent by God. Muhammad was the final prophet with the final revelation from God, and so it is the duty of all people to follow Islam. Those who fail to do so are "infidels" (nonbelievers). The Islamic "Hadith" (non-Koranic sayings of Muhammad) offer teachings that cause many Muslims to believe that Jesus will return to Earth one day with a person called Imam Mahdi to establish Islam as the world religion. When that happens, Christ will "break the cross", which some Muslims interpret to mean that he will destroy Christianity. Some Hadith seem to say that Christ's imposition of Islam on humankind will be violent, and non-Muslims will be forced to convert or die. In spite of this, most Christians have failed to convert to Islam just as most Jews do not pick up the banner for Christianity. The result has been centuries of hard feelings and war.}

According to *Koran* 4:157, "They slew him not but it appeared so to them." In other words, Jesus' executioners only thought that they had killed him. The *Koran* follows with a statement about Jesus being "raised by" or "ascended to" Allah, which is interpreted by the orthodox majority to mean that Christ was physically taken up to heaven *before* the crucifixion, and someone else was crucified in his place. Others interpret the passage to mean that Jesus survived the crucifixion because Allah decided to let him die a natural death later. In the *Kanz-ul-Aimal* — a collection of teachings attributed to Muhammad that were not included in the *Koran* — Jesus was saved from the cross, did some traveling, and avoided further persecution. A famous thirty-volume commentary about the *Koran* by Ibn-i-Jarir at-Tibri says that Jesus and his mother left Jerusalem after the crucifixion and traveled from one country to another. A belief held by Muhammad's wife is that Christ lived to a very old age, perhaps as old as 120.

The evidence of survival suggests that Jesus came extremely close to perishing, maybe to the point where his vital signs had stopped resulting in temporary clinical death, but he was revived in the nick of time before his coma or loss of vital functions became irreversible. In those days, such a recovery might have seemed like a miracle wrought by God, perhaps even to Jesus who must have felt elated to wake up alive. Today we understand it as a medical phenomenon that happens regularly in hospitals around the world. Despite this, Christians are still taught that salvation requires them to hold firm to a belief that Christ's survival was a miraculous resurrection from death.

If Jesus survived the cross, where did he end his mortal life?

[COMMENT: And that brings us back to the life of Apollonius of Tyana. Bramley follows his question with concise discussions of the various legends that "Jesus" died in England, France, Kashmir, Japan or at Masada (Palestine) during the Roman siege of 70 CE. RS]

Page 237

Some critics contend that Christianity has been the most devastating religion ever created. The centuries are witness to millions of people dying at the hands of Christian crusaders, Inquisitors, and zealots. Slavery in America was justified by Christian dogma. The world has still not recovered from the extraordinary intellectual cloud brought on by the Catholic Inquisition centuries ago. Christians are responsible for the wholesale destruction of irreplaceable historical and spiritual records. All the while, hypocrisy seems to run rampant. Christian fundamentalists are often staunch supporters of high military budgets that lead to terrible human slaughter, but they promote Christianity as a religion of peace. Many Christians loudly condemn various forms of sexual conduct while the media feed us with lurid sex scandals involving Catholic priests and fundamentalist preachers. Some Christian "Reconstructionists" and "Dominionists" with surprising influence in American politics advocate a return to literal Old Testament law including legalization of slavery, death by stoning, and forceful eradication of all contrary religion. It has been argued that Christianity attracts people with promises of love and fellowship, but then it quietly turns many of its believers into merchants of hypocrisy, intolerance and division.

[**COMMENT:** In his absorbing book *The Gods Of Eden*, William Bramley discusses how "ancient astronauts" and extraterrestrial aliens may be exerting control over earthly events. He traces this alien involvement from the Nibiruan origin of Cro-Magnon Sapiens until the present-day. In *Jesus Goes To Hollywood* this idea is only briefly addressed in Chapter 20, primarily in connection with the matter of Jesus Christ. However, Bramley includes some insightful comments about the possible motives for this extraterrestrial control of Earth. RS]

Pages 271-273

Why would a space-age race even bother?

This is a compelling question when we consider that many modern-day UFO abductees and contactees return from their experiences bearing religious and/or apocalyptic messages allegedly given to them by their captors. In other words, these ETs seem to have been engaged in religious indoctrination for thousands of years. Most of the sects they inspire remain small and marginalized, but a few make it big like Judaism, Christianity, Islam, and the Latter Day Saints (Mormons).

But to what end?

Many theories have been proposed, and here are just a few:

1. The ETs operate under a directive not to openly interfere with our primitive and violent society, so they remain forever in the background as an enigma to us. All the while, they covertly civilize us by implanting apocalyptic/resurrection religions. Their objective is to help the human race graduate from the "cosmic kindergarten" of Earth to join the more spiritually and technologically advanced peoples of the cosmos. Perhaps the ETs feel a responsibility to do so because they had originally created *Homo sapiens* as a slave race but do not use us for that purpose any more. One day they will openly reveal themselves to guide human society, and that is the true meaning of their apocalyptic messages.

2. An opposing theory is that the ETs have been using classic Machiavellian techniques to covertly manipulate their slave race to keep it spiritually imprisoned on Earth. Ancient Mesopotamian texts state that the "gods" possessed secret knowledge of how to "divide and conquer", and that their intention has been to prevent their slave race from gaining true spiritual wisdom. The apocalyptic religions they plant on Earth have exactly that effect: they keep the human race in a chronic state of division and strife, and many people feel that such religions perpetuate superstition, fear, and ignorance. The ETs will probably continue this activity indefinitely.

3. Earth is a prison where we have been sent by the ET race to work out our spiritual salvation. Once we do that, we can leave.

4. Earth is targeted for invasion by an extraterrestrial race that operates on a much longer time frame than we do (*i.e.*, a thousand years to us is like a decade to them). Earth's population is being kept in a state of disunity to make it "soft" for takeover when the invasion actually occurs. (This technique is described in Suntzu's famous two thousand year old treatise *The Art of War*.) "Judgment Day" and "Second Coming" teachings are methods to throw target populations off-guard by causing them to accept a sudden extraterrestrial presence without resisting it. The Bible says

that cataclysms and massive death will be part of that scenario, *i.e.*, the human race is being promised a rather nasty genocide sometime down the road, but we as a target population will not prepare or resist because of the religious belief we have been taught.

5. In his book, *The Threat*, Dr. [David M.] Jacobs finds evidence of a large-scale effort by ETs to breed half-human half-ET hybrids. Does this mean that *Homo sapiens* will be eradicated to make room for a new hybrid race on Earth? By some estimates, modern *Homo sapiens* may have appeared abruptly on Earth only ten to thirty thousand years ago, and our forerunners seemingly vanished at the same time. Will the same thing happen between us and the ET hybrids?

Postscript

It has been only about 55 years since Dr. Immanuel Velikovsky published *Worlds In Collision* and *Ages In Chaos*, and only about 30 years since Zecharia Sitchin's *The Twelfth Planet* and Robert K.G. Temple's *The Sirius Mystery.* In the grand scheme of the historical evolution and acceptance of radical new ideas, along the lines of those of Copernicus and Galileo, the sciences of Cosmic Catastrophism and Nibiruology are still relatively new. We who are at the forefront of these ideas are intellectually privileged indeed. Future people will look back upon us as revolutionary visionaries. And let me take this opportunity to thank those tireless people who are members of my Internet discussion forum for all of the URLs and other input which they have contributed to my research!

In the Postscript to the Prologue of *Slow-Motion Doomsday* I proposed that Springfield, Missouri, be selected as a centrally located North American gathering hub following Crossover. Subsequently to writing that in October 2003, my Springfield contact is no longer available. Although people in that particular locality could seek safety and information in Springfield, it may be impossible for those from other regions to travel there. Thus, all interested parties should plan their own regional responses to Crossover. Texarkana will probably be where I myself try to organize people from the Ark-La-Tex. If any reader might wish to make suggestions about this idea, you can go to one of my websites and contact me by email.

Crossover Dreamtime Is Coming Again!
Where Will You Be?

Rob Solàrion
Northeast Texas
1 March 2006
Current Date & Time, Planet X — New Year's Eve, 07:40 AM

About The Author

Robertino "Rob" Solàrion was born and educated in Texas. He was a 1964 Phi Beta Kappa graduate of the University of Texas at Austin, majoring in Russian and minoring in French. He joined the Peace Corps that same year and was assigned to Massaua, Eritrea (then a part of the Empire of Ethiopia), where he taught English, science and mathematics at the local high school. Subsequently he attended the Foreign Service Institute in Washington, D.C., and was assigned to the Refugee Program in South Vietnam during the war. Fluent in both French and Vietnamese, Rob also served as a translator for the U.S. Embassy in Saigon. He has studied a number of other foreign languages, has graduate credits in linguistics from UCLA and has widely travelled across 50 countries in North America, Europe, Africa, Asia and the Middle East.

Rob's interest in catastrophism and ancient history began in 1972, and he avidly pursued the ideas of Dr. Immanuel Velikovsky. He became a part of the so-called "Velikovskian School" as propounded by the former *KRONOS* Journal. Following the death of Dr. Velikovsky in November 1979, this "school" fragmented into smaller, often rival groups. After a hiatus from 1980 through 1993, during which time he worked as a professional photographer and writer, Rob became fascinated by the books of Zecharia Sitchin concerning Planet X Nibiru. His original website was uploaded in 1996, after he had made the assumption, now shared by many, that the "cosmic cataclysm" described by Dr. Velikovsky was caused by the "comet-planet" of Mr. Sitchin. Subsequently Rob developed his theory of "The Cosmic Tree" by integrating both these men's ideas with those of Giorgio de Santillana and Hertha von Dechend in *Hamlet's Mill.* Rob's theory of "The Cosmic Tree" is unique and has not been explored by other researchers on the Internet or elsewhere.

In 2004 Rob published *Planet X Nibiru: Slow-Motion Doomsday*, describing his theory of The Cosmic Tree. Rob Solàrion has also emerged as one of the world's leading authorities on the life and times of Apollonius of Tyana, and in 2005 he published *Apollonius Of Tyana & The Shroud Of Turin.* Both of these books are also available from Author House.

http://www.slowmotiondoomsday.com/osiris.html
http://www.slowmotiondoomsday.com/cosmictree.html
http://www.apollonius.net/turinshroud.html

www.ingramcontent.com/pod-product-compliance
Lightning Source LLC
Chambersburg PA
CBHW031813170526
45157CB00001B/41

* 9 7 8 1 4 2 5 9 2 6 2 1 2 *